ELEMENTARY ENGINEERING HYDROLOGY

ELEMENTARY ENGINEERING HYDROLOGY

M. J. DEODHAR

Emeritus Fellow

AICTE, New Delhi

India

Pearson

ISBN 978-81-317-0805-7

First Impression, 2009

Published by Pearson India Education Services Pvt. Ltd, CIN: U72200TN2005PTC057128

Head Office: 1st Floor, Berger Tower, Plot No. C-001A/2, Sector 16B, Noida – 201 301,Uttar Pradesh, India.

Registered Office: Featherlite, 'The Address' 5th Floor, Survey No 203/10B, 200 Ft MMRD Road,Zamin Pallavaram, Chennai - 600044
Website: in.pearson.com; Email:companysecretary.india@pearson.com

Digitally printed in India by Trinity Academy for Corporate Training Ltd, New Delhi in the year of 2024.

In memory

of

one of my dearest friends,

the Late Shri M. J. alias Achyut Karmarkar.

BRIEF CONTENTS

CONTENTS

PREFACE

I am very happy to present this book for the benefit of students and practising engineers.

I have taught this subject for over 35 years and sincerely felt the necessity of a good textbook explaining the subject in a simple language making the fundamentals clear and also indicating its practical application. I have, therefore, made the text comprehensive and covered almost the entire gamut of related topics and not to suit the syllabi of any university.

The sequence of the chapters followed in the book is similar to the processes of the hydrological cycle. Each chapter is divided into sub-topics, which will help to bring clarity to the subject matter and make it easy for students to grasp the concepts. A number of appendices are added to provide all the necessary allied information. The text includes a number of numerical problems with solutions to help clarify the concepts and demonstrate their practical utility. At the end of each chapter 'multiple choice questions' are given to help students prepare for competitive examinations.

A number of my students have helped me in writing this book. It is not possible for me to mention all the names here but I thank them all. The help and guidance extended by Professor P. R. Bhave, Dr C. D. Thatte, Professor S. Y. Kute and Professor R. D. Mahadeokar for checking the text and numerical problems is thankfully acknowledged. I also thank Vasant Vaze, my business partner, for all the help extended by him.

The data provided in this book have been procured from the following institutes:

1. Government College of Engineering, Pune.
2. Bharati Vidyapeeth's College of Engineering, Pune.
3. Sinhgad College of Engineering, Pune.
4. Central Water and Power Research Institute, Pune.
5. National Water Academy, Pune.
6. Bhandarkar Institute of Oriental Research, Pune.
7. Dnyan Prabhodhini, Pune.

I thank the authorities of these institutes for their support.

I was awarded Emeritus fellowship by the All India Council for Technical Education (AICTE), New Delhi, for writing this book. It was an encouragement to me. I express my deep sense of gratitude for the honour conferred on me.

I have made a sincere attempt to make this book a valuable reference book. Any suggestions to enhance the quality of this book are welcome.

M. J. Deodhar

1, Amraphal Apartments,
34, Bharatkunj Society No. 1,
Erandavana
Pune 411038

ABOUT THE AUTHOR

Professor M. J. Deodhar graduated in 1957 and completed his postgraduate studies in Hydraulics and Dam Engineering in 1959 from the University of Pune. He started his career as Assistant Research Officer in Maharashtra Engineering Research Institute, Nasik, and later became Professor and Head of the Civil Engineering Department in the Government College of Engineering at Karad, Amaravati and Pune. He retired as Principal of K. K. Wagh College of Engineering, Nasik.

He has been Visiting Professor and also Emeritus Fellow, AICTE, New Delhi. He has worked as technical expert for UPSC, MPSC, UP PSC and Himachal Pradesh PSC, and as Chairman, Board of Studies in Civil Engineering. He has also been a member of Faculty of Engineering and Technology and the Academic Council of universities in Maharashtra.

He has published technical articles and also presented papers at the international conferences, the prominent being the International Association of Hydraulic Engineering and Research held at San Francisco and Beijing. He has also conducted and attended short-term courses on behalf of the Indian Society for Technical Education, New Delhi, and UNESCO.

With over 50 years of research and teaching experience behind him, Professor Deodhar was awarded the 'best teacher' award and 'outstanding engineer' award by the Institution of Engineers, Nasik Local Centre. Presently, he is working as Technical Advisor for Coastal Power Consultants, Pune.

1 Introduction

Chapter Outline

1.1 DEFINITION

Hydrology today denotes a field of science that covers many more branches of science than it did earlier. Till 1906, hydrology referred to the study of underground water only, as distinct from the present usage that covers both groundwater as well as surface water. In a much broader sense, hydrology refers to the study of water.

The term 'hydrology' is derived from two Greek words *hydor* and *logas* meaning 'water' and 'science', respectively. So, in simple terms, hydrology is a science related to water.

The Merriam-Webster dictionary describes hydrology as *a science dealing with the properties, distribution and circulation of water on the surface of land, in the soil, in the underlying rocks and in the atmosphere, particularly with respect to evaporation and precipitation.*

C. O. Wisler and E. F. Brater define it as *The science that deals with the processes governing the depletion and replenishment of the water resources of the land areas of the earth.*

The ad-hoc panel on hydrology of the Federal Council for Science and Technology (established by the president of the United States of America in 1959) recommended the following definition for hydrology:

> *Hydrology is the science that treats waters on the earth, their occurrence, circulation, distribution, their chemical and physical properties and their environment including their relation to living things.*

In short, *what happens to the rain* is the basis of the definition of the science of hydrology. It should not be confused with hydraulics, which deals with the mechanics of water. Hydrology has got a variety of practical applications. Therefore, it should not be treated as a pure science.

1.1.1 Importance of Hydrology

Water is, indeed, the most valuable natural resource because human race or life will not survive in its absence. It is also described as the *free gift of God.* Without it, no form of life is possible. It not only supports the animal and plant kingdom for its daily subsistence but also serves as a valuable source of energy and a means of transportation. It also serves many other useful purposes.

However, this natural source, at times, assumes the form of a very destructive agent destroying valuable property, taking a heavy toll of life and eroding and carrying thousands of tons of rich and fertile soil into the sea. With a rapid increase in population, the demands for this vital resource are becoming more and more acute. So also the destructive effects of floods are increasing and becoming more devastating. It is, therefore, necessary that an attempt be made to gain a better understanding of the occurrence and behaviour of water on the earth.

1.1.2 Sub-branches of Hydrology

Depending on the applications of hydrology, the hydrological study can be divided into different sub-branches.

- To emphasize its importance in practical application, the term *applied hydrology* is used.
- For the numerous applications of hydrology in the field of engineering, the title *engineering hydrology* is applied.
- The expression *scientific hydrology* is used to distinguish it from practical application.
- The term *synthetic hydrology* is used for the study of hydrology that involves synthesis and simulation techniques by field plots, or physical and mathematical models, or electrical analogy.

1.1.3 Supporting Sciences

For the study of hydrology, knowledge of other sciences is required. These sciences are:
(1) chemistry, (2) physics, (3) biology, (4) geology, (5) fluid mechanics, (6) statistics, (7) mathematics and (8) operation research.

1.1.4 Overlapping Areas of Hydrology

There are other fields and branches of science of which hydrology forms a significant part. The significance of overlapping areas can be seen from the development of the new branches such as: '(1) hydrometeorology, (2) hydrogeology, (3) geohydrology, (4) hydrobiology and (5) biohydrology.'

1.1.5 Influence of Hydrology

Since hydrology is a science also deployed in dealing with the development and regulation of water resources, it influences the following areas:

(1) agriculture, (2) forestry, (3) geography, (4) watershed management, (5) political sciences (water law and policy), (6) hydro-economics, (7) sociology and (8) ecology.

1.1.6 Practical Use of Hydrology

Hydrology has got practical applications in other fields too. These fields are: (1) water supply, (2) waste-water disposal, (3) irrigation, (4) drainage, (5) hydropower, (6) flood control, (7) navigation, (8) erosion and sediment transport and (9) hydroponics.

1.1.7 Embodying Sciences

The science of hydrology embodies various fields of study, as shown in the following:

- Potamology—Science of surface streams
- Limnology—Science of surface lakes
- Cryology—Science of surface snow and ice
- Glaciology—Science of surface glaciers
- Oceanology—Science of surface oceans
- Hydrometry—Science of measurement of water
- Hydrography—Science that describes the physical features of all waters on the earth's surface

1.2 QUANTITY OF WATER AVAILABLE ON THE EARTH

The total quantity of water available on the earth is estimated as 1348.25305 × 10^6 km^3, and this may cover the earth to an average depth of 2.73 km—assuming earth a uniform sphere of 12,800 km in diameter.

1.2.1 Global Water Budget

The water available on the earth in its various forms is given in Table 1.1.

1.3 HISTORY OF HYDROLOGY

Man understood some hydrological phenomena such as precipitation, runoff, and so on. Studies were done related to these in the prehistoric times also, and were based on logical thinking, but as a science, hydrology was developed only recently. Those studies and investigation of nature's processes were initially the tasks of philosophers, and were explained by them based on the assumptions and deductions therefrom.

Table 1.1 Various forms of water available on the earth

Serial no.	Water occurrence	Estimated water (km^3)	Estimated water (%)
A. Salt water			
1	Oceans	1307.410×10^6	
2	Salt water lakes and island seas	0.100×10^6	
Total sea water		$\mathbf{1307.510 \times 10^6}$	**96.9781**
B. Freshwater			
3	Glaciers, Polar ice caps	30.4300×10^6	
4	Atmospheric moisture	0.0140×10^6	
5	Hydrated earth minerals	0.0040×10^6	
6	Water content in plants and animals	0.0011×10^6	
B1.	**Non-utilizable freshwater**	$\mathbf{30.4491 \times 10^6}$	**2.2584**
7	Freshwater lakes	0.1246×10^6	
8	Rivers	0.00115×10^6	
B2.	**Surface resources of freshwater**	$\mathbf{0.12575 \times 10^6}$	**0.0093**
9	Soil moisture	0.0375×10^6	
10	Groundwater up to 800-m depth	4.998×10^6	
11	Groundwater below 800-m depth	5.6309×10^6	
B3.	**Groundwater resources**	$\mathbf{10.1682 \times 10^6}$	**0.7542**
B4.	**Total freshwater resources**	$\mathbf{40.74305 \times 10^6}$	**3.0219**
Total water resources of the earth		$\mathbf{1348.25305 \times 10^6}$	**100**

Men, in earlier days, noticed that rainfall annually occurred occasionally for a short span of time. They also saw there were some perennial rivers and springs. The problem for the early thinkers was 'Wherefrom these rivers and springs received their supply?' The solution of these problems was based on the following two assumptions:

1. Rainfall was inadequate to account for all the surface water in rivers and springs.
2. The earth was impervious below a certain depth from the surface.

Homer (1000 BC), Thales (650 BC) and Plato (400 BC) put forth the idea of subterranean flow from sea to land. By subterranean water, they meant that there is a continuous underground supply from sea to land, and during the flow the sea water loses its salinity. They also believed that the subterranean water flow supplied water to springs. Aristotle (350 BC) also assumed that there was an atmosphere below the ground. According to him, water vapour came from the seas and condensation took place, and this was the major source for the springs and rivers during the non-rainy days. Lucretius (90 BC), Pliny (30 BC) and Seneca (65 AD) accepted this theory. Vitruvius, who lived during the time of Christ, had mentioned that the source of rivers and springs was rain or ice only. But nobody believed this, and the theories proposed by the early thinkers were believed.

Leonardo da Vinci (1442–1519) and Bernard Palissy (1509–1589) correctly understood the hydrological process, and proposed the infiltration theory. Vitruvius, Vinci and Palissy are, therefore, considered as the pioneers in advocating correct hydrological principles.

Quantitative measurements of the hydrological processes were started by Pierr Perrault (1608–1680), Edme Mariotte (1620–1684) and Edmund Halley, the English astronomer (1656–1742). Perrault observed the discharge of river Seine near Burguendy for three years, and noticed that it was only one-sixth of the precipitation. Mariotte made observations on the same river near Paris and accepted Perrault's observations. Halley conducted experiments on salt-water evaporation and stated that the evaporation from the Mediterranean Sea was adequate to supply all waters to rivers discharging into the sea. The study by these three scientists put an end to the subterranean flow theory.

These three scientists are, therefore, considered as founders of hydrology. Measurement of water is the basic requirement in hydrology. Pitot (1732) proposed the flow velocity measurement tube, while Chezy (1775) proposed the flow formula. Dalton's evaporation measurements (1801) and infiltration measurements (1802) were important for the development of this science.

Until the 19th century, slow progress was made in the development of hydrology. The important contributions in this century were as follows:

- Francis formula (1856)
- Venturimeter by C. Herschel (1886)
- Currentmeter by Ellis
- Ganguillet and Kutter's formula to determine Chezy's coefficient (1869)
- Contribution by Darcy, Dupuit, Gabriel, Daubree, Abbe Parramelle, Prinz Adolph Thiem, Gunther, Theiss Forcheimer, Slitcher and Hazen in groundwater flow
- The first book *Manual of Hydrology* was published in 1850 by a civil engineer Nathaneil Bardmore.

Mead published *Notes on Hydrology* in 1904 and modified it in 1909 under the title *Hydrology*. Meyer published *Elements of Hydrology* in 1927. These books are important contributions for the development of this science.

In 1919, in Brussels and then in 1922, in Rome, the International Union of Geology and Geophysics was held. This led to the development of scientific hydrology. In 1932, L. K. Sherman proposed the unit hydrograph theory, and laid an important milestone in the development of the science and all further developments and research in this science, namely, *instantaneous unit hydrograph* leading to conceptual models, and so on, and are based on the unit hydrograph theory proposed by Sherman.

In order to focus the attention of all countries on the importance of the hydrology, the International Hydrological Decade, as proposed by UNESCO, started from 1965, and its activities are continued beyond 1975 till today.

1.3.1 Phases of Development of Hydrology

Professor Ven Te Chow divided the development of hydrology in eight phases. The division is not exact and the phases do overlap. The phases are as follows:

1. **Period of speculation (up to 1400 AD)** The concept of hydrological cycle was speculated by many scientists and philosophers. However, all these ideas were incorrect.
2. **Period of observations (1400–1600)** In this period, scientists started taking observations and deriving theories based on these observations rather than speculating ideas.
3. **Period of measurement (1600–1700)** Scientists started measurements based on these observations and were able to make some conclusions.
4. **Period of experimentation (1700–1800)** Some understandings and discoveries were established in this period, based on the experiments done.

5. Period of modernization (1800–1900) The experiments were modernized, and the foundation of modern science of hydrology was laid.
6. Period of empiricism (1900–1930) In this period, empiricism in hydrology became more evident. A number of empirical formulae were suggested.
7. Period of rationalization (1930–1950) In this period, rational analysis of hydrological problems was used instead of empiricism.
8. Period of theorization (1950–till date) In this period, theoretical approaches have been used extensively in hydrological problems. Many hydrological principles have been subjected to mathematical analysis. Use of sophisticated instruments and computer techniques has been developed.

These phases indicate that hydrology is a young science with many important problems only imperfectly understood, and research in this field is still going on.

1.3.2 Hydrology in Ancient India

During the Vedic period, it was believed that water is not lost during the various processes of hydrological cycle, namely, evaporation, condensation, rainfall, stream flow, and so on, but gets converted from one form to another. Water intake by plants, division of its particles into minute particles by the Sun's rays and wind, different types of clouds, their heights, their rainfall capacity, and so on, along with the prediction of rainfall quantity in advance, by observations of natural phenomena, is illustrated in the Puranas, the *Varah Sanhita, the Meghmala* and other literature.

The references of rain gauges are available in Kautilya's *Arthashastra* and Panini's *Astadhyayi*. The distribution of rainfall in various parts of India was also known to Kautilya. The cyclonic and orographic effects on rainfall were also known. Various other phenomena such as infiltration, interception, stream-flow, geomorphology and erosive action of water were well understood.

Groundwater development and quality consideration were getting sufficient attention, as evidenced by Varah Sanhita. Water management and conservation, well-managed water-pricing system, construction methods and materials for dams, bank protection, spillways, and so on, mentioned in the ancient literature, reflect the high-stage development of *water resources and hydrology in ancient India.*

1.4 PROPERTIES OF WATER

Water consists of two hydrogen and one oxygen molecules strongly bonded together. It is an inorganic liquid, and the only substance occurring naturally as solid, liquid and gas. Water has the highest surface tension compared to other liquids, except mercury. Some chemical properties of water are mentioned below:

- Water is the universal solvent and many compounds and salts dissolve in it.
- Pure water boils at 100 °C and freezes at 0 °C.
- Water by itself is a bad conductor of electricity; however, dissolved ions like Ca, Mg, Na, Cl, and so on, have the ability to conduct electricity.
- Water has the greatest thermal conductivity except mercury.
- The physical properties of water are as follows:

$$\text{Mass density} = \rho = 1000 \text{ kg/m}^3, \quad \text{weight density} = \gamma = 9.81 \text{ kN/m}^3.$$

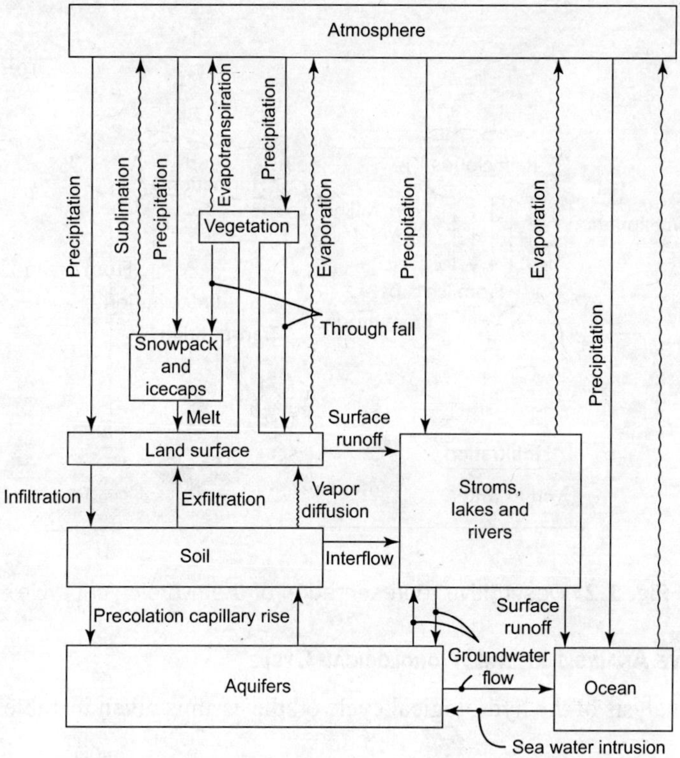

Fig. 1.1 Engineering representation of the hydrological cycle

1.5 HYDROLOGICAL CYCLE

The cyclic movement of water from sea and back to the sea is known as *hydrologic cycle, hydrological cycle or water cycle*. This cyclic order of events may be explained as follows:

The water from the sea evaporates due to solar radiation. It becomes lighter than air and hence moves up in the atmosphere. Here, under favourable conditions, clouds are formed. These clouds move upwards and then over the land. Precipitation from the clouds occurs over the land area. Water, thus, precipitated gets collected in streams and rivers and then flows back to the sea. All these events are repeated continuously. Thus, the hydrological cycle is a case of circular infinity.

The hydrologic cycle has no beginning. Nobody knows for sure when it started and from which stage it started. But it is surmised that it started some 3500 million years ago. It is the most distinctive and important process on this planet. It is driven by the energy of the Sun, influenced by the gravity of the earth and the physical, chemical and biological properties of water.

The cyclic order of events does occur in nature, but there are some short cuts also at different stages. Figure 1.1 shows the engineering representation of the hydrological cycle.

Figure 1.2 shows the descriptive representation of the hydrological cycle.

The circulation of water penetrates the three phases of the earth system:

1. Hydrosphere: Bodies of water that cover the surface of the earth
2. Atmosphere: The gaseous envelop above the hydrosphere
3. Lithosphere: Rocks below the hydrosphere

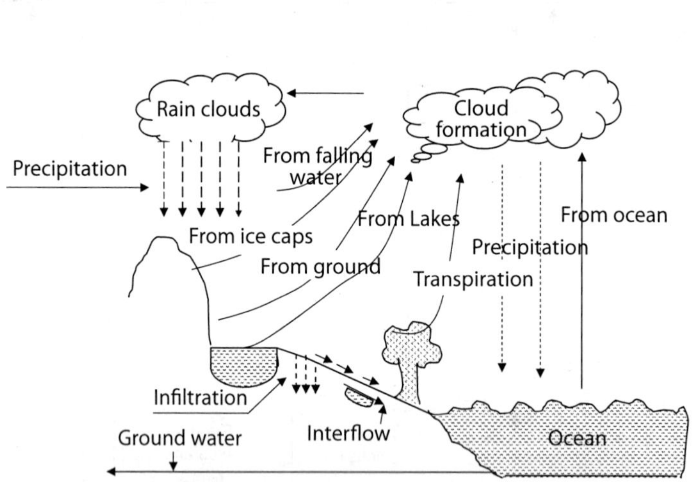

Fig. 1.2 Descriptive representation of the hydrological cycle

1.5.1 Quantitative Analysis of the Hydrological Cycle

The quantitative analysis of the hydrological cycle on the earth is given in Table 1.2.

Table 1.2 The quantitative analysis of the hydrological cycle on the earth

Serial no.	Item	Symbol	Area (km^2)	Total quantity (km^3)	Annual average per unit area (mm)
A. Precipitation					
1	Precipitation on ocean	P_O	360×10^6	346,000	961
2	Precipitation on land	P_L	150×10^6	99,000	660
Total precipitation		P	510×10^6	445,000	872
B. Evaporation					
3	Evaporation from oceans	E_O	360×10^6	383,000	1063
4	Evaporation from land	E_L	150×10^6	62,000	413
Total evaporation		E	510×10^6	445,000	872
C. Runoff		R	150×10^6	37,000	247

$$P = P_O + P_L$$
$$E = E_O + E_L$$
$$R = P_L - E_L$$

The average global precipitation is 872 mm, which is equivalent to 445,000 km^3 of water. The average atmospheric moisture is 14,000 km^3. This means that the atmospheric moisture is replaced 32 times in a year, or the residence time of atmosphere moisture is 10 days. Figure 1.3 shows the disposition of global annual average precipitation.

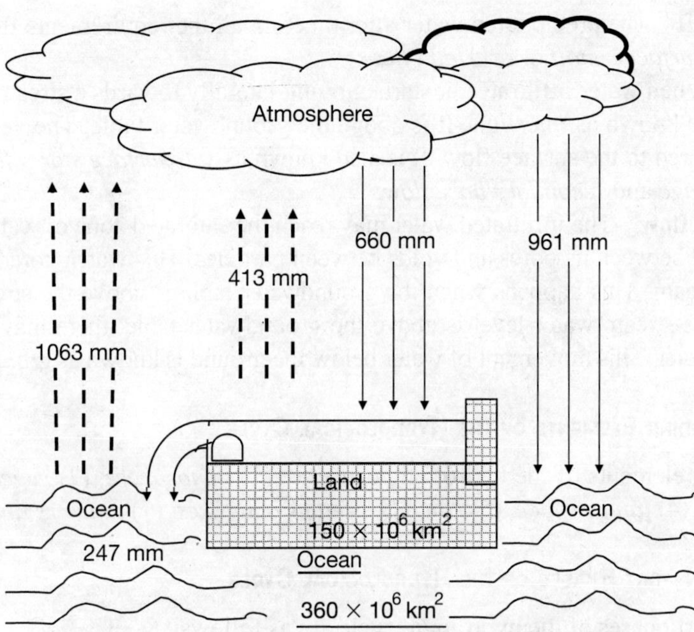

Fig. 1.3 Disposition of global annual average precipitation

1.5.2 Processes in Hydrological Cycle

The various processes in a hydrological cycle are as follows:

1. Evaporation It is a process by which water from water bodies or land mass passes into vapour state and is diffused into the atmosphere. Evaporation is observed from oceans, lakes, streams; land; vegetation; glaciers and ice and during rainfall.

2. Transpiration It is the process by which water passes from the liquid state to the vapour state through plant metabolism. Evaporation and transpiration are sometimes combined, as often their separate assessment is difficult and are termed as *evapotranspiration*.

3. Precipitation It is the general term for all the moisture emanating from the clouds and falling on the ground. Precipitation can occur in many different forms, including rain, hail, mist, snow, ice and so on.

Through fall is that part of precipitation, intercepted by vegetation, which then falls on the ground.

4. Interception It is that part of precipitation, which is received and retained by vegetation and evaporates later.

5. Infiltration It is that part of precipitation, which enters into the ground and then flows downwards.

6. Vapour diffusion It is that part of water retained by the soil, which flows in the form of vapour towards the ground surface.

7. Surface detention When a river flows, a significant volume of water is contained in the river channel. This temporary storage in the river channel is called *surface detention*.

8. Depression storage When there is precipitation over a catchment, and before the river starts flowing, part of water is stored in ditches, small ponds, and so on, depending upon the topography of the catchment. This quantum of water thus stored is known as *depression storage* or *surface retention*.

9. Surface runoff The precipitated water after meeting all the requirements flows in a stream. This is known as *surface runoff* or *overland flow*.

10. Interflow When water infiltrates, it starts moving laterally towards a stream and appears on the surface. This is known as *interflow*. It is above the groundwater table. The velocity of flow is very low as compared to the surface flow. It is also known as *sub-surface storm flow*, *sub-surface runoff*, *storm seepage* and *secondary base flow*.

11. Groundwater flow The infiltrated water may reach the saturated zone of water below ground and may get stored between its pores and voids between particles. This water stored eventually may flow towards a stream. This happens when the groundwater table is above the stream water level. However, when the stream water level is above the groundwater table, there may be a flow from stream to groundwater. This movement of water below the ground is known as *groundwater flow*.

1.5.3 Water-Holding Elements of the Hydrological Cycle

The water-holding elements of the hydrologic cycle are (1) *atmosphere*, (2) *vegetation*, (3) *snow caps and ice caps*, (4) *land surface*, (5) *soil*, (6) *streams and lakes*, (7) *aquifers and* (8) *oceans*.

1.5.4 Liquid-Transport Phases of the Hydrological Cycle

The liquid-transport phases of the hydrologic cycle are as follows:

1. Precipitation on the land surface and vegetation
2. Through fall from vegetation to land surface
3. Melting of snow and ice onto land surface
4. Surface runoff
5. Infiltration from land surface to subsoil
6. Ex-filtration from subsoil to land surface
7. Interflow from subsoil to streams
8. Percolation from subsoil to aquifers
9. Capillary rise from aquifers to soil
10. Groundwater flow from aquifer to streams and from streams to aquifer

1.5.5 Vapour-Transport Phases of the Hydrological Cycle

The vapour-transport phases of the hydrologic cycle are as follows:

1. Evaporation from land, stream, lakes, oceans and atmosphere
2. Evapotranspiration from vegetation to atmosphere.
3. Sublimation from snow-packs and ice-caps to atmosphere
4. Vapour diffusion from subsoil to land surface

1.6 HYDROLOGICAL PROCESS

Any phenomenon that undergoes continuous changes is generally termed as *process*. All hydrological phenomena change with time and thus they are termed as *hydrological processes*. These hydrological processes are complex and involve many variables that are interdependent and can be classified as follows:

1. Deterministic
2. Stochastic or probabilistic

1.6.1 Deterministic Process

The study in a deterministic process is based on exact law, and the chance of occurrence of variables is ignored. For example,

$$Q = A \times V$$

Discharge can be calculated with the help of this identity and there is no possibility of any chance.

1.6.2 Stochastic Process

In the case of stochastic or probabilistic study, the chance of occurrence of the variables is taken into consideration. For calculating the flood discharge, the chance or probability of the maximum precipitation is required to be taken into account.

1.6.3 The Hydrological Equation

All the hydrologic studies are mainly based on the continuity equation, which is known as *hydrologic equation*. This equation is as follows:

$$\text{Inflow} - \text{Outflow} = \text{Change in storage}$$

1.7 MODERN TECHNIQUES

In the study of hydrology, the following modern developments and techniques are used:

- Remote sensing
- Artificial neural network
- Geographical information system
- Genetic algorithm
- Fuzzy logic

1.7.1 Remote Sensing

Remote sensing (RS) is an art and the science of obtaining information about an object, area or phenomenon through the analysis of the data acquired by a device that is not in contact with an object, area or phenomenon under consideration.

The electromagnetic energy sources are normally installed on satellites and these sensors acquire data from the earth. Various features of the earth surface exhibit typical spectral reflectance patterns. This spectral reflectance is measured by the sensor and the digital images are stored. The image processing provides the detailed information.

1.7.2 Artificial Neural Network

The idea of artificial neural network (ANN) was proposed by McCulloch and Pitts in 1943. Its development is inspired by a desire to understand the human brain and emulate its functioning. An ANN is a massively parallel-distributed information-processing system that resembles biological neural networks of human brain. The individual units in neural network perform computations in parallel. The ANN has an ability to extract the relation between the input and the output process, without the physics being explicitly provided to them. It has a potential to learn from experience.

A neural network is an interconnected assembly of simple processing elements, units, nodes or neurons whose functionality is loosely based on the biological neuron. The processing ability of the network is stored in the inner unit connection. Strengths or weights, obtained by a process of adaptation to or learning form, a set of learning form, a set of training patterns.

The major strength of neural network lies in the way knowledge is represented by it. Knowledge refers to stored information or models used by a person or machine to interpret, to predict and appropriately respond to the outside world. The functioning of the ANN consists of the following basic steps:

1. Information is passed between nodes through connection links.
2. Signals are passed between nodes through connection links.
3. Each connection link has an associated weight that represents its connection and strength.
4. Each node typically applies a non-linear transformation called activation function to its net to determine its output signal.

1.7.3 Geographical Information System

Geographical information system (GIS) is a very effective digital tool that handles and analyses the spatially referred data. The digitalized data forms a framework for regional map coverage for easier comparison with other categories of data, which is amenable for computer aided analysis.

GIS originated in the middle of 1960. It is a powerful set of tools for collecting, storing and retrieving at will, transforming and displaying spatial data from the real world for a particular set of purposes. It acts as a decision-support system, involving the integration of spatially referenced data in problem solving. GIS consists of four basic operations:

1. Data gathering and input processing
2. Geographic data bases
3. Data analysis and modelling
4. Output presentation

The data is presented in a number of layers laid one over the other in GIS while presenting the output. The applications of GIS have increased many folds in various fields. It is used widely as a standard tool in agriculture, environmental management, forestry, hazard monitoring, watershed management, land analysis and hydrology.

1.7.4 Genetic Algorithm

Genetic algorithm (GA) is a stochastic global-search method based on the mechanics of natural selection and neutral genetics. It works on the principles of the *survival of the fittest* and inheritance of the characteristics of the parent populations. The problems based on non-convex functions can be easily solved by GA, where other conventional optimization methods fail to do so. GA is often viewed as function optimizer. It is a process of a number of solutions simultaneously.

The GA approach consists of the following features:

- All parameters are encoded in binary digits in the form of a string.
- When operated on such population of string, the GA operations act randomly to explore search space from different points. This global search gives an optimum result.

- No gradient information is required that avoids the mathematical complexity of conventional non-liner optimization methods.
- GA reduces the computational time during evolution process due to the possibility of parallel computation.

Following are the key steps in GA:

1. Generating an initial population
2. Describing a coding scheme for all the variables
3. Running the flow simulation finite element model for all the sets of variables
4. Computing the fitness function from objective functions
5. Performing the operations with genetic operators
6. Termination condition

1.7.5 Fuzzy Logic

Fuzzy logic (FL) was first developed for solving imprecise or vague problems in the field of artificial intelligence, especially for imprecise reasoning and modelling linguistic terms. When the information is incomplete and vague, the precise mathematics will not be sufficient to model a complex system. Such problems can be solved by FL.

A fuzzy can be defined mathematically by assigning each possible individual its grade of membership in the fuzzy set. This grade corresponds to the degree to which that individual is similar with the concept represented by the fuzzy set. Such individuals may belong to the fuzzy set to a greater or lesser degree as indicated by a large or smaller membership grades. These membership grades are very often expressed by real numbers values between 0 and 1. Fuzzy system can serve different purposes such as modelling, data analysis, prediction or control.

REVIEW QUESTIONS

1. Define hydrology. Also discuss its importance in modern times.
2. Is hydrology a pure science? Discuss.
3. What are the sub-branches of hydrology?
4. State and discuss the supplementary sciences for the study of hydrology.
5. What are the overlapping areas of hydrology?
6. Explain the practical use of hydrology in other fields.
7. State and explain the influence of hydrology in other sciences.
8. Write a note on water budget of the world.
9. Write a detailed note on the history of hydrology.
10. Write a note on hydrology in ancient India.
11. Explain the different phases of development of hydrology as suggested by Prof. Ven Te Chow.
12. Explain with a neat sketch hydrological cycle.
13. State and explain the different processes in a hydrological cycle.
14. Analyse the hydrological cycle quantitatively.
15. What are the water-holding elements in a hydrological cycle?
16. What are the liquid-transport phases of a hydrological cycle?
17. What are the vapour-transport phases of a hydrological cycle?

18. Write short notes on
    a. Definition of hydrology
    b. Total quantity of water on the earth
    c. Hydrological equation
    d. Hydrological processes
    e. Hydrology in ancient India
    f. Interflow
    g. Water budget of the world
    h. Remote sensing
    i. Fuzzy logic
    j. Artificial neural network
    k. Geographical-information system
    l. Genetic algorithm
19. Differentiate between
    a. Surface detention and surface retention
    b. Deterministic process and stochastic process
    c. Synthetic hydrology and applied hydrology
    d. Geohydrology and hydrogeology
    e. Biohydrology and hydrobiology
    f. Overland flow and interflow
    g. Hydraulics and hydrology

MULTIPLE CHOICE QUESTIONS

1. The science of surface streams is
   (a) Potamology
   (b) Limnology
   (c) Cryology
   (d) Morphology
2. The science of measurement of water is
   (a) Hydrography
   (b) Hydrometry
   (c) Oceanography
   (d) Hydraulics
3. The total quantity of water on the earth is
   (a) 1348×10^6 km^3
   (b) 2348×10^6 km^3
   (c) 3348×10^6 km^3
   (d) 4348×10^6 km^3
4. Assuming earth to be a uniform sphere, the total quantity of water may cover to an average depth of
   (a) 1.73 km
   (b) 2.73 km
   (c) 3.73 km
   (d) 4.73 km
5. The per cent of salt water on the earth is
   (a) 67
   (b) 77
   (c) 87
   (d) 97
6. The total area covered by oceans on the earth is
   (a) 260×10^6 km^2
   (b) 360×10^6 km^2
   (c) 460×10^6 km^2
   (d) 560×10^6 km^2
7. The total area covered by land on the earth is
   (a) 150×10^6 km^2
   (b) 250×10^6 km^2
   (c) 350×10^6 km^2
   (d) 450×10^6 km^2

8. The average annual precipitation on the earth is
   (a) 672 mm (b) 772 mm
   (c) 872 mm (d) 972 mm

9. The average annual runoff from the land to the sea is
   (a) 147 mm (b) 247 mm
   (c) 347 mm (d) 447 mm

10. If the earth is assumed to be a uniform sphere, its diameter is
    (a) 10,800 km (b) 11,800 km
    (c) 12,800 km (d) 13,800 km

11. The phases of development of hydrology were suggested by
    (a) L. K. Sherman (b) Prof. Ven Te Chow
    (c) Leonardo da Vinci (d) Dalton

12. Science of occurrence, distribution and circulation of water is known as
    (a) Hydrography (b) Hydrometry
    (c) Hydraulics (d) None of the above

13. The full form of GA is
    (a) General algorithm (b) Genetic algorithm
    (c) Geologic algorithm (d) Geographic algorithm

14. The full form of RS is
    (a) Regular sensing (b) Remote surveying
    (c) Regular surveying (d) Remote sensing

15. The full form of FL is
    (a) Fuzzy logic (b) Finite logic
    (c) Fuzzy location (d) Finite location

16. The full form of ANN is
    (a) Artificial normal network (b) Automatic neural network
    (c) Artificial neural network (d) Automatic normal network

17. The full form of GIS is
    (a) Geographical information system (b) Graphical information system
    (c) Geographical information source (d) Graphical information source

ANSWERS TO MULTIPLE CHOICE QUESTIONS

1. a 2. b 3. a 4. b 5. d 6. b 7. a 8. c
9. b 10. c 11. b 12. d 13. b 14. d 15. a 16. c
17. a

*** ——— ***

2 Hydrometeorology

Chapter Outline

2.1 DEFINITION

The word *atmosphere* is derived from two Greek words, *atmos* and *spheria.* Atmos means *vapour* and spheria means *sphere* or spherical earth as its extended meaning. Atmosphere, therefore, means the gaseous envelope surrounding the earth.

Hydrometeorology is the study of that atmospheric process, which affects the hydrologic characters of a region.

2.1.1 Atmosphere and Its Constituents

The permanent constituents of air, in gaseous form, are: (1) nitrogen—78%, (2) oxygen—21% and (3) inert gases 1%. The variable constituents may be in the solid, liquid or gaseous form.

Water is one of these variable constituents, which may comprise as much as 4% of the atmosphere and may exist in all the three forms. When water exists in the form of gas, it is invisible and is termed as *humidity*. Besides water, ozone (O_3) and carbon dioxide (CO_2), exist in the atmosphere in varying proportions.

2.1.2 Existence of the Atmosphere

The atmosphere has got no upper limit. It is densest near the earth surface and gradually decreases in density and practically does not exist beyond 600 km from the earth surface. About 90% of the atmosphere is within 20 km from the earth surface.

2.1.3 Density and Pressure of the Atmosphere

At the equator, the normal atmosphere density is 1.02 kg/m^3 and pressure is 1.02×10^5 N/m^2 equivalent to 10.33 m of water.

2.1.4 Vertical Structure of the Atmosphere

The vertical structure of the atmosphere is divided into five layers.

Figure 2.1 shows the vertical structure of the atmosphere.

1. Troposphere: It roughly extends up to a height of about 10 km, 8 km over the poles and 16 km near the equator. The entire phenomenon related to weather is confined to this layer. Around 75% of mass of the moisture, dust, etc. lies in this layer. Its temperature reduces at a rate approximately of 6.5 °C/km. The top of troposphere is called *tropopause*.

2. Stratosphere: This layer lies above the troposphere and is considered to be of a height ranging from 16 to 30 km. In this layer, major portion is ozone and the temperature is more or less constant. The top of the stratosphere is called *stratopause*.

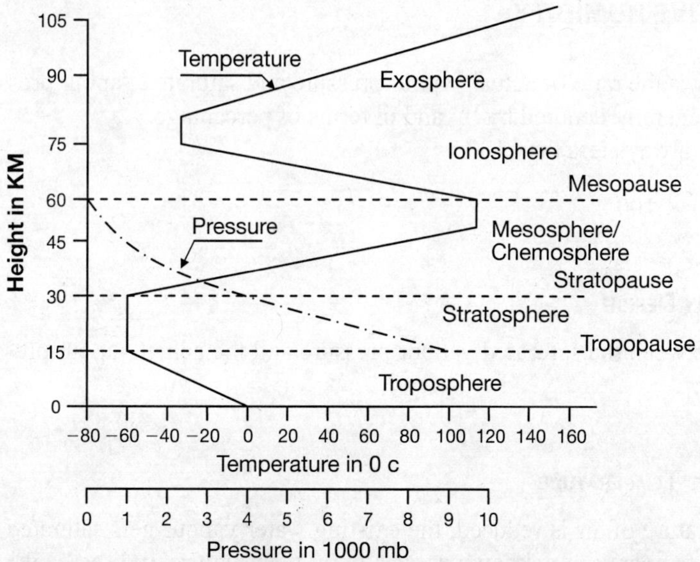

Fig. 2.1 Vertical structure of atmosphere

3. Mesosphere: This is a warm layer and is above the stratosphere and the thickness is roughly 30 km. It is also known as *chemosphere*. The region at the top of the mesosphere is called *mesopause*.

4. Ionosphere: This layer is above the mesosphere; the pressure in this layer is very low. It consists of a number of layers with different electrical characteristics.

5. Exosphere: The ionosphere gradually merges into the outermost shell called *exosphere*.

2.2 VAPOUR PRESSURE

Water exists in the form of vapour in the atmosphere. Vapour is lighter than air.

The partial pressure exerted by this vapour is known as *vapour pressure*. The more the vapour content, the more will be the vapour pressure. The vapour pressure is normally denoted by 'e'. For any temperature, there is a limit for the water vapour that can be held in the atmosphere. For a given temperature, when the air holds the vapour to its maximum capacity, the vapour pressure is said to be *saturated vapour pressure* at that temperature and is generally denoted by 'e_s'.

Saturated vapour pressure increases with temperature.

Air with vapour is lighter than dry air.

Vapour pressure is normally expressed in millibars (mb)

$$1 \text{ bar} = 10^5 \text{ N/m}^2$$

$$1 \text{ mb} = 10^2 \text{ N/m}^2$$

$$\text{and} \quad 1.33 \text{ mb} = 1 \text{ mm of mercury.}$$

2.3 RELATIVE HUMIDITY

Relative humidity is the ratio of actual vapour pressure and saturated vapour pressure for that temperature, and is generally denoted by 'h' and in terms of percentage.

Naturally, it is always less than 100.

Thus, $h = (e/e_s) \times 100$

2.3.1 Saturation Deficit

The difference between the saturated vapour pressure and the actual vapour pressure is known as *saturation deficit*.

2.3.2 Dew-Point Temperature

When the temperature of air is reduced, the existing water vapour gets saturated at a certain temperature and this temperature is known as *dew-point temperature*. Indirectly, it is the temperature at which the saturated vapour pressure equals the existing vapour pressure.

2.3.3 Wet-Bulb Temperature

If evaporation is allowed in the air without controlling the temperature, then the temperature of air goes on reducing, reaching a saturation point. This temperature, when the air gets saturated, is known as *wet-bulb temperature*.

2.3.4 Measurement of Humidity

The instrument used to measure humidity is known as *psychrometer*. It consists of two glass thermometers: (a) a dry bulb thermometer and (b) a wet-bulb thermometer. The dry bulb thermometer records the ambient air temperature normally denoted as 'T'. The wet-bulb thermometer has its mercury bulb wrapped in a wick. The other end of the wick is submerged in a container having distilled water. With this arrangement, the mercury bulb is supplied with moisture continuously due to surface tension. The water in the wick, wrapping the mercury bulb, evaporates reducing the temperature recorded by this thermometer. For this reason, this thermometer is known as wet-bulb thermometer and the temperature recorded as wet-bulb temperature normally denoted by 'T_w'.

From the observed atmospheric pressure 'P' and from these two temperatures, the relative humidity can be calculated. For this purpose psychometric tables are used.

2.3.5 Hygrograph

It is noticed that human or animal hair reacts to the change in the humidity. It expands as humidity increases and contracts with a decrease in humidity. This variation is very uniform. This property is used to measure humidity and is automatically recorded on a graph. This instrument is known as *hygrograph*. Sometimes hygrograph is also equipped with a thermograph and is called a *thermohygrograph*.

2.4 WIND

Air in motion is called wind. The horizontal component parallel to the earth surface is called *wind* and the vertical component is referred to as *air current*.

2.4.1 Wind Velocity

The wind velocity is measured by an instrument known as *anemometer*.

A cup-type anemometer is the standard instrument. It consists of three or four hemispherical cups mounted on a sleeve that rotates freely about the vertical axis. Because of the wind velocity, the cups rotate along the vertical axis and the number of rotations is recorded by a counter. The more the velocity, the more will be the rotations. The rotations are recorded automatically on a chart so that the wind velocity during the 24 hours is recorded.

Figure 2.2 shows an anemometer. In recent years, electrical/electronic anemometers are available.

2.4.2 Direction of Wind

The direction of the wind is measured and recorded by *wind vanes*. The wind vanes are normally installed along with the anemometer.

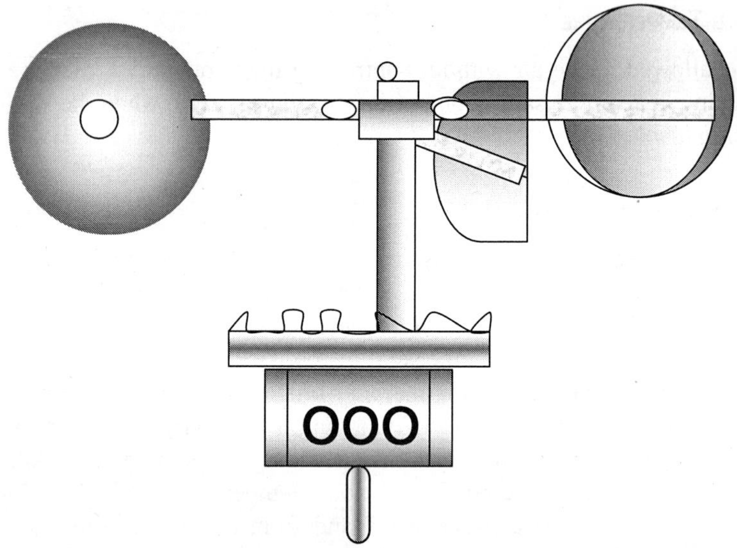

Fig. 2.2 Anemometer with a counter

2.4.3 Level of Anemometer

No standard has been specified for the height of an anemometer. The wind speed varies with the height above the ground and it follows the logarithmic profile or power law profile. It can be expressed as

$$V/V_0 = (Z/Z_0)^{0.15}$$

where, V = Speed of wind required at a height Z

V_0 = Speed of wind observed at a height Z_0

2.4.4 Effect of Wind

Wind transports vapour and heat and hence is responsible for precipitation, evaporation and temperature changes at a location.

2.5 TEMPERATURE

The temperature indicates the heat energy in soil, water and air mass. It affects the hydrologic processes like evaporation, precipitation, infiltration, etc.

2.5.1 Factors Affecting Temperature

Temperature depends on the following factors:

(1) altitude, (2) latitude, (3) nature of surface, (4) condition of sky, (5) wind speed (6) time after sunrise, (7) vegetation and (8) orographic influence.

2.5.2 Measurement of Temperature

Temperature is normally measured by the length of mercury column in a glass bulb-type thermometer kept in a wooden shelter known as *Stevenson screen*. Standards have been laid down for the installation of a Stevenson screen.

An automatic recording instrument known as *thermograph* obtains a continuous record of temperature with time.

2.5.3 Average Daily Temperature

The average daily temperature is calculated by either of the following two methods:

1. Average of maximum and minimum temperatures.
2. Average worked out from thermograph chart.

These two values differ slightly.

2.5.4 Normal Average Temperature

The normal average temperatures required in the study of hydrology are: (1) mean daily temperature, (2) mean monthly temperature and (3) mean annual temperature.

These average values are worked out from 30-years period and are revised every decade, that is, 10 years.

2.5.5 Lapse Rate

Lapse rate is the change in temperature with height above ground.

2.6 AIR MASS

Air mass is a vast and deep body of air. The temperature, humidity, etc. are homogeneous in the air mass. The air mass may move because of the pressure gradient. It can be categorized into the following two types:

- When the air mass moves over a warmer surface, it is called *cold air mass*.
- When the air mass moves over a cooler surface, it is called *warm air mass*.

2.6.1 Air Front

The surface of contact between two air masses or with the surrounding atmosphere is called *air front*. The air front is of the following three types:

- The air front with the earth surface is called *surface front*.
- When a moving cold air mass comes in contact with a warmer one, the warm air mass moves over the cold one and then it is called a *cold front*.
- Similarly, when a moving warm air mass comes in contact with a cold one, it is called *warm front*.

2.6.2 Cyclone

When a low atmosphere pressure develops in an area, the air in the adjoining area revolves moving spirally towards the low-pressure belt with a high velocity. The storm thus formed is known as *tropical cyclone*.

The cyclone may cover a circular area of 500 km^2 and its centre is called *eye of the storm.* A cyclone in the northern hemisphere revolves always anticlockwise, while in the southern hemisphere it revolves clockwise. The wind speed may be to the extent of 300 km/h and may pick up moisture from the land and ocean surface and may cause heavy precipitation.

2.6.3 Causes of Cyclone

The tropical cyclones normally occur at low latitudes over the seas because of the development of low pressure. The reason may be a combination of the following: (a) intense sunshine, (b) high humidity, (c) low surface friction, (d) deflation due to earth rotation and (e) lapse rate.

2.6.4 Extratropical Cyclone

When air masses of contrasting properties come in contact with each other, the storm formed is known as *extratropical cyclone.* It covers a very large area to the extent of 2000 km^2 and is mostly associated with heavy precipitation.

2.6.5 Anticyclone

When high pressure is developed at a location, the air mass moves away spirally outwards and then it is called *anticyclone* The spiral movement is clockwise in the northern hemisphere and clockwise in the southern hemisphere.

REVIEW QUESTIONS

1. Discuss the permanent and variable constituents of the atmosphere.
2. Describe the vertical structure of the atmosphere.
3. Discuss the factors that affect temperature at a location.
4. Explain the working of an anemometer.
5. What is humidity? How is it measured?
6. Discuss the effect of wind on various hydrological processes.
7. Explain the effect of temperature on various hydrological processes.
8. Write short notes on
   a. Vapour pressure
   b. Saturation deficit
   c. Relative humidity
   d. Measurement of temperature
   e. Eye of a cyclone
   f. Lapse rate
   g. Variation of wind with the height above ground
   h. Dew-point temperature
   i. Wet-bulb temperature
   j. Measurement of humidity
9. Differentiate between
   a. Warm front and cold front
   b. Average temperature and normal temperature
   c. Cold air mass and warm air mass
   d. Cyclone and anticyclone
   e. Tropical cyclone and extratropical cyclone
   f. Dry bulb thermometer and wet-bulb thermometer
   g. Wind and air current
   h. Isobar and isotherm
   i. Hygrograph and thermograph

MULTIPLE CHOICE QUESTIONS

1. Hydrometeorology is the science dealing with
   (a) Water in the atmosphere (b) Water below the earth surface
   (c) Water in the oceans (d) Water in surface streams

2. A pressure of 1 mb is equal to
   (a) 100 N/m^2 (b) 10 N/m^2
   (c) 1 N/m^2 (d) 0.1 N/m^2

3. The atmosphere has no existence from the earth surface beyond
   (a) 6000 km (b) 600 km
   (c) 60 km (d) 6 km

4. At the equator, the normal atmospheric density is
   (a) 100.02 kg/m^3 (b) 10.02 kg/m^3
   (c) 1.02 kg/m^3 (d) 0.102 kg/m^3

5. At the equator, the atmospheric pressure is
   (a) 100.33 m of water (b) 10.33 m of water
   (c) 1.03 m of water (d) 0.103 m of water

6. The vertical structure of the atmosphere is divided into
   (a) Six layers (b) Five layers
   (c) Four layers (d) Three layers

7. Most of the weather phenomenon takes place in
   (a) Troposphere (b) Stratosphere
   (c) Mesosphere (d) Ionosphere

8. The permanent constituents of air in gaseous form are
   (a) 1. Nitrogen 50% 2. Oxygen 30% 3. Inert gases 20%
   (b) 1. Nitrogen 70% 2. Oxygen 10% 3. Inert gases 20%
   (c) 1. Nitrogen 60% 2. Oxygen 20% 3. Inert gases 20%
   (d) 1. Nitrogen 78% 2. Oxygen 21% 3. Inert gases 1%

9. Water exists in the atmosphere in the form of
   (a) Solid (b) Liquid
   (c) Gas (d) All of the three

10. The temperature in the atmosphere is fairly constant in
    (a) Troposphere (b) Stratosphere
    (c) Mesosphere (d) Ionosphere

11. Isobar is the line joining the points of equal
    (a) Temperature (b) Humidity
    (c) Atmospheric pressure (d) Wind velocity

12. Wind velocity is measured by
    (a) Current meter (b) Atmometer
    (c) Anemometer (d) Aerometer

13. Atmospheric humidity is measured and recorded by
    (a) Barograph (b) Hygrograph
    (c) Thermograph (d) Pantograph

14. Atmospheric temperature is measured and recorded by
    (a) Pantograph (b) Thermograph
    (c) Barograph (d) Hygrograph

15. Thermohygrograph records
    (a) Temperature and humidity (b) Temperature and wind velocity
    (c) Temperature and pressure (d) Humidity and pressure

16. In the cyclone in the southern hemisphere, wind flows
    (a) Clockwise inward (b) Anticlockwise inward
    (c) Clockwise outward (d) Anticlockwise outward

17. In the anticyclone in the southern hemisphere wind flows
    (a) Clockwise inward (b) Anticlockwise inward
    (c) Clockwise outward (d) Anticlockwise outward

18. Wind speed varies with the height above ground and the profile is
    (a) Elliptical (b) Logarithmic
    (c) Straight line (d) Circular

19. Wind transports vapour and heat and is responsible for
    (a) Precipitation (b) Evaporation
    (c) Temperature changes (d) All the three

20. The p eriod after which the normal average temperature values are revised is
    (a) 10 years (b) 20 years
    (c) 30 years (d) 40 years

ANSWERS TO MULTIPLE CHOICE QUESTIONS

1. a	2. a	3. b	4. c	5. b	6. b	7. a	8. d
9. d	10. b	11. c	12. c	13. b	14. b	15. a	16. a
17. c	18. b	19. d	20. c				

*** ——— ***

3

Evaporation and Transpiration

Chapter Outline

3.1 DEFINITION

Evaporation is an important process in the hydrologic cycle preceding precipitation. It is the process by which water in the liquid form transforms into vapour through the transfer of energy. The exact laws governing this phenomenon are not completely known.

The transformation from solid to the vapour state without passing through the usual intermediate liquid state is known as *sublimation*. In the atmosphere, evaporation occurs from free water surfaces like (1) seas, (2) lakes and rivers, (3) moisture in solid surfaces of (4) land and soil, (5) vegetation, (6) snowfields, (7) glaciers and (8) even from the falling rain drops.

3.1.1 Process of Evaporation

Any water body, regardless of its size, is made up of a number of molecules. Each one of them is constantly in motion at varying velocities and in different directions. This velocity of movement is dependent on temperature. So also, in a given mass, every molecule is attracted by another by a force inversely proportional to the square of the distance between them and directly proportional to the product of their masses. The molecules underneath attract the molecules which are near the top surface. Thus, some molecules near the surface, which have more kinetic energy than the attracting force—quite enough to overcome the attraction of the molecules in the lower part of the water body, escape into the atmosphere. This process is known as *evaporation*.

The molecules that escape into the atmosphere carry with them some heat energy. Thus evaporation is a cooling process. Some molecules are continuously moving into the atmosphere and some are returning from the atmosphere to the water surface. There may be a collision of molecules moving from and to the water surface. Thus, the rate of evaporation is the net rate of movement of molecule to and from the water surface.

When the rate of molecules coming from the atmosphere to the water surface is more than the rate of molecules going out from the water surface, then the process is known as *condensation*.

Evaporation is expressed in terms of the depth of water per unit area per unit time (e.g. mm/m^2/h).

3.1.2 Dalton's Law of Evaporation

John Dalton was the first scientist to describe this process, scientifically.

Immediately adjacent to the water surface, there is a thin layer of air that is saturated with vapour and its temperature is the same as that of water.

Evaporation depends upon the difference of vapour pressure in the atmosphere and the saturated vapour pressure of this air film.

Mathematically

$$E = C \times \frac{de}{dz}$$

where, E = Evaporation rate

C = A constant

$\frac{de}{dz}$ = Vapour pressure gradient in the vertical direction.

If the air above the film has vapour pressure same as that of the film, i.e. when $de = 0$, then there will be no evaporation.

However, if the air above the film has vapour pressure less than that of the film, i.e. $de \neq 0$, then there will be evaporation.

Dalton suggested the formula

$$E = C\,(e_s - e_a)$$

where, E = Evaporation rate

C = A constant

e_s = Saturated vapour pressure at the water surface temperature

e_a = Vapour pressure in the air above

Dalton's Law of evaporation is fundamental and all further work is based on this law.

3.2 NECESSITY OF ESTIMATION OF EVAPORATION

Large amounts are spent in constructing a dam and creating a reservoir to store water. Canals are also constructed to carry water. Some water stored in these reservoirs and flowing in canals is lost due to evaporation. The amount of the loss of water due to evaporation can be substantial (about 30%).

It is, therefore, necessary to estimate the loss due to evaporation to assess the water available for use in different seasons.

3.3 FACTORS AFFECTING EVAPORATION

The factors affecting evaporation are (1) temperature, (2) solar radiation, (3) wind, (4) barometric pressure and altitude, (5) dissolved solids, (6) turbidity, depth of water, (7) shape of surface, (8) extent, i.e., total area of water surface, (9) colour of water, (10) velocity of water, and (11) waves at water surface. These are discussed below in detail.

3.3.1 Temperature

Evaporation is directly affected by the temperature of water as well as that of the air above. The kinetic energy of the molecules of water increases with temperature. Similarly, the saturated vapour pressure of the air film near the water surface and the vapour pressure of the air above increases. Since evaporation depends upon the difference of these two vapour pressures, change in temperature may not affect evaporation in the same proportion and, as such, effect of temperature on evaporation is somewhat complicated.

Evaporation observed in a lake for one complete year indicates that for the same mean temperature observed in different months, the evaporation figures differ. However, evaporation increases with temperature, even though a close relation between evaporation and temperature cannot be established.

3.3.2 Solar Radiation

About 590 calories are required for converting one gram of water into water vapour. This heat energy is directly or indirectly derived from solar radiation. It is quite evident that evaporation rate changes during day, night and during different months in a year, since incidence of the Sun's radiation on the earth is different. Air and water temperatures depend upon solar radiation, and one can expect a close relation between the Sun's radiation and evaporation.

3.3.3 Wind

The air above the water surface receives the evaporated water molecules, and the rate of evaporation depends upon the moisture content in the air. If the water molecules in the air above are removed quickly, then the rate of evaporation increases. Removal of water molecules in the air above water surface is done by the wind and naturally depends on the velocity of wind. The more the velocity of wind, the faster will be the process of removal of water vapour particles and more will be the evaporation rate.

However, there is a limiting velocity of wind up to which the rate of evaporation increases with the increase in the wind velocity. Further increase has no effect on the evaporation rate. This higher limit of wind velocity depends on other factors also.

The velocity of wind referred above is naturally near the water surface, because the variation in the velocity of wind in the vertical direction is not a straight-line relation. Wind speed observed by an anemometer at a higher level will have to be modified considering the variation in the vertical direction.

So also, the velocity of wind will affect the evaporation rate if the air movement is turbulent. The water vapour will be removed quickly by turbulent diffusion. If the airflow is laminar, it will have no significant effect on the removal of water vapour and indirectly on the evaporation rate.

3.3.4 Barometric Pressure and Altitude

It is evident that the barometric pressure has an influence on the rate of evaporation. As the barometric pressure reduces, the rate of evaporation increases. However, the change in barometric pressure brings about changes in other variables that influence evaporation, and, as such, it is very difficult to establish an exact relation between the change in barometric pressure and evaporation rate.

Similarly, as altitude increases, the barometric pressure reduces and so the temperature. However, evaporation increases with the reduction of barometric pressure, but reduces with the reduction in temperature. The result will be the combined effect.

However, field observations indicate that evaporation goes on reducing with altitude.

3.3.5 Quality of Water

When water contains some dissolved solids, the vapour pressure reduces as compared to the pure water. Thus, the evaporation rate reduces with an increase in soluble solids. Evaporation is about 2–3% less from sea water as compared to fresh water. It is also noticed that the evaporation rate reduces by 1% with 1% increase in the specific gravity due to the dissolved solids.

Turbidity of water has no direct effect on the evaporation rate. The reflection of the solar radiation may increase and thus turbidity may indirectly affect the evaporation rate.

3.3.6 Depth of Water

The depth of water in the lake or reservoir has definitely an effect on the evaporation rate. The heat energy received by water from the Sun's radiation is utilized in three different ways:

1. Reflecting it back into atmosphere
2. Increasing the temperature of water
3. For evaporation

In the case of water of high depth during summer season, heat is utilized to increase the temperature of water at lower depths and thus evaporation reduces slightly. In cold season for higher depths, heat is available for evaporation from water and hence the evaporation rate increases. In the case of shallow waters, it is exactly the opposite way. The evaporation rate observed in two different lakes having different depths is shown in Fig. 3.1.

3.3.7 Other Factors

The other factors that have an effect on evaporation are as follows:

- Shape of the water surface
- Size, i.e. total area of the water surface

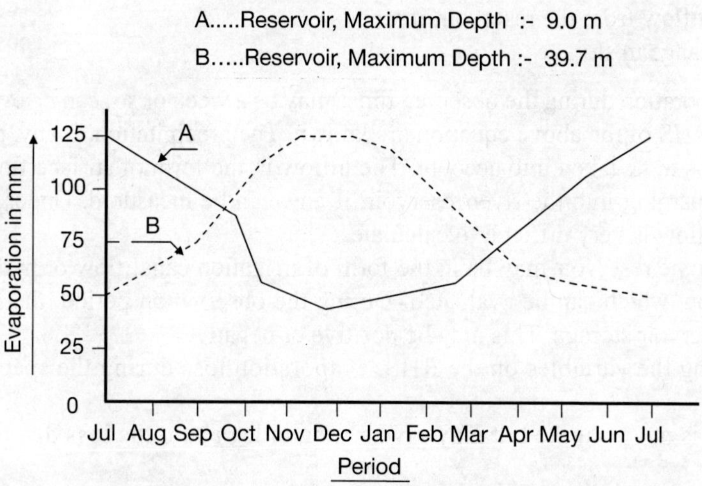

Fig. 3.1 Evaporation in lakes with different depths

- Colour of water surface
- Flow velocity of water
- Surface waves

However, efforts are still on to evaluate the effect of these factors on the rate of evaporation, though it is quite certain that these factors affect the evaporation rate. No exact relation has been concluded so far.

3.4 ESTIMATION OF EVAPORATION

The following methods are generally adopted to evaluate the rate of evaporation from a reservoir.

- Water budget method
- Energy budget method
- Mass transfer method
- Actual observations
- Evaporation formulae

3.4.1 Water Budget Method

This method is the application of continuity equation to the water content in a reservoir which is expressed as

$$E = P + I - (U_i - U_o) - O \pm dS$$

where, E = Evaporation
P = Precipitation
I = Surface inflow
U_i = Underground inflow into the reservoir
U_o = Underground outflow from the reservoir

O = Outflow from the reservoir, if any
dS = Change in storage

The total evaporation during the observed time (may be a week or so) can be evaluated provided all terms on the RHS of the above equation are known. The precipitation, if any, during the period of observation has to be taken into account. The inflow in the form of surface flow from the various streams discharging into the river/reservoir, if any, can be measured. Underground inflow or underground outflow is very difficult to calculate.

Outflow from the reservoir may be in the form of irrigation canal flow or hydroelectric power generation outflow which can be evaluated. During the observation period, there might be some change in the reservoir storage. This may be positive or negative.

Thus, knowing the variables on the RHS, evaporation loss during the specified period can be estimated.

This method is not accurate but may give a rough idea about the loss due to evaporation.

Example 3.1

The catchment area of an irrigation tank is 70 km^2. The constant water spread during October 2006 was 2 km^2. During that month, the uniform precipitation over the catchment was recorded to be 100 mm. 50% of the precipitation reaches the tank. The irrigation canal discharges at a uniform rate of 1.00 m^3/s in the month of October.

Assuming seepage losses to be 50% of the evaporation losses, find out the daily rate of evaporation for October 2006.

Solution:

$$\text{Total inflow} = 70 \times 10^6 \times \frac{100}{1000} \times 0.5 = 3.50 \times 10^6 \text{ m}^3$$

$$\text{Outflow from canal} = 1 \times 3600 \times 24 \times 31 = 2.68 \times 10^6 \text{ m}^3$$

$$\text{Loss of water} = 3.50 \times 10^6 - 2.68 \times 10^6 = 0.82 \times 10^6 \text{ m}^3 = \text{Seepage loss} + \text{evaporation loss}$$

Since seepage loss is 50% that of evaporation loss,

$$\text{Loss of water} = 1.5 \times \text{evaporation loss}$$

$$\text{Therefore, evaporation loss} = \frac{0.82 \times 10^6}{1.5} = 0.55 \times 10^6 \text{ m}^3$$

$$\text{Rate of evaporation} = \frac{0.55 \times 10^6 \times 1000}{2 \times 10^6 \times 31} = 8.8 \text{ mm/day/m}^2$$

Example 3.2

The surface area of a reservoir in m^2 is given by $A = 100\,y^2$, where y is the depth of water in metres in the reservoir. In one week, the water depth in the reservoir has reduced from 10 to 9 m. Find the average hourly rate of evaporation. Assume seepage loss to be 40% of the evaporation loss.

Solution:

Therefore, A_1 = Area at depth 10 m

$$= 100 \times 10^2 = 10{,}000 \text{ m}^2$$

$$A_2 = 100 \times 9^2 = 8100 \text{ m}^2$$

$$\text{Average surface area} = \frac{1}{3}\left[A_1 + A_2 + \sqrt{(A_1 \times A_2)}\,\right]$$

$$= \frac{1}{3}\left(10{,}000 + 8100 + \sqrt{10{,}000 \times 8100}\,\right)$$

$$= 9033.33 \text{ m}^2$$

$$\text{Loss of water} = 9033.33 \times 1$$

$$= 9033.33 \text{ m}^3$$

Since seepage loss = 0.4 × evaporation loss

Therefore, evaporation loss + seepage loss = 1.4 × Evaporation loss

$$\text{Therefore, rate of evaporation} = \frac{9033.33 \times 10^3}{1.4 \times 9033.33 \times 7 \times 24}$$

$$= 4.25 \text{ mm/h/m}^2$$

Example 3.3

The catchment area of a reservoir is 10.0 km^2. A uniform precipitation of 0.5 cm/h for 2.0 h was observed on 7th of July. 50% of the runoff reached the reservoir. A canal carrying a discharge of 1.25 m^3/s is taken from the reservoir. The rate of evaporation observed was 0.7 mm/m^2/h. The seepage loss was observed to be 50% of the evaporation loss. Find the change in the reservoir level on 7th July from 8:00 a.m. to 6:00 p.m. if the water spread of the reservoir was 0.476 km^2.

Solution:

$$\text{Inflow into the reservoir} = 10.0 \times 10^6 \times \frac{0.5}{100} \times 2.0 \times 0.5 = 5.0 \times 10^4 \text{ m}^3$$

$$\text{Canal outflow} = 1.25 \times 10 \times 3600 = 4.5 \times 10^4 \text{ m}^3$$

$$\text{Evaporation and seepage losses} = 0.476 \times 10^6 \times 10 \times \frac{1}{1000} \times 0.7 \times 1.5$$

$$= 0.5 \times 10^4 \text{ m}^3$$

$$\text{Total outflow from the reservoir} = 4.5 \times 10^4 + 0.5 \times 10^4 = 5.0 \times 10^4 \text{ m}^3$$

$$= \text{inflow.}$$

Hence, there will not be any change in the reservoir water level.

Example 3.4

A trapezoidal channel of bed width 4.0 m and side slopes 1:1 carries water at a depth of 2.0 m. The rate of evaporation observed was 0.35 mm/m^2/h. Find the daily loss due to evaporation from the canal in a length of 10 km in ha m.

Solution:

$$\text{Top width of water level} = 4.0 + 2.0 + 2.0 = 8.0 \text{ m}$$

$$\text{Total water spread in the canal in 10.0 km} = 8.0 \times 10.0 \times 1000 = 8 \times 10^4 \text{ m}^2$$

Rate of evaporation from the canal = 0.35 mm/m²/h

$$\text{Daily loss of canal water} = 8.0 \times 10^4 \times \frac{0.35}{1000} \times 24 = 672 \text{ m}^3$$
$$= 672 \times 10^{-4} \text{ ha m}$$
$$= 0.0672 \text{ ha m}$$

3.4.2 Energy Budget Method

In this method, in a specific time, the energy received by water from the Sun or from the atmosphere is accounted and then the energy utilized for evaporation is calculated; thus, the actual evaporation is worked out.

Thus,
$$Q_I = Q_R \pm Q_s + Q_E$$

where, Q_I = Total energy received from Sun's radiation
Q_R = Energy reflected back into the atmosphere by water
Q_s = Change in the energy of the stored water
Q_E = Energy required for evaporation

Q_s may be positive or negative depending on energy utilized by the water or energy received from the water. This will depend upon the season. Q_E can be calculated if all other factors in the equation above are evaluated. However, this requires a lot of instrumentation. Once Q_E is calculated, the actual evaporation can be calculated knowing the following:

- Latent heat of evaporation
- Temperature of water
- Barometric pressure

This method too is not accurate, as it requires very accurate instrumentation. However, it may give a rough idea.

Example 3.5

A reservoir having a water spread of 0.5 km² and water storage of 1 million m³ receives the Sun's radiation at the average rate of 0.025 Langley per second. 5% of this radiation is reflected back into the atmosphere. The rise in the average temperature of the water is 2 °C during 9:00 a.m. to 5:00 p.m.

Assuming no exchange of heat between water and ground as well as the atmosphere, find the average rate of evaporation.

Solution:

1 Langley = 1 cal/cm²
Latent heat of evaporation: 539.55 cal/g
Specific heat of water: 1 cal/g/°C

(A) Energy received from the Sun $= 0.5 \times 10^6 \times 10^4 \times 0.025 \times 8 \times 3600$
$= 3.6 \times 10^{12}$ cal

(B) Energy received by water $= 0.95 \times 3.6 \times 10^{12}$
$= 3.42 \times 10^{12}$ cal

(C) Energy utilized by water $= 2 \times 1 \times 10^6 \times 10^6$ cal
$= 2.0 \times 10^{12}$ cal

(D) Energy utilized by water for evaporation $= (3.42 - 2.0) \times 10^{12}$ cal
$= 1.42 \times 10^{12}$ cal

(E) Volume of water evaporated $= \dfrac{1.42 \times 10^{12} \times 10^{-6}}{539.55} = 2.63 \times 10^3\ m^3$

(E) Rate of evaporation $= \dfrac{2.63 \times 10^3 \times 1000}{0.5 \times 10^6 \times 8}$
$= 0.66\ mm/m^2/h$

3.4.3 Mass Transfer Method

This method is based on the determination of the mass of water vapour transferred from the water surface to the atmosphere. It is also known as *vapour flow approach* or *aerodynamic approach.* The concept of boundary layer theory and continuous mixing, are applied. It is assumed that wind velocity in the vertical is logarithmic and atmosphere is adiabatic.

The evaporation is expressed as:

$$E = \frac{46.08(e_1 - e_2)(v_2 - v_1)}{(T + 273)\log_c(z_2/z_1)^2}$$

where, E = Evaporation in mm/h
z_1, z_2 = Arbitrary levels above water surface levels in metres
e_1, e_2 = Vapour pressures at z_1, z_2 in mm Hg
v_1, v_2 = Wind velocity at z_1, z_2 in km/h
T = Average temperature in °C between z_1 and z_2

3.4.4 Methods of Actual Observations

Evaporation can be measured by an instrument known as *atmometer*.

In an atmometer, there is a continuous supply of water to a surface. This surface is kept constantly wet and evaporation from this surface is measured. Atmometers are not common because of their small size. They do not have sufficient exposure. The rate of evaporation observed from an atmometer is on higher side than that from any other method.

The following are normal atmometers: (1) Livingstone atmometer, and (2) Piche atmometer

(1) Livingstone atmometer It consists of a 50-mm diameter spherical surface of 2.5-mm thick porous material as shown in Fig. 3.2.

The bottle is filled with distilled water that is supplied continuously to the porous bulb. The loss of water from the bottle is due to evaporation.

(2) Piche atmometer It consists of a graduated glass tube of 15-mm diameter and 300-mm in length. It is filled with water and covered with a filter paper. The tube is kept in an inverted position so that there is continuous supply of water to the filter paper. The loss of water from the glass tube is the loss due to evaporation. The Piche atmometer is shown in Fig. 3.3.

3.4.5 Pan Observations

The estimation of evaporation can be done by taking actual observations from a pan and correlating these results to a reservoir.

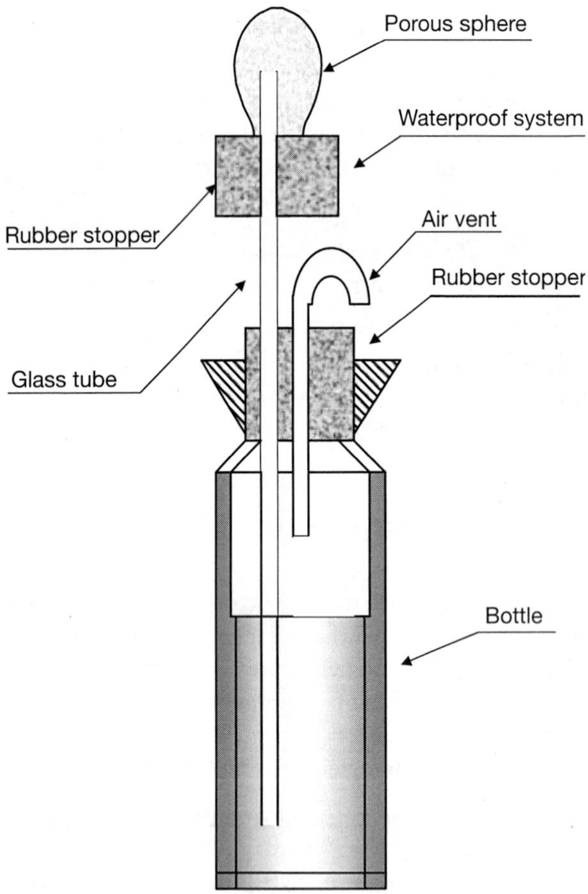

Fig. 3.2 Livingstone atmometer

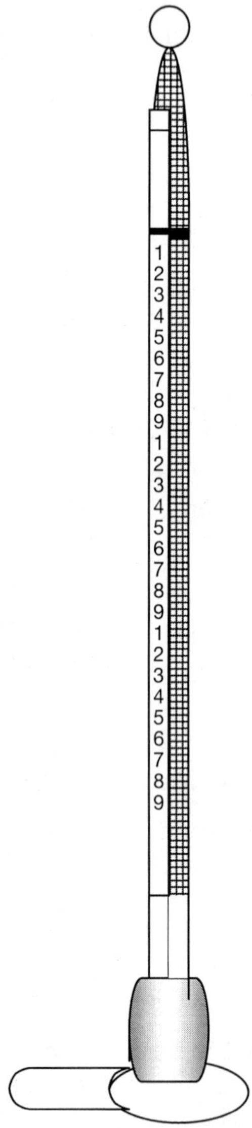

Fig. 3.3 Piche atmometer

3.4.5.1 Pan

A pan is a metal container, square or circular, and of uniform cross-section. Normally it is circular. The diameter may range from 300 to 1500 mm.

This pan is filled with water and the loss of water from this pan in a specified period is measured. The rate of evaporation observed here is correlated to the evaporation from a reservoir.

3.4.5.2 Observations on a Pan

The observations over a specified time are taken by two methods.

1. In the first method, the pan is filled up to a specific level and the loss of water is calculated by observing the level of water over a specified time. Knowing the cross-section of the pan and the reduction in the water level, the loss due to evaporation over the specified time can be calculated. The water levels in the pan are taken accurately by a hook gauge.
2. In the second method, the pan is filled up to a specific level. This level is maintained constant by adding water to the pan periodically to meet the evaporation loss. The loss due to evaporation over that period is the quantity of water added to the pan to maintain a constant level.

Example 3.6

The drop in water level in the case of a 1.5-m diameter evaporation pan during 24 h is 10 mm. The precipitation recorded during this period was 15 mm. Find the rate of evaporation from the pan.

Solution:

$$\text{Area of pan} = \frac{\pi}{4}(1.5)^2 = 1.767\ \text{m}^2$$

$$\text{Evaporation in 24 h} = \frac{E + P}{\text{Area}}$$

$$\text{Therefore, Rate of evaporation} = \frac{10 + 15}{1.767 \times 24} = 0.59\ \text{mm/h/m}^2$$

Example 3.7

The amount of water added to a 1.0-m diameter circular land pan over a period of 10 h to maintain the water level constant is 9.56 lit. Find the hourly rate of evaporation from the pan.

Solution:

$$\text{Area of pan} = \frac{\pi}{4}D^2 = \frac{\pi}{4}(1)^2 = 0.785\ \text{m}^2$$

$$\text{Rate of evaporation per hour} = \frac{9.56 \times 10^{-3} \times 10^3}{0.785 \times 10}$$

$$\text{Rate of evaporation} = 1.21\ \text{mm/h/m}^2$$

Example 3.8

Observations taken on a 1.0-m diameter circular land pan on 7 July 2006 from 8:00 a.m. to 6:00 p.m. were as follows:

1. Quantity of water taken out of the pan = 5.0 lit
2. Precipitation during this time period = 20 mm

If there was no change in the water level in the pan, find the rate of evaporation.

Solution:

1. Area of pan $= \frac{\pi}{4} \times 1^2 = 0.785\ m^2$
2. Amount of water received due to precipitation $= 0.785 \times 10^3 \times 20 \times 10^{-3} = 15.706$ lit
3. Evaporation from the pan: $15.706 - 5.0 = 10.706$ lit

   Therefore, the rate of evaporation $= \dfrac{10.706 \times 10^{-3} \times 1000}{0.785 \times 10} = 1.36\ mm/m^2/h$

Example 3.9

In the case of a 1.5-m diameter circular pan, following observations were taken from 8:00 a.m. to 6:00 p.m.

1. Quantity of water added to keep the water level in the pan constant is 5.0 lit
2. Precipitation during the observation period is 10 mm
3. Leakage from the pan is 2.0 lit

Find the rate of evaporation from the pan.

Solution:

Area of the pan $= \frac{\pi}{4} \times 1.5^2 = 1.767\ m^2$

Quantity of water received by the pan $= 1.767 \times 10^3 \times 10 \times 10^{-3}$

$= 17.67$ lit by way of precipitation

Evaporation from the pan $= 17.67 + 5.0 - 2.0 = 20.67$ lit

$$\text{Rate of evaporation} = \frac{20.67 \times 10^{-3} \times 10^3}{1.767 \times 10} = 1.17\ mm/m^2/h$$

3.4.5.3 Pan Coefficient

The observations from a pan are extrapolated to a reservoir by using a pan coefficient. A pan coefficient may be defined as follows:

$$\text{Pan coefficient} = \frac{\text{Rate of evaporation from the reservoir}}{\text{Rate of evaporation from a pan}}$$

The value of pan coefficient is always less than 1.0 and is dimensionless because, in the case of a pan, a very small area is exposed to atmosphere and, secondly, the metal container absorbs more energy and in turn is utilized for evaporation.

Example 3.10

The rate of evaporation from a 1.25-m pan having a pan coefficient of 0.8, was 1 mm/m²/h. Find the total evaporation from a reservoir in a week, having a water spread of 2.0 ha.

Solution:

Evaporation from the reservoir during a week $= \dfrac{1}{1000} \times 0.8 \times 2 \times 10^4 \times 7 \times 24$

$= 2688\ m^3$

Example 3.11

The monthly evaporation (in mm) observed on a pan from January to December 2006 is as follows:

111, 126, 127, 132, 141, 146, 148, 143, 138, 126, 118, 114

The water spreads of reservoir in January 2006 and December 2006 were 2.56 and 2.69 km^2, respectively. Assuming a pan coefficient of 0.8, find the total loss of water due to evaporation in 2006 from the reservoir. Neglect all other losses.

Solution:

$$\text{Mean spread of reservoir} = \frac{1}{3}\left(A_1 + A_2 + \sqrt{A_1 \times A_2}\right)$$
$$= \frac{1}{3}(2.56 + 2.69 + \sqrt{2.56 \times 2.69})$$
$$= 2.624 \text{ km}^2$$

$$\text{Annual loss due to evaporation} = 111 + 126 + 127 + 132 + 141 + 146 + 148 + 143 + 138 + 126 + 118 + 114 = 1570 \text{ mm}$$

$$\text{Therefore, evaporation loss in 2006} = 2.624\ 10^6 \times \frac{1570}{1000} \times 0.8$$
$$= 3.295 \times 10^6 \text{ m}^3 = 3.295 \text{ million m}^3$$

3.4.5.4 Factors Affecting Pan Coefficient

The following factors affect the pan coefficient.

- Diameter of pan
- Depth of water in the pan
- Height of rim of pan above the water level
- Location of pan
- Colour of pan
- Material used for pan

The pan coefficient can be evaluated by taking observations on the different sizes of pans and extrapolating the observations as shown in Fig. 3.4.

The different types of pans are: land pan, floating pan and sunken pan

3.4.5.5 Land pan

This type of pan is located on land above the ground normally on the bank of the lake or the reservoir. There are different types and sizes. The standard one is shown in Fig. 3.5.

It is made of ungalvanized iron sheet 1200 mm in diameter and 250 mm in depth. The bottom is supported on a wooden frame placed 150 mm above ground level, and the water level is kept 50–75 mm below the rim. The pan coefficient varies between 0.6 and 0.7.

3.4.5.6 Floating pan

In this type, the pan is kept floating in the reservoir. Thus with this arrangement, similar surrounding conditions are maintained. Initially, the water level in the pan is kept the same as that of reservoir and observations are taken regularly by approaching it from the bank.

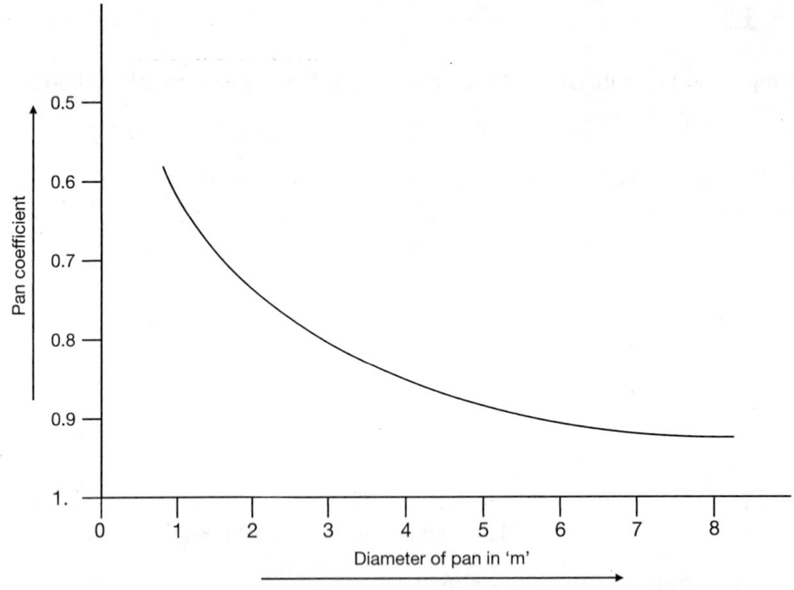

Fig. 3.4 Variation of pan coefficient w.r.t. pan diameter

The standard Floating Pan of 900-mm^2 area and 450-mm depth is as shown in Fig. 3.6. The pan coefficient varies from 0.7 to 0.82.

3.4.5.7 Sunken pan

In this type, the pan is sunk in the ground on the banks of the reservoir. This arrangement avoids splash of rain, drifting of dust or trash and obstruction to the wind. It also eliminates boundary effects such as the radiation of the side walls and heat exchange between the pan and the atmosphere.

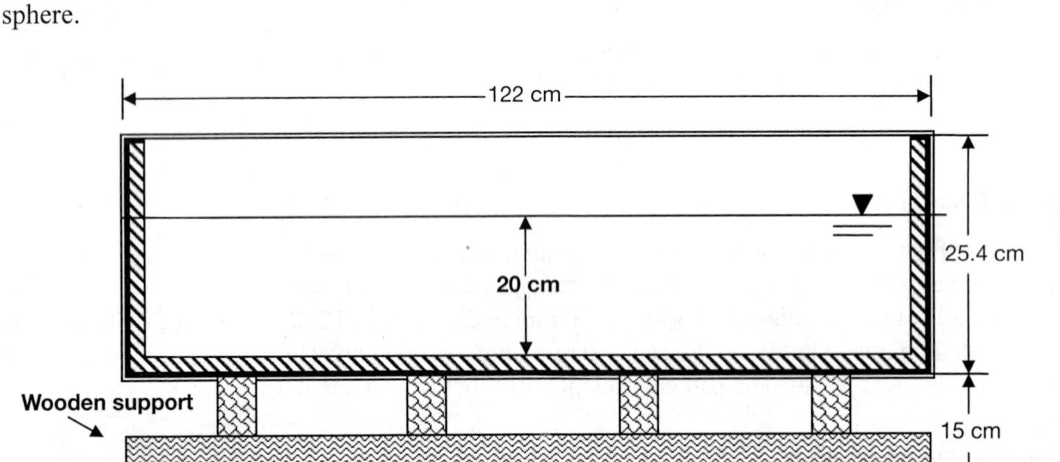

Fig. 3.5 Land pan

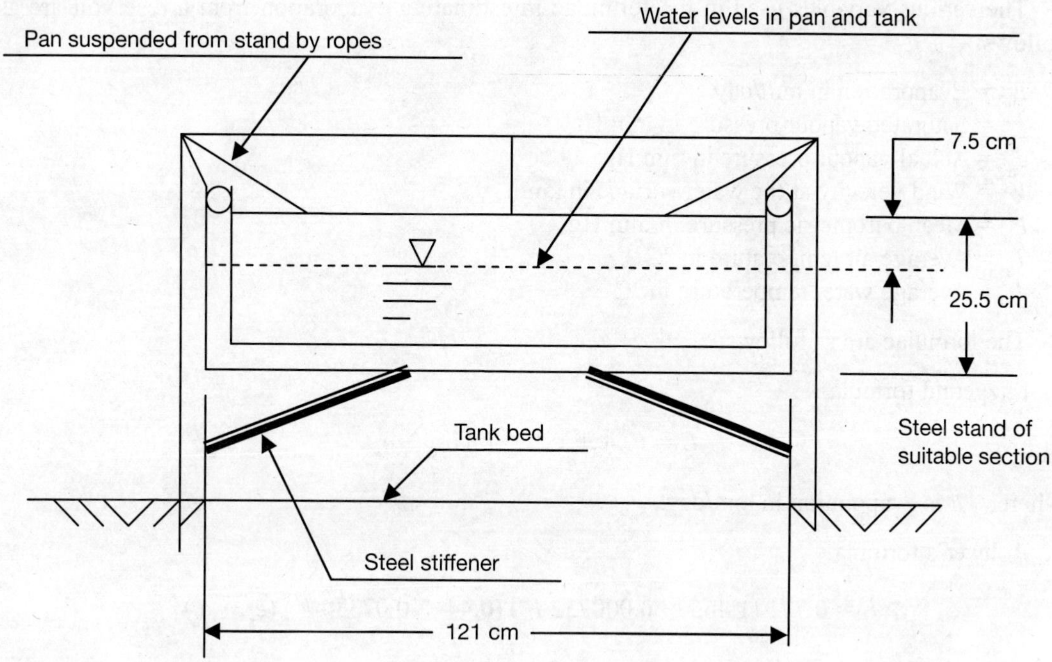

Fig. 3.6 Floating pan

The standard type is made of ungalvanized unpainted iron sheet with an area of 900 mm^2 and 450 mm in depth. The rim is kept 100 mm above water level and the water level same as that of ground, as shown in Fig. 3.7. The pan coefficient varies between 0.75 and 0.86.

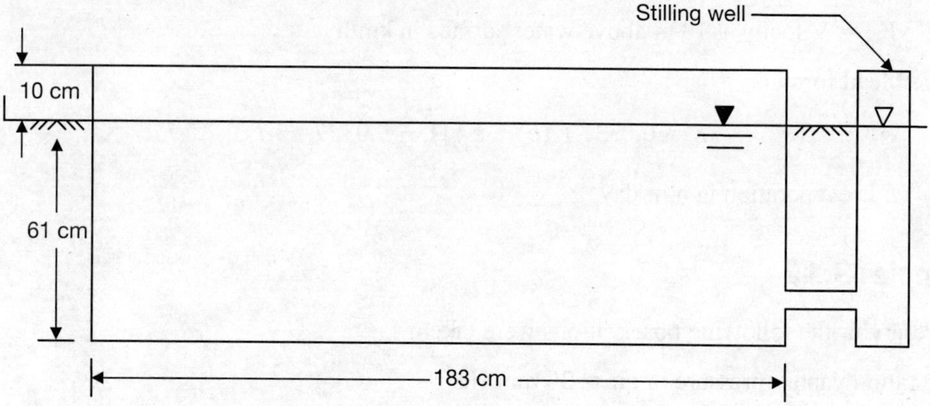

Fig. 3.7 Sunken pan

3.4.6 Formulae to Estimate the Evaporation from a Reservoir

There are several formulae to estimate evaporation from a reservoir. All of them are empirical.

The formulae normally used are: Fitzgerald's formula, Rohwer's formula, Meyer's formula and Lake Mead formula.

The various variables used in the formulae in estimating evaporation from a reservoir are as follows:

E = Evaporation in mm/day
e_s = Saturated vapour pressure in mm Hg
e_a = Actual vapour pressure in mm Hg
V = Wind velocity at the water surface in km/h
P_a = Mean barometric pressure in mm Hg
T_a = Average air temperature in °C
T_w = Average water temperature in °C

The formulae are as follows:

1. Fitzgerald formula

$$E = (0.4 + 0.124\,V)\,(e_s - e_a)$$

where, E is evaporation in mm/day.

2. Rohwer's formula

$$E = 0.771\,(1.465 - 0.000732\,P_a)\,(0.44 + 0.07334\,V)\,(e_s - e_a)$$

where, E is evaporation in mm/day.

3. Meyer's formula

$$E = C\,(e_s - e_a)\,(1 + 0.06215\,V_{10})$$

where, E is evaporation in mm/month
C a constant having a value of 11 for deep water bodies and 15 for shallow water bodies.

V_{10} = Velocity at 10 m above water surface in km/h

4. Lake Mead formula

$$E = 0.0331\,V\,(e_s - e_a)\,[1 - 0.03\,(T_a - T_w)]$$

where, E is evaporation in mm/day.

Example 3.12

For a reservoir the following observations were taken:

1. Saturation vapour pressure of air = 30 mm Hg
2. Relative humidity = 0.5
3. Velocity of wind at the rate of 0.5 m above ground = 25 km/h
4. Water spread of reservoir = 10 km^2
5. Atmospheric pressure = 10.4 m of water
6. Average air temperature = 29 °C
7. Average water temperature = 27 °C

Evaluate monthly evaporation by using different formulae.

Solution:

The monthly evaporation rate was calculated by four empirical formulae.

1. Fitzgerald' s equation

$$E = (0.4 + 0.124\ V)\ (e_s - e_a)$$

$$\text{Relative humidity} = \frac{\text{Actual vapour pressure of air}}{\text{Structure vapour pressure of air}}$$

$$0.5 = \frac{e_a}{e_s} = \frac{e_a}{30} \quad \text{since } e_s = 30 \text{ mm Hg}$$

Therefore, $e_a = 15$ mm Hg

Assuming average velocity of wind

V = Velocity of wind at the rate of 0.5 m above ground

$$E = (0.4 + 0.124\ V)\ (e_s - e_a)$$
$$= (0.4 + 0.124 \times 25)\ (30 - 15)$$
$$= (0.4 + 3.1)\ (15)$$
$$E = 52.5 \text{ mm/day}$$

Therefore, monthly evaporation loss $= E \times 30 \times$ water spread of reservoir

$$= \frac{52.5}{1000} \times 30 \times 10 \times 10^6$$
$$= 15.75 \times 10^6 \text{ m}^3 = 15.75 \text{ million m}^3$$

2. Meyer's equation

$$E = C(e_s - e_a)(1 + 0.0621\ V_{10})$$

where, E = Evaporation in mm per month

V = Velocity at 10 m above ground

$C = 15$

Velocity at the rate of 10 m above ground $= V_{10}$

Velocity at the rate of 0.5 above ground $= V_{0.5} = 25$ km/h

Therefore, $\dfrac{V_{10}}{V_{0.5}} = \left(\dfrac{10}{0.5}\right)^{0.15}$

$$V_{10} = V_{0.5}\left(\frac{10}{0.5}\right)^{0.15}$$
$$V_{10} = V_{0.5} \times 1.567$$
$$V_{10} = 25 \times 1.567$$
$$V_{10} = 39.18 \text{ km/h}$$
$$E = C(e_s - e_a)\ (1 + 0.0621\ V_{10})$$
$$= 15\ (30 - 15)\ (1 + 0.0621 \times 39.18)$$
$$= 15 \times 15 \times 3.433$$
$$= 772.42 \text{ mm/month}$$

Therefore, monthly evaporation loss $= E \times$ water spread of reservoir

$$= \frac{772.42}{1000} \times 10 \times 10^6$$
$$= 7.724 \times 10^6 \text{ m}^3 = 7.724 \text{ million m}^3$$

3. Lake Mead equation

$$E = 0.0331\, V (e_s - e_a)\, [1 - 0.03\, (T_a - T_w)]$$

where, T_a = Average air temperature = 29 °C

T_w = Average water temperature = 27 °C

$E = 0.0331 \times 25\, (30 - 15)\, [1 - 0.03\, (29 - 27)]$

E = 11.66 mm/day

Therefore, monthly evaporation loss $= E \times 30 \times$ water spread of reservoir

$$= \frac{11.66}{1000} \times 30 \times 10 \times 10^6$$

$$= 3.498 \times 10^6 \text{ m}^3 = 3.498 \text{ million m}^3$$

4. Rower's equation

$$E = 0.771\, (1.465 - 0.000732\, P_a) \times (0.44 + 0.07334\, V)\, (e_s - e_a)$$

where, P_a = Pressure in mm Hg

$$P_a = \frac{10.33}{13.6} \times 1000 = 764.70 \text{ mm Hg}$$

$$E = 0.771\, (1.465 - 0.000732 \times 764.70) \times (0.44 + 0.07334 \times 25)\, (30 - 15)$$

$$= 0.771\, (1.465 - 0.559)\, (0.44 + 1.833)\, 15$$

$$= 23.81 \text{ mm/day}$$

Therefore, monthly evaporation loss $= \frac{23.81}{1000} \times 30 \times 10 \times 10^6$

$$= 7.14 \times 10^6 \text{ m}^3$$

$$= 7.14 \text{ million m}^3$$

3.5 CONTROL OF EVAPORATION FROM A RESERVOIR

Evaporation from a reservoir can be reduced by as much as 30%. The different methods to control evaporation from a reservoir are as follows.

3.5.1 Surface Area Reduction

Evaporation depends mostly on the exposed surface area. If by any means the surface area exposed is reduced, the evaporation from a reservoir can be reduced.

This can be achieved by the following ways:

- Reducing the meandering length of the streams
- Selecting the dam site such that the ratio of surface area to storage is the minimum
- Storing water underground

This method has a limited scope, but evaporation can be controlled by this method.

3.5.2 Mechanical Covers

If a reservoir is covered, completely or even partially, the movement of water vapour can be controlled and thus evaporation will reduce.

This is a very costly method and can be practised for small reservoirs only.

3.5.3 Wind Shields

Wind velocity near the water surface affects the evaporation. Less the velocity, less will be the evaporation. The wind velocity near the water surface can be reduced by providing obstruction to wind. This obstruction can be achieved by providing plants and bushes all around the reservoir. However, the exact effect due to this method cannot be estimated.

This method can be adopted for small lakes and reservoirs only.

3.5.4 Surface Films

Reduction in evaporation from a reservoir is achieved by spreading a film over the water surface. This can be done by means of some chemicals. This film will retard the movement of water molecules from the water surface to the air above.

This film is one-molecule thick and hence is termed as *monomolecular film* (one millionth of a millimetre). The reduction in evaporation due to such films is observed to be 40–70% in the laboratory and 25% in the lakes and the reservoirs.

The chemical to be used should have the following properties:

- It should prevent movement of water molecules into the atmosphere.
- It should allow movement of oxygen and carbon dioxide.
- It should not get disturbed due to wind or rain.
- It should not be harmful to plants and animals.
- It should be cheap and available in the required quantity.

3.5.4.1 Chemicals normally used

Cetyl alcohol ($C_{16}H_{33}OH$) known as hexadecanol or stearyl alcohol ($C_{18}H_{37}OH$) known as octadecanol are the chemicals normally used. These are available in powder form and a solution is prepared by using turpentine, petrol, kerosene, or so.

3.5.4.2 Application of chemicals

The chemical can be applied by one of the following methods:

- Hand spreading from the banks
- Spraying from a boat
- Aerial application using aeroplanes
- By automatic dispensing units mounted on well-spaced anchored barges

3.5.5 Floating Covers

Evaporation from a reservoir can be controlled by providing floating covers over the water surface. This retards the movement of water molecules from the water surface to the air above.

These floating covers may be any of the following:

- Wooden planks or sheets
- Plastic balls
- Polyethylene sheets

This method is very costly. However, it has definitely some positive result.

This method can be adopted for very small lakes and reservoirs.

3.6 TRANSPIRATION

Plants absorb water from the soil through their roots. Mineral salts are also absorbed in dilute solution using water as its vehicle. This solution is transported through roots and stems to the leaves where plant food is produced from this sap, CO_2 from atmosphere and using energy from the Sun through chlorophyll. The plant food thus produced is again distributed using water as a vehicle for the cell growth and tissue building.

Most of the water thus absorbed by the plants is discharged back into the atmosphere in the form of vapour. This process is known as *transpiration*. It is observed that only 1% of water sucked in by the roots is retained by the plants.

3.6.1 Factors Affecting Transpiration

Practically, all the meteorological factors that affect evaporation, influence transpiration also. In addition, the following factors affect transpiration:

Sunlight: The growth of plants depends on sunlight. About 95% of the transpiration occurs during the day time.

Moisture available: Transpiration is limited by the rate of moisture that becomes available to the plants. The plants can extract water between field capacity and the wilting point. If the water available is less than the wilting point, then the plants may not be able to suck up water from the soil. Thus, it may affect transpiration.

Stage of plant development: Transpiration depends upon the plant growth. The growth of plant varies (1) diurnal, (2) seasonal and (3) annual. Also, the growth of the plants is more during its early stage. Transpiration varies accordingly.

3.6.2 Measurement of Transpiration

Transpiration can be measured in terms of the depth of water transmitted daily or annually.

A phytometer provides a practical method for measuring transpiration. This is a large vessel filled with soil in which one or two plants are rooted. The soil surface is sealed to prevent evaporation from the soil. The initial weight of the container along with the soil and the plants is recorded. The transpiration loss can be worked out by observing the weight of the container over a known period. A small phytometer containing water only is known as *potometer*.

3.6.3 Transpiration Reduction

Water conservation can be achieved to some extent through transpiration reduction. The following are the different methods to conserve water:

- Use of chemicals to inhibit water consumption (similar to mono molecular films in the case of evaporation)
- Harvesting of plants
- Improved irrigation methods
- Removing unwanted, unproductive and useless trees, bushes and grasses

Transpiration ratio (*TR*) is the ratio of the total weight of water transpired by a plant during its complete development to the weight of the dry matter produced. It is dimensionless.

The TR for rice is between 300 and 600 and that for wheat is between 600 and 800.

3.6.4 Evapotranspiration

In studying water balance from an area, it is difficult to separate evaporation and transpiration and hence these two processes are combined and treated as a single and termed as *evapotranspiration*.

3.6.4.1 Potential Evapotranspiration

When adequate quantity of water is available for evaporation and transpiration, the combined process is known as *potential evapotranspiration*. It is normally known as PET. Actual evapotranspiration, normally denoted by AET is, therefore, always less than potential evapotranspiration (AET < PET).

3.6.4.2 Measurement of Evapotranspiration

Evapotranspiration can be measured by the following methods:

- Lysimeter
- Inflow–outflow measurements

Lysimeter: It consists of a circular tank, its diameter ranging from 600 mm to 3000 mm. It is buried in the ground so that its top is flush with the ground. It is filled with the soil similar to the field conditions, and the crop or the tree of which AET is to be found is grown in the lysimeter. A typical lysimeter is shown in Fig. 3.8.

Water is added to the lysimeter and an account of water added, etc. is kept. Then AET is calculated as follows:

$$P + W = O + \text{AET} + \Delta S$$

where, P = Precipitation during the period of observations
W = Water supplied
O = Water drained
AET = Actual evapotranspiration
ΔS = Change in soil moisture in lysimeter

Lysimeters are expensive and the process of observation is time-consuming.
Lysimeter is also called Evapotranspirometer.

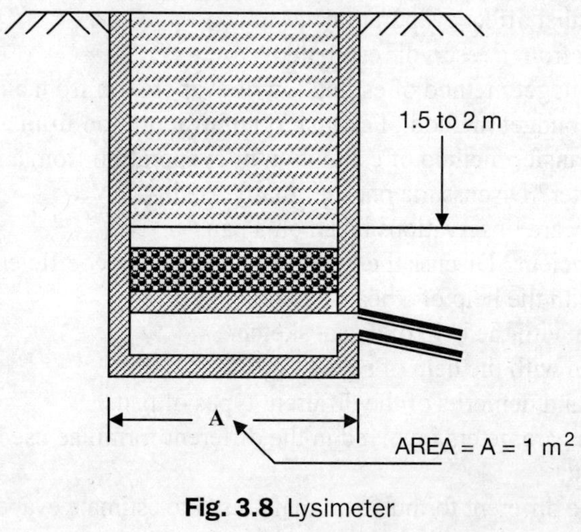

Fig. 3.8 Lysimeter

Inflow–outflow measurements: This method is applicable to large areas, such as up to 100 ha. An account of the inflow of water is maintained for a specific period. The inflow may be precipitation or surface inflow or even groundwater inflow. During the same period, an account of outflow is maintained. The outflow may be the surface outflow or the groundwater outflow. Naturally, the difference between these two will be the loss due to evapotranspiration.

If the observation period is small, then it is assumed that the groundwater inflow and outflow are equal. In this method, it is very difficult to measure inflow, outflow and groundwater storage to the desired accuracy.

3.7 SOIL EVAPORATION

Immediately after the rain stops, evaporation from the soil starts. Evaporation from a saturated soil surface is approximately the same as that from a water surface at that temperature. As the soil begins to dry, the evaporation goes on reducing. Finally it virtually stops as there is no supply of water. If the groundwater table is within 1.0 m from the ground, evaporation from soil is noticed since there is some supply of water.

Evaporation opportunity is defined as 'the ratio of evaporation from soil and evaporation from equivalent water surface'. It is dimensionless and is generally expressed in terms of percentage. Thus,

$$\text{Evaporation opportunity} = \frac{\text{Evaporation from soil}}{\text{Evaporation from equivalent water surface}} \times 100$$

Evaporation from soil is measured by using a lysimeter of size 1 m × 1 m × 1 m. The soil is filled flush with the tank edge and the groundwater table is maintained at a specific level. The evaporation from the soil can be determined by weighing the tank at known intervals and from the quantity of water added to maintain the groundwater table.

REVIEW QUESTIONS

1. Define evaporation. Explain the process of evaporation.
2. Explain Dalton's law of evaporation.
3. Discuss the factors that affect evaporation.
4. How is evaporation from a reservoir estimated?
5. Explain the water budget method of estimation of evaporation from a reservoir.
6. Explain the energy budget method of estimation of evaporation from a reservoir.
7. Explain the mass transfer method of estimation of evaporation from a reservoir.
8. What is an atmometer? Discuss the ones normally used.
9. What is a pan? How are observations taken on a pan?
10. What is a pan coefficient? Discuss the factors that affect pan coefficient.
11. Explain land pan with the help of a neat sketch.
12. Explain sunken pan with the help of a neat sketch.
13. Explain floating pan with the help of a neat sketch.
14. Discuss the merits and demerits of the different types of pans.
15. Discuss the various parameters involved in the different formulae used to estimate evaporation from a reservoir.
16. State and explain the different formulae normally used to estimate evaporation from a reservoir.

17. Can evaporation from a reservoir be reduced? Explain the different methods.
18. Write a detailed note on the use of surface films to reduce evaporation from a reservoir.
19. Explain the process of transpiration. Discuss the different methods to reduce transpiration.
20. Discuss the different methods to measure transpiration.
21. What is evapotranspiration? Explain the different methods to estimate it from a catchment.
22. Discuss soil evaporation.
23. Write a note on the pattern of evaporation over India.
24. Write short notes on
    a. Potential evapotranspiration.
    b. Factors affecting transpiration.
    c. Livingstone atmometer.
    d. Pitch atmometer.
    e. Transpiration ratio.
    f. Phytometer.
    g. Potometer.
    h. Lysimeter.
    i. Pan coefficient.
    j. Factors affecting transpiration.
    k. Annual evaporation in India.
    l. Atmometer.
    m. Evaporation opportunity.
25. Differentiate between
    a. Evaporation, sublimation and transpiration.
    b. Phytometer and potometer.
    c. Land pan, sunken pan and floating pan.
    d. Evapotranspiration and potential evapotranspiration.

NUMERICAL QUESTIONS

1. The uniform precipitation over a catchment area covering 80 km^2 received a total precipitation of 80 mm in a month. 60% of the precipitation reached the reservoir. The spread of reservoir during that month was 2.5 km^2. The irrigation canal discharges at a constant rate of 1.1 m^3/s. If the seepage losses were 60% of the evaporation losses, find the daily rate of evaporation.

   Ans: 8.24 mm/h/m^2

2. In the case of a reservoir, the surface area in m^2 is given by $A = 90\,y^2$, y is the average depth of water in the reservoir in metres. In a week, the depth of water reduced from 9.0 m to 8.5 m. If the evaporation loss is double the seepage loss, find the rate of hourly evaporation loss.

   Ans: 2 mm/h/m^2

3. The bed width of a trapezoidal channel is 5.0 m and the side slopes are 1:1. It carries water at a depth of 1.8 m. The rate of evaporation observed on a 1.5-m diameter floating pan was 0.25 mm/h/m^2. Find the loss of evaporation in a day from the canal in a length of 5 km. Assume the pan coefficient to be 0.8.

   Ans: 0.0206 ha m

4. During a test, the drop in water level in the case of a land pan of diameter 1.0 m in 24 h was observed to be 15 mm. The precipitation recorded during this period was 10 mm. Find the rate of evaporation from the pan.

   Ans: 1.32 mm/h/m^2

5. A test was conducted on a 0.5-diameter sunken pan. The quantity of water added to maintain the water level constant over a period of 10 h was 950 cc. Find the rate of evaporation.

Ans: 0.5 mm/h/m^2

6. The following observations were taken on a 1.0-m diameter land pan from 8 a.m. to 5 p.m.
   (i) Precipitation during the observation period = 15 mm
   (ii) Leakage from the pan = 1.5 l
   (iii) Quantity of water added to keep the level constant = 4.0 l

   Find the rate of evaporation.

Ans: 2.02 mm/h/m^2

7. The following observations were taken from a reservoir:
   (i) Velocity of wind at 0.5 m above ground = 24 m/s
   (ii) Atmospheric pressure = 10.4 m of water
   (iii) Average air temperature = 30 °C
   (iv) Average water temperature = 27 °C
   (v) Relative humidity = 0.5
   (vi) Water spread of reservoir = 15 km^2
   (vii) Saturation vapour pressure of air = 30 mm Hg

   Find the monthly evaporation using empirical formulae.

Ans: (i) 22.50×10^6 m^3
(ii) 11.25×10^6 m^3
(iii) 4.87×10^6 m^3
(iv) 10.36×10^6 m^3

MULTIPLE CHOICE QUESTIONS

1. The pan coefficient is given by
   (a) $\dfrac{\text{Rate of evaporation from reservoir}}{\text{Rate of evaporation from pan}}$
   (b) $\dfrac{\text{Rate of evaporation from pan}}{\text{Rate of evaporation from reservoir}}$
   (c) Rate of evaporation from reservoir − Rate of evaporation from pan
   (d) Rate of evaporation from reservoir × Rate of evaporation from pan

2. The dissolved salts in water
   (a) Reduce the rate of evaporation
   (b) Increase the rate of evaporation
   (c) Have no effect on the rate of evaporation
   (d) Double the rate of evaporation

3. Lysimeter is an instrument used to measure
   (a) Evaporation
   (b) Transpiration
   (c) Evapotranspiration
   (d) Sublimation

4. The unit of evaporation is
   (a) mm/h
   (b) mm/m^2/h
   (c) mm/m/h
   (d) mm/m^3

5. The pan coefficient is always

(a) More than 1 (b) Equal to 1
(c) Less than 1 (d) Less than 0.5

6. Dalton's law of evaporation is given by

(a) $E = C\,(e_a - e_s)$ (b) $E = C\,(e_s - e_a)$
(c) $E = C\dfrac{e_a}{e_s}$ (d) $E = C\dfrac{e_s}{e_a}$

7. Transpiration ratio is given by

(a) $\dfrac{\text{Weight of dry matter produced}}{\text{Weight of water transpired}}$ (b) $\dfrac{\text{Weight of water transpired}}{\text{Weight of dry matter produced}}$
(c) Weight of dry matter produced + Weight of water transpired (d) Weight of dry matter produced − Weight of water transpired

8. Transpiration is confined to

(a) Day time hours (b) Night hours
(c) All the day (d) Any of the three

9. The reduction of evaporation by using monomolecular surface is approximately

(a) 2% (b) 10%
(c) 25%. (d) 50%

10. Out of the water sucked by the roots, the percentage retained by plants is

(a) 1 (b) 5
(c) 10 (d) 15

11. The transpiration ratio for rice is between

(a) 50 to 300 (b) 300 to 600
(c) 600 to 900 (d) 900 to 1200

12. PET means

(a) Practical evapotranspiration (b) Progressive evapotranspiration
(c) Perfect evapotranspiration (d) Potential evapotranspiration

13. AEP means

(a) Additional evapotranspiration (b) Accumulated evapotranspiration
(c) Actual evapotranspiration (d) Allowable evapotranspiration

14. Evapotranspiration means

(a) Evaporation − transpiration (b) Evaporation + transpiration
(c) Evaporation ÷ transpiration (d) Evaporation × transpiration

15. Evaporation is a

(a) Cooling process (b) Heating process
(c) Combined process (d) None of the three

16. Annual evaporation loss from a reservoir is about
    (a) 5% (b) 10%
    (c) 30% (d) 60%

17. From a reservoir evaporation loss can be evaluated from
    (a) Land pan (b) Floating pan
    (c) Sunken pan (d) Any of the three

18. The chemical notation of cetyl alcohol is
    (a) $C_{16}H_{32}OH$ (b) $C_{10}H_{20}OH$
    (c) $C_{20}H_{40}OH$ (d) $C_{16}H_{33}OH$

19. The chemical symbol for stearyl alcohol is
    (a) $C_{18}H_{38}OH$ (b) $C_{18}H_{40}OH$
    (c) $C_{18}H_{37}OH$ (d) $C_{19}H_{38}OH$

20. The thickness of a monomolecular film is
    (a) One tenth of a millimetre (b) One hundredth of a millimetre
    (c) One thousandth of a millimetre (d) One millionth of a millimetre

21. In the case of a crop TR means
    (a) True ratio (b) Transit ratio
    (c) Transpiration ratio (d) Technical ratio

22. The dimensions of evaporation opportunity are
    (a) mm/m^2/h (b) Dimensionless
    (c) mm/h (d) mm/m^2

ANSWERS TO MULTIPLE CHOICE QUESTIONS

1. a	2. a	3. c	4. b	5. c	6. b	7. b	8. a
9. c	10. a	11. b	12. d	13. c	14. b	15. a	16. c
17. d	18. d	19. c	20. d	21. c	22. b		

*** ——— ***

4 Precipitation

Chapter Outline

4.1 DEFINITION

All water flowing or stored on the land surface or subsurface is derived directly or indirectly from precipitation, i.e. rainfall including snowfall. It is one of the important basic processes in the hydrological cycle.

Precipitation is the general term for all the moisture emanating from the clouds and falling on the ground. Precipitation data are of utmost importance to hydrologists as they form the basis of all hydrological studies. Variation of rainfall distribution over time and space creates extreme problems like droughts and floods. Study of precipitation, therefore, requires great attention.

4.2 DIFFERENT FORMS OF PRECIPITATION

The different forms of precipitation are as follows:

- Drizzle: It consists of water drops less than 0.5 mm in diameter and intensity less than 1 mm/h.
- Rain: It consists of water drops of 0.5 mm and above.
- Cloudburst: It is rainfall, which is exceptionally of very high intensity.
- Hail: It is precipitation in the form of balls or lumps of ice with diameter from 0.5 mm up to 5.0 cm.
- Snow: It is precipitation in the form of ice crystals or thin flakes of ice resulting directly from the water vapour.
- Sleet: It is a mixture of ice and rain.
- Dew: It forms on the ground directly by condensation during the night when the surface has been cooled due to outgoing radiation.
- Glaze: It is the ice coating when drizzle or rain freezes as it comes in contact with cold objects at the ground.
- Fog: It is a low-level cloud that touches the ground.
- Smog: It is a mixture of smoke and fog.

Precipitation of high intensity occurring over a substantial time covering a large area is normally termed as a *storm*.

4.2.1 Terminal Velocity of a Raindrop

As a raindrop starts falling under gravity, the velocity is low. The air resistance is a function of velocity. As the raindrop starts falling under gravity, the velocity of fall of the raindrop goes on increasing so also the air resistance. A stage is reached when the air resistance and the weight of the raindrop are equal and opposite. When this stage is reached, the raindrop starts falling with a constant velocity.

This constant velocity is called the *terminal velocity* of the raindrop and is a function of the diameter of the raindrop. The terminal velocity for different diameters of the raindrop is given in Table 4.1.

When the raindrop diameter is more than 5.5 mm, because of the resistance offered by air during its fall, it deforms and splits. The air resistance of a falling body depends on the area exposed to air. The volume contained by the surface area per unit weight is maximum in case of a sphere. Hence to have minimum resistance, it is nature's tendency to have a spherical shape for a raindrop.

Table 4.1 The terminal velocity for different diameters of raindrops

Serial no.	Diameter of raindrop (mm)	Terminal velocity (m/s)
1	1.0	4.4
2	2.0	5.9
3	3.0	7.0
4	4.0	7.7
5	5.0	7.9
6	5.5	8.0

4.3 PROCESS OF PRECIPITATION

Four conditions are necessary for the formation of precipitation. These are:

1. Mechanism to cool the air
2. Mechanism to produce condensation
3. Mechanism to produce growth of droplets
4. Mechanism to produce accumulation of moisture of sufficient quantity

4.3.1 MECHANISM TO COOL THE AIR

Water vapour is the most essential for the formation of precipitation. Generally, it is available in the atmosphere.

Air near the ground surface and the oceans carries a lot of moisture. When this air is heated, it becomes lighter and rises up in the atmosphere along with the water vapour and, naturally, as it rises it gets cooled.

Cooling of water vapour is necessary so that the vapour gets saturated and helps in condensation.

4.3.2 MECHANISM TO PRODUCE CONDENSATION

When the air gets cooled it condenses. Condensation of the water vapour in the high atmosphere takes place on a *hygroscopic nucleus*.

Hygroscopic nuclei may be defined as very small particles that have an affinity for water. These are very small particles of dust of the size 10^{-3} to 10 micron (1 micron = 10^{-3} mm). Sodium chloride, sulphur trioxide and cement particles have affinity for water. These hygroscopic nuclei are available in atmosphere in abundance. The condensation may take place even before the air is saturated with vapour.

Condensation depends on the curvature of the nucleus. When, due to condensation, the raindrop size increases, the curvature is reduced and thus the affinity to attract water vapour also reduces. When the raindrop size equals to 10^{-3} mm, the effect of curvature and hygroscopicity becomes negligible.

4.3.3 MECHANISM OF DROPLET GROWTH

Clouds can be considered as colloidal or in suspension. A tendency of the droplet thus formed to remain small and not to fall on the earth is termed as *colloidal stability*. On the other hand, if these droplets tend to combine to form bigger size droplets sufficient to overcome the air resistance to fall on the earth then it is termed as *colloidal instability*.

Therefore, for precipitation colloidal instability is necessary. The small droplets, of size less than 10^{-3} mm formed due to condensation coalesce (combine to form a bigger one), which is due to the following two reasons:

1. **Gravitational coalescence process** The small droplets formed due to condensation are in suspension and move in a haphazard manner in any direction. The speed of movement of different particles is different and thus the particles collide with each other. During this collision, the larger particles grow at the cost of the smaller ones. When the size of the droplet is sufficient to overcome the air resistance, it falls down.

   This process is more predominant in tropical countries where the cloud temperature is well above 0 °C.

2. **Ice crystal theory** In the high atmosphere where condensation takes place, the temperature is 0 °C or even lower. The droplets thus formed are in the form of ice crystals or even water droplets. There is a difference in saturation pressure over the ice crystals and the water droplets. Thus evaporation of water droplets and condensation over the ice crystals occurs.

   This process is comparatively slow and occurs in very cool (super cooled) clouds, i.e. having temperatures lower than 0 °C.

4.3.4 Mechanism to Produce Accumulation of Moisture

If a vertical column of atmosphere from land or from sea is considered, the amount of water vapour remains the same or slightly more than during and after rains. Thus to have rains there should be continuous inflow of water vapour into the vertical column from the sides. This flow is known as *convergence* and is observed during precipitation.

4.4 FACTORS AFFECTING PRECIPITATION

The factors affecting precipitation at a specific location are as follows:

- Height of the station
- Presence of mountains and their relative position
- Nearness of large lakes or oceans
- Forestation or urbanization
- Prevalent wind direction
- Sunspots

4.4.1 Height of Station

Normally temperature decreases with the altitude, and hence the moisture holding capacity of the air decreases. So also, the temperature reduces with the altitude and less moisture is available for precipitation. Thus, generally, higher the altitude lesser will be the precipitation. There are some exceptions to this tendency, but these exceptions have special conditions associated with them.

4.4.2 Presence of Mountains and their Relative Position

Mountains having their ridges across the direction of the flow of wind cause lifting of air. As the air containing moisture is lifted up, the temperature reduces and causes precipitation on the windward side. This is the case of *orographic precipitation.*

Of course, precipitation reduces on the leeward side of the mountains.

If the mountain range is oriented parallel to the movement of air, it does not affect the precipitation.

4.4.3 Nearness of the Oceans and Seas

A place near a large water body will have a higher relative humidity because of continuous supply of water vapour. If other favourable conditions exist, then it will ensure higher precipitation. So also, as the air moves towards land, precipitation will occur due to vertical movement of moisture and due to cooling. Normally coastal areas receive higher precipitation as compared to the interior ones.

4.4.4 Forestation and Urbanization

In the forest area, transpiration is more and the temperature is less. Thus, due to the forestation, precipitation tends to increase. Higher precipitation helps in an increase in forestation. The African and Amazon forests are good examples. The heat from the urban area is more and naturally the vapour-holding capacity of the air increases.

Precipitation due to urbanization, therefore, reduces.

4.4.5 Prevalent Wind Direction

Areas that receive winds from equatorial or warm oceans receive higher precipitation. Winds travelling from the cold areas receive little precipitation. The coasts near the equator facing east or west are rich in precipitation, whereas coasts facing north or south poles receive less precipitation.

The precipitation over India is named according to the direction from which the wind is blowing.

4.4.6 Sunspots

There are spots on the Sun and scientists have observed that these sunspots have a cycle of variation of 11 years, i.e. the nature of the sunspots are repeated after every 11 years. Some scientists believe that there is a correlation between the hydrological cycle and the sunspots cycle, but have failed to prove any exact correlation.

4.4.7 Precipitation Cycle

It was formerly assumed that the precipitation cycle in India repeats after 35 years. However, after scanning the observed data, it is now noticed that no such cycle of precipitation after 35 years is prevalent.

4.5 TYPES OF PRECIPITATION

The different types of precipitation are as follows:

- Cyclonic precipitation
- Convective precipitation
- Orographic precipitation

In nature, the effect of these types is interrelated and hence precipitation of any specific type cannot be exactly identified.

4.5.1 Cyclonic Precipitation

Very often, low-pressure belts are developed as a result of thermal variations in some regions, and hence air from the surrounding area flows towards these low-pressure belts. The air rushing from the surroundings changes into a whirling mass because of the rotary motion of the earth.

The movement of such whirling air results in cyclone formation. Some features of a cyclone are mentioned below:

- A cyclone may be defined as *a whirling mass of air at the centre of which the barometric pressure is low.*
- It is also known as *hurricane, blizzard, typhoon* and *tornado* according to the whirling speeds. An anticline similarly whirls round a high-pressure centre.
- A cyclone comprises of a very large mass of air spread over an area from 10 km^2 to as high as 1500 km^2 moving with a velocity of 100 km/h or even more.
- The whirling air mass carries water vapour with it. The central portion of the cyclone acts as a chimney through which the air gets lifted, expanded and cooled, and the water vapour gets condensed causing precipitation.

Thus, precipitation caused due to a cyclone is known as *cyclonic precipitation* and may result into a drizzle or into heavy precipitation covering a large area.

4.5.2 Convective Precipitation

On a hot day, the ground surface is heated unequally and so is the air near the ground surface. This causes air that is heated more to rise in atmosphere, to cool and then to condense resulting into precipitation. Such type of precipitation is called *convective precipitation* and normally covers a small area for a short duration but has a high intensity.

4.5.3 Orographic Precipitation

When the winds carrying sufficient water vapour are obstructed by a range of hills or mountains, they are mechanically lifted up. During this lifting, the air is cooled and condensation takes place. This results in heavy precipitation on the windward side and the precipitation on the leeward side reduces substantially.

The precipitation caused due to the obstruction by mountains is called *orographic precipitation.* In India, precipitation is mostly due to this type.

Figure 4.1 shows a specific case of orographic precipitation.

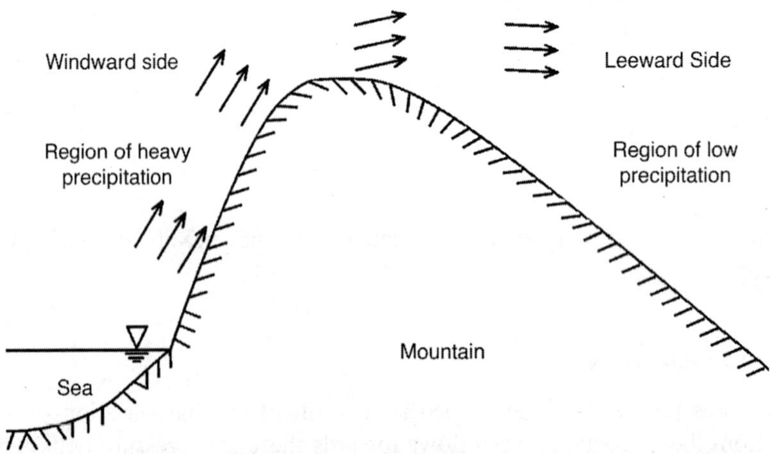

Fig. 4.1 Orographic precipitation

4.5.4 Artificial Precipitation

Precipitation can also be achieved artificially. Many times it happens that there is sufficient water vapour in the atmosphere and all other conditions are also favourable for precipitation, but there may be a deficiency of the hygroscopic nuclei, which are essential for the formation of raindrops.

A nucleus is essential for the formation of a raindrop. Artificial precipitation can be produced by providing the required hygroscopic nuclei. The hygroscopic nuclei for such artificial precipitation used so far are:

- Dry ice (solid CO_2)
- Silver iodide
- Sodium chloride
- Portland cement

This process of artificial formation of clouds is also known as *cloud seeding*. Cloud seeding is accomplished by the following methods:

- Emitting the hygroscopic nuclei with the help of special guns or through high-level chimneys
- Using aeroplanes for spreading the nuclei
- Using rockets by using suitable delivery systems (these nuclei are taken to the desired height followed by explosions)

Artificial rain-inducing system is not much prevalent because it has the following three major disadvantages.

1. It is a very costly process.
2. It may affect rainfall in the adjoining area.
3. The success rate is not very high.

4.6 MEASUREMENT OF PRECIPITATION

The instrument used to measure rainfall is known as *rain gauge*. All forms of precipitation are measured on the basis of the vertical depth of water that would accumulate on a plain surface, if the precipitation remains where it falls. It is measured in millimetres or tenths of millimetres.

In olden days, a rain gauge was known as *hyetometer, ombrometer or pluviometer*.

4.6.1 History of Precipitation Measurement

The earliest systematic hydrological measurements ever made were probably those of precipitation. There is evidence that precipitation was measured as far back as 400 BC. However, precipitation measurements done by Mr Townley in 1671 are available even now.

The measurement of precipitation is never subjected to check either by repetition or by duplication and hence care should be taken during its measurement. So also, rainfall measurement forms the basic data on which all the designs are based and hence utmost accuracy has to be observed during its measurement. The precipitation per unit time is the *intensity of precipitation.*

It is measured as mm/h or cm/day.

4.6.2 Considerations for Selecting Site for a Rain Gauge

The following considerations should be followed while installing a rain gauge:

- The site should be an open-level ground.
- Clear distance between the obstruction and the rain gauge should be at least twice the height of the obstruction. In no case, it should be nearer to the rain gauge than 30 m.
- The site should be representative of the area.
- Roof installation and windward slopes should be avoided.
- The serious factor that affects the rainfall measurement is wind velocity, its vertical as well as horizontal components. Obstruction to either should be avoided.
- The rain gauge should be installed upright in a vertical mode.

A gauge inclined at 10° with vertical will catch 1.5% less rainfall as compared to that in a vertical direction. Some engineers consider that the rain gauge should be kept perpendicular to ground and the rainfall thus calculated may be multiplied by the cosine of the angle of inclination of the ground surface. However, this is not practicable and hence avoided.

The best site would be a level ground with trees and bushes all around, which serve as windbreak. However, care should be taken that these trees and bushes do not affect the catch of the rain gauge and the distance between the two is more than twice their height.

4.6.3 Hyetograph

A bar chart of time versus precipitation is known as *hyetograph. The ordinate graph* presents the rainfall in a year drawn to some scale at the corresponding year.

Specific and ordinate graphs are shown in Figs. 4.2 and 4.3, respectively.

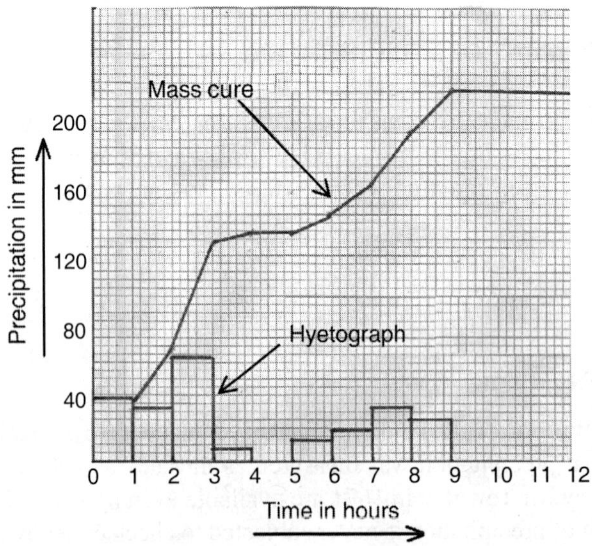

Fig. 4.2 Mass curve and hyetograph of precipitation

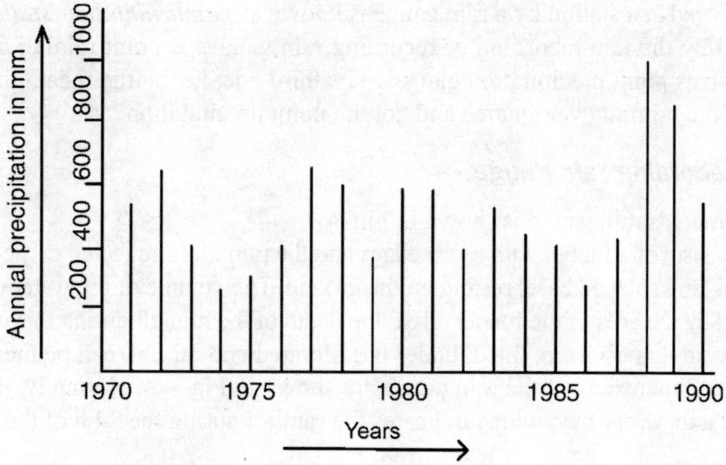

Fig. 4.3 Ordinate graph of precipitation

4.6.4 Mass Curve of Precipitation

A mass curve of precipitation is a cumulative plot of the accumulated precipitation.

A mass curve is shown in Fig. 4.2.

The slope of the mass curve is the intensity of rainfall at that time. When there is no rainfall, the mass curve is horizontal.

Example 4.1

The hourly precipitation data during a storm are as follows:

Time (h)	0	1	2	3	4	5	6	7	8	9	10
Precipitation (mm)	0	30	25	50	5	0	10	15	25	20	0

Plot 1. Hyetograph
2. Mass curve

Solution:

The mass curve coordinates will be as follows:

Time (h)	0	1	2	3	4	5	6	7	8	9	10	11	12
Precipitation (mm)	0	30	55	105	110	110	120	135	160	180	180	180	180

The hyetograph and the mass curve are shown in Fig. 4.2.

4.6.5. Types of Rain Gauges

The different types of rain gauges are as follows:

- Non-recording rain gauge
- Recording or automatic rain gauge
- Radar

The rainfall observed at a station by a rain gauge is known as *point rainfall or station rainfall*. The rainfall observed by the non-recording or recording rain gauges is point rainfall and hence these gauges are known as point precipitation gauges. The third one, i.e. by the radar, gives an instantaneous picture of the rainfall over an area and not the point precipitation.

4.6.5.1 Non-recording rain gauge

The standard Symons's rain gauge is shown in Fig. 4.4.

The gauge consists of a funnel with sharp edges and the rainfall is collected in the cylinder, which has a narrow neck and splayed base, resting on the ground. The volume of the water collected in specific time divided by the area of the funnel, gives the depth of the rainfall during this period.

In order to avoid this division, the cylinder is calibrated and rainfall can be measured immediately. The various dimensions of the rain gauge are shown in Fig. 4.4. Normally, the observations are taken at 8:30 a.m. every day, which indicates the rainfall during the 24 h of the previous day.

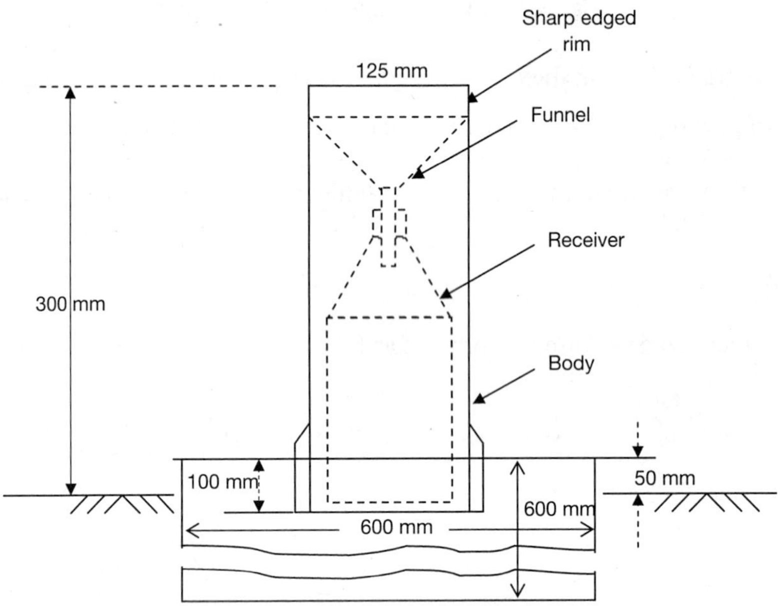

Fig. 4.4 Non-recording rain gauge

Disadvantages in the non-recording rain gauge

The disadvantages in the non-recording rain gauge are as follows:

- It records the precipitation only during the observation period and not the intensity of precipitation.
- If the rainfall is heavy, the cylindrical collecting bottle may overflow and thus may not record the entire quantum of precipitation.

4.6.5.2 Automatic rain gauges

In this case the rainfall is recorded automatically using different mechanisms. The different types of automatic rain gauges are as follows:

- Siphon bucket-type rain gauge
- Tipping bucket-type rain gauge
- Weighing bucket-type rain gauge

All these automatic rain gauges record not only the rainfall but also its intensity.

Siphon bucket-type rain gauge

The rainfall is collected in a float chamber through a funnel. The funnel has a sharp rim. The float chamber contains a light float that rises in the chamber as rainfall is collected in the chamber.

Thus, the level of the float is indirectly a measure of the rainfall.

The vertical movement of the float is recorded by a mechanism on a paper chart fixed on a drum, by a pen. This pen rests on the chart. This chart is kept rotating at a uniform speed on the drum by a spring wound mechanism. The drum makes one rotation a day.

Thus, the chart records time on the x-axis and rainfall on the y-axis.

There is a small compartment by the side of the float chamber. This is connected by a small opening at the bottom. This is known as the *siphon chamber*. A small vertical pipe is used that works as a siphon. This siphon is used to empty the chamber to avoid the increase in size of the float chamber. As the level of water in the float chamber reaches a specific level, the siphon starts working and water in the float chamber is drained. The float moves down to its lowest position. Along with the float, the pen moves down and draws a vertical line on the chart. As the rain water is again collected in the float chamber due to rainfall, the float moves and with it the pen too.

Normally, the paper on the drum is replaced everyday at 8:30 a.m. Figure 4.5 shows the details of a siphon bucket-type rain gauge and the rainfall chart. The chart is the mass curve of precipitation.

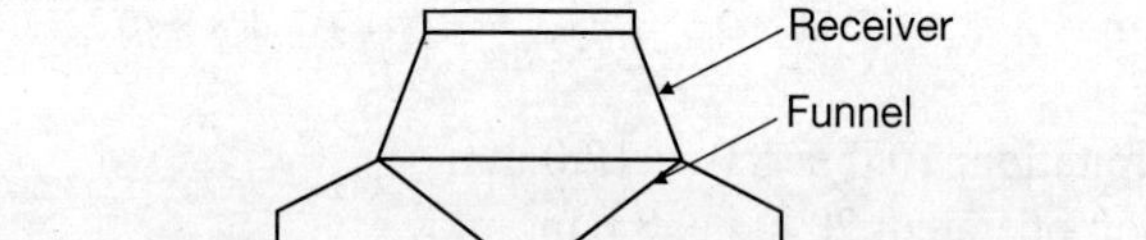

Fig. 4.5 Siphon bucket-type rain gauge

The vertical line indicates that the siphon has worked and a horizontal line indicates no precipitation. The slope of the chart indicates the intensity of rainfall.

The siphoning action requires 15 seconds and record of rainfall during this period is lost. So also the rainfall record may be lost during the time taken in changing the paper, if there is rainfall during that time.

Example 4.2

The chart fixed to an automatic float-type rain gauge is shown in Fig 4.5

Find
1. Hourly precipitation
2. Daily precipitation
3. Time when the siphon was operated
4. Period of no precipitation
5. Maximum intensity of precipitation

Solution:

1. The hourly precipitation as read from the chart will be as follows:

Time	8	9	10	11	12	13	14	15	16	17	18	19
Cumulative ppt (mm)	0	0.25	0.5	1	2.5	4	4	4	4.9	6	7.4	8.5
Precipitation (mm)	0	0.25	0.25	0.5	1.5	1.5	0	0	0.9	1.1	1.4	1.1

Time	20	21	22	23	24	1	2	3	4	5	6	7	8
Cumulative ppt (mm)	10	0.5	1	1.8	2.8	4.2	6	6	6	6.5	7	7.5	8
Precipitation (mm)	1.5	0.5	0.5	0.8	1	1.4	1.8	0	0	0.5	0.5	0.5	0.5

2. Total daily precipitation: 10.0 + 8.0 = 18.0 mm
3. Time when siphon operated: 20 h, i.e. 8 p.m.
4. Time of no precipitation: 13–14
   14–15
   2–3
   3–4
   Total 4 h
5. Maximum intensity: 1.8 mm/h from 1 to 2 h

Tipping bucket-type rain gauge

The details of a tipping bucket-type rain gauge are shown in Fig. 4.6.

In this type, the rainwater received through the funnel is collected alternately in twin buckets. When one bucket receives water and the water level reaches a specific level, it tips along a pivot and discharges into a measuring jar. As the bucket tips, the other twin bucket receives the rainwater. Both the buckets receive the rainwater alternately and discharge into the same measuring bucket.

The funnel diameter, the bucket size and shape are so designed that the bucket may tip after it has received a rainfall of 25 mm. The tipping of the bucket is recorded by an electric circuit or by some mechanism on a drum. The measuring jar is calibrated to give the depth of rainfall.

This rain gauge gives the total precipitation during the specific period. The intensity of rainfall can be worked out from the time required to tip the bucket as well as the number of tippings of each

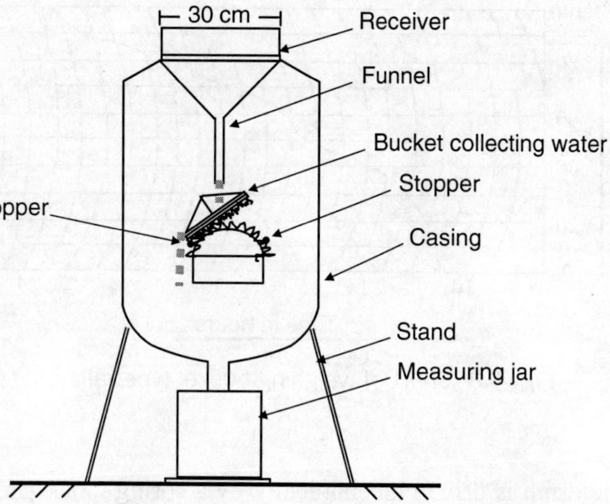

Fig. 4.6 Tipping bucket-type rain gauge

bucket. However, for low intensity of rainfall, the results may not be very accurate since no-rainfall period is not recorded in this rain gauge.

Weighing bucket-type rain gauge

The details of a weighing bucket-type rain gauge are shown in Fig. 4.7.

The rainfall is collected through a funnel into a bucket resting on a platform. The weight of water thus collected is transformed through a mechanism to a pen, which makes a trace on a paper

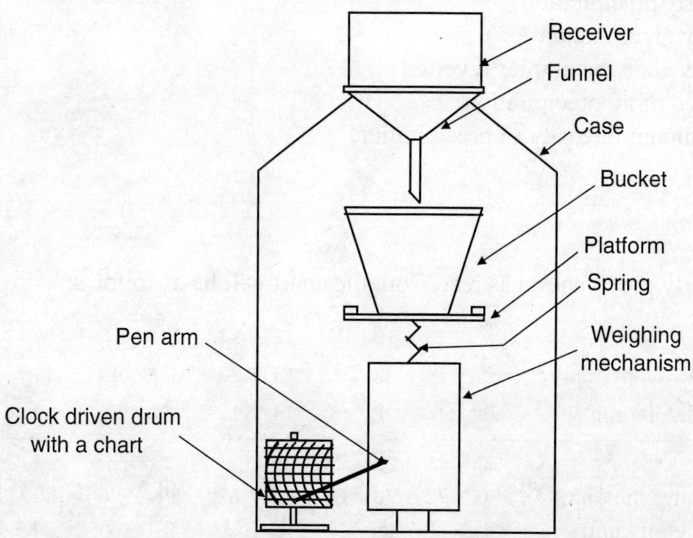

Fig. 4.7 Weighing bucket-type rain gauge

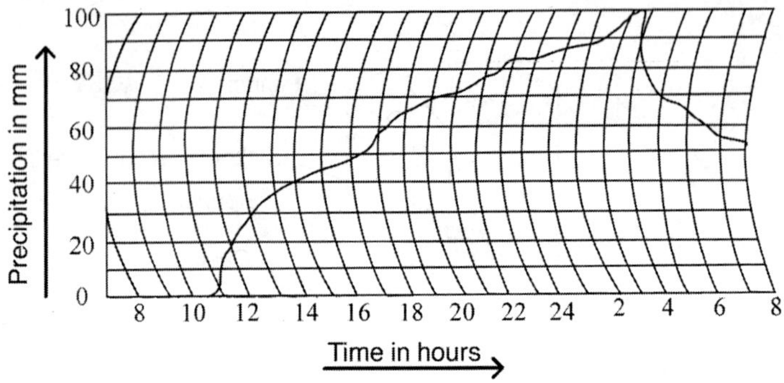

Fig. 4.8 Chart of weighing bucket-type rain gauge

fixed on a drum. The drum is driven mechanically by a spring clock and makes one revolution a day at a uniform speed. Thus, the trace of the pen on the paper records the weight of rainwater and indirectly the rainfall.

After a certain amount of precipitation, the mechanism reverses the travel of the pen. However, the movement of the drum remains unaltered. Figure 4.8 shows a typical rainfall chart of a weighing bucket-type rain gauge.

The horizontal line indicates the no-rainfall period. Reversal of line has to be taken into account. With this, the total precipitation as well as the intensity of precipitation can be calculated accurately.

This type of rain gauge can be used for the measurement of snow, rainfall, hail, and so on.

Example 4.3

The chart fixed to an automatic weighing bucket-type rain gauge reads as in Fig. 4.8:

Find
1. Hourly precipitation
2. Daily precipitation
3. Time when the pointer reverted
4. Period of no precipitation
5. Maximum intensity of precipitation

Solution:

1. The hourly precipitation as read from the chart will be as follows:

Time	8	9	10	11	12	13	14	15	16	17	18	19	
Cumulative ppt (mm)	0	0	0	5	18	30	37	40	44	48	56	66	
Precipitation (mm)	0	0	0	5	13	12	7	3	4	4	8	10	
Time	20	21	22	23	24	1	2	3	4	5	6	7	8
Cumulative ppt (mm)	70	72	78	83	84	87	89	96	100	70	65	55	52
Precipitation (mm)	4	2	6	5	1	3	2	7	4	30	5	10	3

2. Total daily precipitation: 100 + (100 − 52) = 148 mm
3. Time when the pointer reverted: 4 h
4. Period of no precipitation: 8–9
   9–10
   Total 2 h
5. Maximum intensity of precipitation: 30 mm/h from 4 to 5 h

Difference between the record graph papers

The difference between the record graph papers of siphon-type rain gauge and weighing-type rain gauge is as follows:

1. The pen, in the case of a float-type, records the vertical movement of float and the graph paper is a squared graph paper. However, in case of weighing-type, the weight of water collected on the platform is transferred to the pen through a point and hence its movement is along the arc of a circle with pivot as the centre.
2. In the case of float-type, when the level of water in the collection chamber reaches the specific limit, the siphon operates and the pen records a vertical line up to the x-axis. On the contrary, when the weight of water collected on a platform reaches the specific limit, the mechanism reverts and the graph records precipitation in the negative direction.

4.6.5.3 Radar measurement of rainfall

Radar is an acronym and its full form is **RA**dio **D**etection **A**nd **R**anging. It was originally designed to detect the location of aircrafts flying in the air.

High frequency electromagnetic waves are radiated in the atmosphere and the reflections of these waves due to the rainfall (which is known as *echo*) are recorded on a radarscope.

The intensity of rainfall, its duration and also the coverage can be estimated by scanning the intensity of radiation of the electromagnetic waves, the echo due to rainfall, the time of travel, and so on. The range of a radar is about 200 km. This procedure is very costly and requires high-level instrumentation.

4.6.6 Difficulties in the Measurement of Precipitation

The difficulties experienced in the measurement of precipitation are as follows:

- The rain gauge itself may cause eddy currents and may affect the catch of the rain gauge.
- There may be some evaporation from the water collected in the rain gauge.
- Some of the precipitation water may be lost in wetting the sides of the funnel or the measuring flask.
- In some case, the splash into or out of the gauge may modify the true value of rainfall.

4.6.6.1 Rainy days

When the rainfall during a day is 2.5 mm or more the day is known as *rainy day*.

4.6.6.2 Classification of intensity of rainfall

The rainfall intensity is classified as follows:

1. Up to 2.5 mm/h: Light rain
2. 2.5–7.5 mm/h: Medium rain
3. 7.5 mm/h and above: Heavy rain

4.6.6.3 Average precipitation

The average precipitation, whether it is annual, seasonal or even daily, is taken as the average of the last 30 years. It is revised after every 10 years by deleting the previous 10-year data and adding the recent 10-year data.

4.6.6.4 Index of wetness

Index of wetness is the ratio of rainfall in a given year and annual average precipitation. When this index is less than one, it is called a *bad year* or *a deficient year* or *a dry year*. When it is more than one, it is called a *good year* or *a surplus year* or *a wet year*. When it is one, it is a *normal year*.

4.7 SUPPLEMENTING RAINFALL DATA

Sometimes there might be breaks or gaps in the rainfall data of some stations. These gaps may be due to the following reasons:

- Absence of an observer
- Instruments failure
- Unapproachable circumstances such as flooding or washing out of the approach road, etc., resulting in non-recording of the rainfall

The missing rainfall data may be supplemented by the following methods:

- Arithmetic average method
- Normal ratio method
- Weighted average method

4.7.1 Arithmetic Average Method

When the data of a rain gauge station, e.g. 'A', is missing for a year, consider the adjacent rain gauge stations—B, C and D. If the rainfall at these adjacent stations B, C and D is within 10% of the average rainfall of A (less or more), then the simple arithmetic average of these adjacent stations of the year may be considered as the missing rainfall of A for that year.

$$P_A \text{ (missing)} = \frac{1}{n}(P_B + P_C + P_D + \ldots + P_n)$$

where, n = Number of adjacent rain gauge stations

P_B, P_C, P_D = Precipitation at B, C and D for the period for which the data is missing at A

Example 4.4

In a catchment area, daily precipitation was observed by 11 rain gauge stations. On 2 August 2005, the observations indicated that one rain gauge was out of order. The observations taken by the 10 rain gauge stations are as follows:

Station	A	B	C	D	E	F	G	H	I	J	K
Precipitation (mm)	21	23	19	20	23	24	19	?	21	22	18

Estimate the missing data at H.

Solution:

Since there is not much variation in the precipitation data, a simple arithmetic average of the precipitation observed at the 10 remaining stations was taken as under.

$$\text{Precipitation at H} = \frac{21 + 23 + 19 + 20 + 23 + 24 + 19 + 21 + 22 + 18}{10}$$

$$= \frac{210}{10} = 21 \text{ mm}$$

4.7.2 Normal Ratio Method

If the rainfall of station A is missing for a year and the variation of the adjacent rain gauge stations B, C and D is more than 10%, then simple principle of linearity is used to evaluate the missing rainfall of A as follows:

$$P_A \text{ (missing)} = \frac{1}{n}\left[\frac{N_A}{N_B} \times P_B + \frac{N_A}{N_C} \times P_C + \frac{N_A}{N_n} \times Pn\right]$$

where, N_A = Average of A excluding the missing period
$N_B\ N_C$ = Average of B and C excluding the missing period
$P_B\ P_C$ = Precipitation at B and C during the missing period
n = number of adjacent rain gauge stations considered

For this method, a minimum of three adjacent rain gauge stations are considered.

Example 4.5

The average annual precipitation at five rain gauge stations in a catchment is as follows:

Station	P	Q	R	S	T
Average annual precipitation (mm)	2400	2332	2431	2207	2231

However, the precipitation at station P was not available for the year 1996 because the rain gauge was out of order. The precipitation observed at the other stations for 1996 was as follows:

Station	P	Q	R	S	T
Precipitation (mm)	?	2113	2200	2028	2095

Evaluate the precipitation at P during 1996.

Solution:

$$\text{Precipitation at } P \text{ (in mm)} = \frac{1}{4}\left(\frac{2400}{2332} \times 2113 + \frac{2400}{2431} \times 2200 + \frac{2400}{2207} \times 2028 + \frac{2400}{2231} \times 2095\right)$$

$$= \frac{1}{4}(2174 + 2172 + 2205 + 2253)$$

$$= 2201 \text{ mm}$$

4.7.3 Weighted Average Method

The station of which the data is missing, e.g. A, is considered as the origin and the adjacent rain gauge stations are considered and their distances from A are calculated or measured on a map.

It is assumed that the rainfall variation between these two stations A and B is inversely proportional to the square of the distance between them.

Thus,

$$P_A \text{ (missing)} = \frac{P_B/r_1^2 + P_C/r_2^2 + P_D/r_3^2}{1/r_1^2 + 1/r_2^2 + 1/r_3^2}$$

where, r_1, r_2 and r_3 = the distances from A of the adjacent stations B, C and D
P_B, P_C and P_D = the rainfall at the adjacent stations B, C and D for the missing period.

This method is also known as United States National Weather Service (USNWS) method.

Example 4.6

The location coordinates in km of the five rain gauge stations w.r.t. X are as follows:

Station	X	A	B	C	D
x and y coordinates (km) w.r.t. X	0,0	20,25	−40,15	−30,−20	25,−15

The annual precipitation at X for the year 2005 is missing. The annual precipitation at the remaining four stations for 2005 is as follows:

Station	A	B	C	D
Precipitation (mm)	2735	2805	2680	2560

Evaluate the missing precipitation at X for the year 2005.

Solution:

The relative positions of the stations A, B, C and D w.r.t. X are shown in Fig. 4.9.

$AX = \sqrt{(20^2 + 25^2)} = 32.01$ km
$BX = \sqrt{(40^2 + 15^2)} = 42.72$ km
$CX = \sqrt{(30^2 + 20^2)} = 36.05$ km
$DX = \sqrt{(25^2 + 15^2)} = 29.15$ km

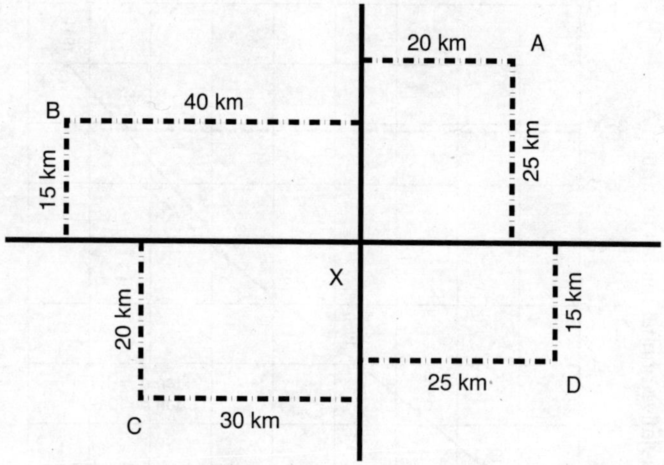

Fig. 4.9 Estimation of missing precipitation data

$$P_X = \frac{\frac{P_A}{AX^2} + \frac{P_B}{BX^2} + \frac{P_C}{CX^2} + \frac{P_D}{DX^2}}{\frac{1}{AX^2} + \frac{1}{BX^2} + \frac{1}{CX^2} + \frac{1}{DX^2}}$$

$$= \frac{\frac{2735}{32.01^2} + \frac{2805}{42.72^2} + \frac{2680}{36.05^2} + \frac{2560}{29.15^2}}{\frac{1}{32.01^2} + \frac{1}{42.72^2} + \frac{1}{36.05^2} + \frac{1}{29.15^2}}$$

$$\text{Precipitation at X} = \frac{2.66 + 1.54 + 2.06 + 3.01}{(9.76 + 5.48 + 7.69 + 11.77) \times 10^{-4}}$$

$$= \frac{9.27 \times 10^4}{34.70} = 2671 \text{ mm}$$

4.8 CONSISTENCY VERIFICATION OF RAIN GAUGE

It may happen that the rainfall recorded by a rain gauge station is doubtful. It then becomes necessary to verify the rainfall record of this station. This is known as *verification of consistency of a rain gauge*. This may by due to the following reasons:

- Change in the location of the rain gauge
- Change in the surroundings, namely, growth of trees, buildings, and so on.
- Change in the instrument
- Fault developed in the rain gauge

The verification can be done by the double mass curve method.

4.8.1 Double Mass Curve Method

On a simple graph paper, the mass curve of the precipitation of the doubtful station, e.g. 'A' versus the mass curve of the average precipitation for the remaining rain gauge stations whose data are available for the corresponding period is plotted.

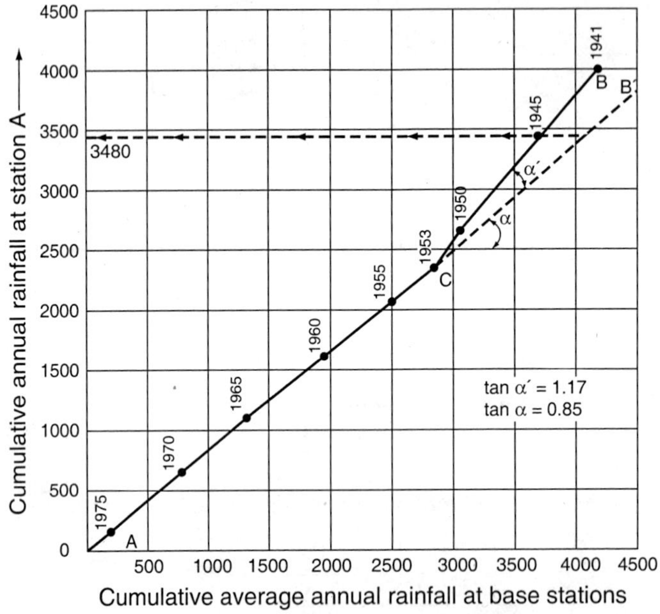

Fig. 4.10 A double mass curve plotted for a specific study

Normally, one should get a straight line through origin if the record at A is correct. If there is inconsistency at A from a particular year, the slope of the straight line may change from that year. It may, therefore, be concluded that the records of A are incorrect from that year and need modification.

The slope of the straight line is maintained and extended and the record of A is corrected accordingly. This procedure cannot be applied for studies for storm rainfall or daily rainfall.

A double mass curve plotted for a specific study is shown in Fig. 4.10.

Example 4.7

The average annual precipitation data of six rain gauge stations in a catchment area during 1991–2000 are as follows:

Station	A	B	C	D	E	F
Year	Precipitation (mm)					
1991	608	704	378	768	576	544
1992	663	768	402	837	628	593
1993	611	707	388	722	519	548
1994	704	815	433	889	668	629
1995	606	702	374	766	575	543
1996	722	836	446	913	685	646
1997	647	748	402	815	612	578
1998	797	924	717	1007	756	713
1999	721	835	700	913	683	646
2000	686	794	680	865	649	612

The data observed at station C were doubtful because of some topographical changes there. Verify whether the data observed at C is consistent and correct it, if necessary.

Solution:

The mass curve coordinates of precipitation at C and also the combined mass curve coordinates of precipitation at A, B, D, E and F will be as follows:

Serial no.	Mass curve for C	Corrected mass for A, B, D, E and F	Corrected curve for C	Corrected C precipitation at C (mm)	Remarks
1	378	3200	—	—	
2	780	6689	—	—	
3	1168	9796	—	—	
4	1601	13501	—	—	
5	1975	16693	—	—	
6	2421	20495	—	—	
7	2823	23895	—	—	←Graph
8	3540	28092	3302	479	changes
9	4240	31890	3756	454	slope
10	4920	35496	4173	417	

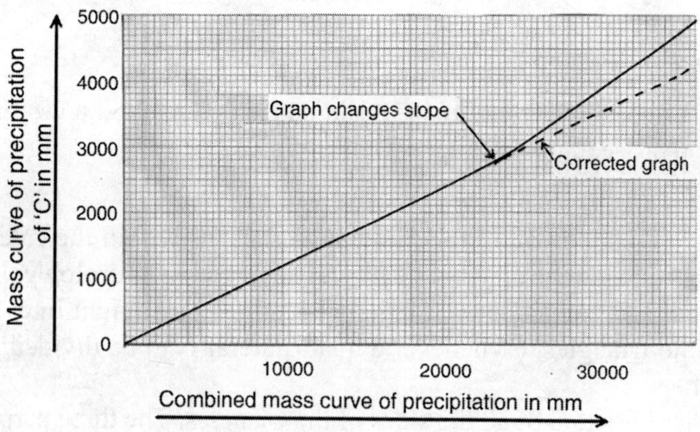

Fig. 4.11 Double mass curve of precipitation

A graph of mass curve of precipitation at A, B, D, E and F versus mass curve of precipitation at C was plotted on a simple graph paper. It was noticed that the curve follows a straight line up to 1997. It changes its slope from 1997.

This means that due to the change in topography at C, the observations from 1997 were incorrect. The straight line graph up to 1997 was extended further and the precipitation at C was corrected (as shown in Fig. 4.11) from this extended graph for the respective years and entered in the table.

4.9 AVERAGE DEPTH OF PRECIPITATION

The precipitation over a catchment area is never uniform. This becomes quite clear from the figures of the average depth of precipitation of the various rain gauge stations in the catchment area. One of the basic requirements in the study of a catchment area is the *average depth of precipitation over the entire catchment.*

This is also known as *equivalent uniform depth of rainfall.* The average depth of precipitation can be calculated by the following methods:

- Arithmetic mean method
- Thiessen polygon method
- Isohyetal method

4.9.1 Arithmetic Mean Method

This method is very simple. The arithmetic mean of average precipitation values of all the rain gauges within the catchment area is determined, namely,

$$P_{av} = \frac{P_A + P_B + P_C + \ldots}{n}$$

where, P_{av} = Average precipitation over the catchment area

P_A, P_B, P_C = Average precipitation at various rain gauge stations A, B, C, ...

This method may be adopted when the catchment area is flat and there is not much variation in the average values of the precipitation of all the rain gauge stations. This method is very simple for calculations. However, no weightage is given neither to the influence area of the individual rain gauge nor the topography. This method is also known as *unweighted mean method.*

4.9.2 Thiessen Polygon Method

A. M. Thiessen suggested this method in 1911 and hence this method takes this name after him. A regular step-by-step procedure is as follows:

1. Mark the catchment area correctly on a sheet of paper.
2. Mark correctly the locations of all the rain gauge stations within the catchment area. Other rain gauge stations outside the area but nearby, if any, having hydrological homogeneity may also be considered. Join the adjacent rain gauge stations by straight lines so as to divide the catchment area into triangles. (Whenever a quadrilateral is to be divided, the shorter diagonal is preferred.)
3. Draw perpendicular bisectors of all the sides of the triangles. The three perpendicular bisectors of a triangle will meet at a point inside the triangle. Near the ridge line of the catchment, the perpendicular bisector lines may be extended beyond the ridge line.

Thus, every rain gauge station will be surrounded by a polygon and it is presumed that the rain gauge station has got an influence on the area of the polygon surrounded by it.

The area of each polygon is measured. The area of influence lying inside the catchment area of the rain gauge, located outside the catchment, should also be considered. The average precipitation over the catchment area is worked out as follows:

$$P_{av} = \frac{P_A A_A + P_B A_B + P_C A_C + \ldots}{A}$$

where, P_{av} = Average precipitation over the catchment area

P_A, P_B, P_C = Average precipitation at various rain gauge stations

A_A, A_B, A_C = Area of the polygon surrounding the rain gauge stations A, B and C

A = Catchment area = $A_A + A_B + A_C + \ldots$

P_{av} can also be mentioned as follows:

$$P_{av} = \frac{A_A}{A} \times P_A + \frac{A_B}{A} \times P_B + \frac{A_C}{A} \times P_C + \ldots$$

and the factors $\frac{A_A}{A}, \frac{A_B}{A}$ and $\frac{A_C}{A}$ are called *Thiessen's weights*.

In this method, weightage to each rain gauge is given depending upon its influence area. Rain gauges having hydrologic homogeneity outside the area, if any, are also considered for more accuracy. However, no consideration is given to the topography.

Example 4.8

Figure 4.12 shows a typical layout of a catchment area ABCDF. Six rain gauge stations are established at A, B, C, D, E and F, as shown in the figure. The precipitation observed at these six stations in July 2004 is as follows:

Station	A	B	C	D	E	F
Precipitation (mm)	100	120	130	180	125	150

Find the average precipitation over the catchment during July 2004 by Thiessen's polygon method.

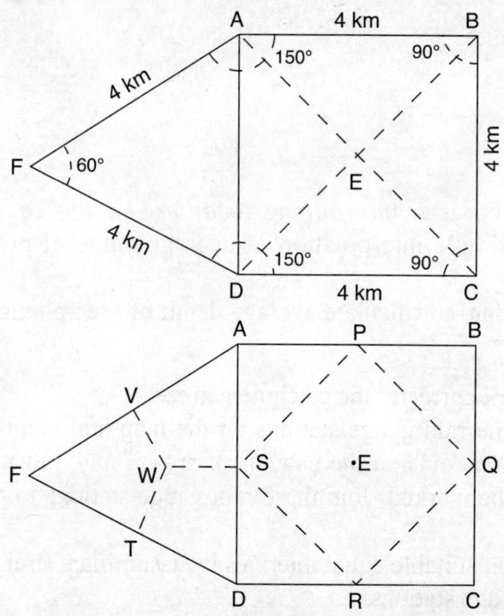

Fig. 4.12 Rain gauge stations surrounded by a polygon

Solution:

Join the locations of the adjacent rain gauge stations, AB, BC, CD, DA, AE, BE, CE, DE, DF and FA, and draw their perpendicular bisectors. The perpendicular bisectors will meet at P, Q, R, S, T, V and W.

Thus, each rain gauge station will be surrounded by a polygon and its area will be as follows:

Serial no.	Station	Polygon	Area (km^2)
1	A	APSWV	$\frac{1}{2} \times 2 \times 2 + \frac{1}{3} \times \frac{1}{2} \times 4 \times 2\sqrt{3} = 4.309$
2	B	BQP	$\frac{1}{2} \times 2 \times 2 = 2.00$
3	C	CRQ	$\frac{1}{2} \times 2 \times 2 = 2.00$
4	D	DTWSR	$\frac{1}{2} \times 2 \times 2 + \frac{1}{3} \times \frac{1}{2} \times 4 \times 2\sqrt{3} = 4.309$
5	E	PQRS	$\frac{4 \times 4}{2} = 8.00$
6	F	FVWT	$\frac{1}{3} \times \frac{1}{2} \times 4 \times 2\sqrt{3} = 2.309$
		Total	22.927 km^2

Therefore, average precipitation over catchment in July 2004 is as follows:

$$= \frac{4.309 \times 100 + 2 \times 120 + 2 \times 130 + 4.309 \times 180 + 8 \times 125 + 2.309 \times 150}{22.927}$$

$$= 133 \text{ cm}$$

4.9.3 Isohyetal Method

An isohyet may be defined as *a line joining locations having equal rainfall*. Isohyets are drawn by the method of simple interpolation of average value of precipitation similar to the level contours.

The step-by-step procedure to calculate average depth of precipitation by isohyetal method is as follows:

1. On a sheet of paper, mark correctly the catchment area.
2. Show the locations of the rain gauge stations on the map and mention their average precipitation. Rain gauge stations outside the catchment area if any, but nearby, having hydrologic homogeneity may also be marked. Join these rain gauge stations to divide the catchment area into triangles.
3. Isohyets may be drawn at suitable equal intervals by assuming a straight line variation between the two adjacent rain gauge stations.
4. The area between two isohyets inside the catchment area should be measured accurately.

5. The average precipitation of the area between two isohyets may be calculated as follows:

$$P_1 = B + \frac{i(2a + b)}{3(a + b)} \quad \text{or} \quad A - \frac{i(a + 2b)}{3(a + b)}$$

where, P_1 = Average precipitation between two isohyets
A = Higher value between the two consecutive isohyets
B = Lower value between the two consecutive isohyets
i = Isohyetal interval (A−B)
a = Length of higher value of isohyet A within the catchment's area
b = Length of lower value of isohyet B within the catchment's area

However, this procedure is complicated.

Alternately, the average precipitation of the area between the two isohyets is calculated by taking a simple mean of the two isohyets, i.e. $\frac{(A+B)}{2}$.

For the area between the higher value isohyet (A) and the ridge line, it is not possible to calculate the average precipitation. In such case, the average is considered as A only. Similarly the average precipitation for the area between the lower value isohyet and the ridge line is considered as only B.

Thus, the average of the catchment area may be calculated as follows:

$$P_{av} = \frac{P_1A_1 + P_2A_2 + P_3A_3 + \ldots}{A}$$

where, P_{av} = Average precipitation over a catchment
P_1 = Average precipitation between two consecutive isohyets
A_1 = Area between two consecutive isohyets within the catchment area
A = Catchment area = $A_1 + A_2 + A_3 + \ldots$

Drawing isohyets is to some extent a skilled job. While drawing isohyets, one may consider the topographical effect, and such other things. Since rain gauges outside the area, but nearby, having hydrologic homogeneity are considered, this isohyetal method is more accurate.

Use of the three methods followed for a specific study is indicated in Fig. 4.13.

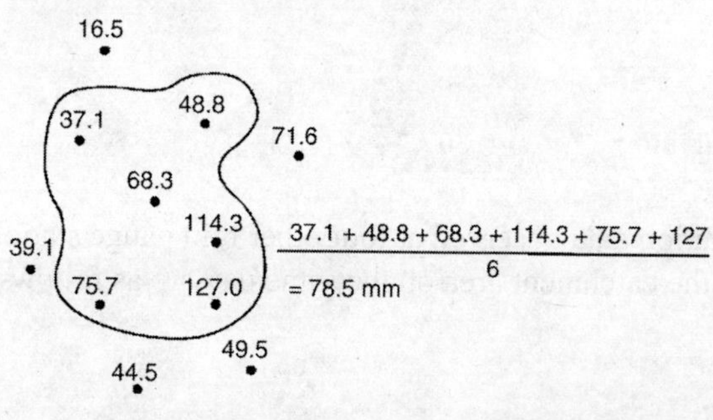

Fig. 4.13a Arithmetic mean method

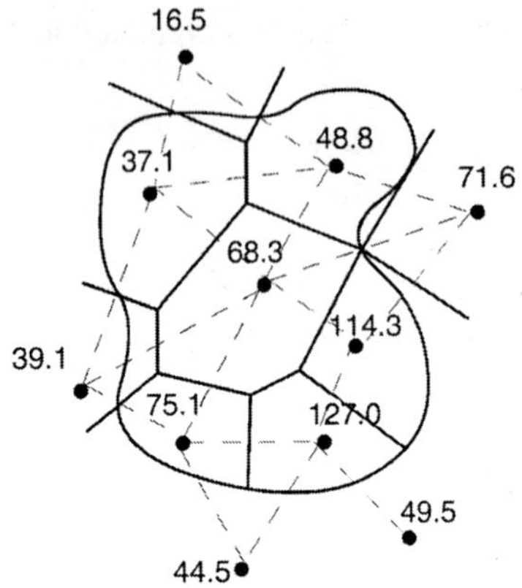

Precipitation	Area	A x ppt
16.5	18	297
37.1	31	11538
48.8	282	13761
68.3	311	21241
39.1	52	2033
75.7	238	18016
127.0	212	26924
114.3	197	22517
Total	1621	116327

$$\text{Average precipitation} = \frac{116327}{1621} = 72 \text{ mm}$$

Fig. 4.13b Thiessen polygon method

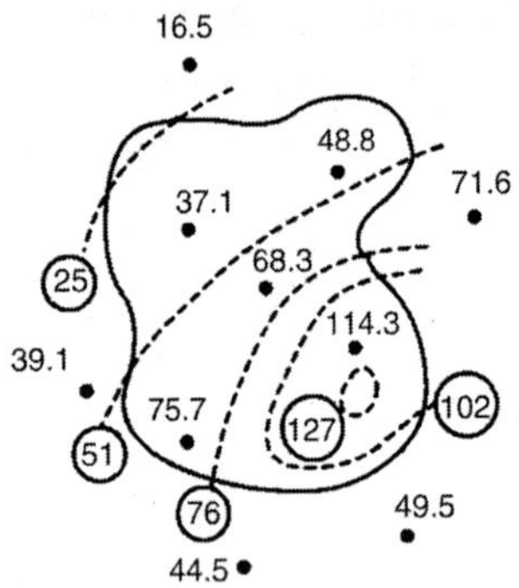

Isohyet	Average	Area	A x Avg
25 & less	25.0	80	2000
25 & 51	38.0	500	19000
51 & 76	63.5	508	32258
76 & 102	89.0	300	26700
102 & 127	114.5	199	22785
127 & above	127.0	34	4318
Total		1621	107061

$$\text{Average precipitation} = \frac{107061}{1621} = 66 \text{ mm}$$

Fig. 4.13c Isohyetal method

Example 4.9

The daily precipitation data observed at four rain gauge stations located inside a catchment area on 2 August 2005 are as follows:

Station	B	D	F	H
Precipitation (mm)	30	35	50	45

So also the daily precipitation data observed at four other rain gauge stations that are meteorologically similar but outside the catchment area on the same day are as follows:

Station	A	C	E	G
Precipitation (mm)	40	25	45	55

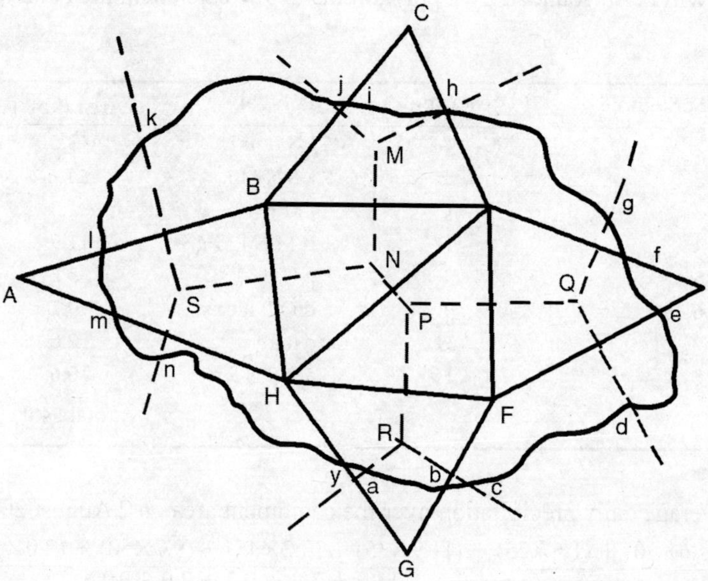

Fig. 4.14 Rain gauge surrounded by a polygon inside the catchment area

Figure 4.14 shows the catchment area and the locations of the rain gauge stations. If the catchment area is 54 km^2, find the daily average precipitation over the catchment.

Solution:

The average daily precipitation over the catchment will be calculated by three different methods as stated below.

1. Arithmetical mean method
2. Thiessen's polygon method
3. Isohyetal method

1. *Arithmetical mean method* The average daily precipitation of the catchment area will be the average of the four rain gauge stations located inside the catchment area.

$$\text{Average daily precipitation of the catchment area} = \frac{30 + 35 + 50 + 45}{4} \text{ mm}$$

$$= \frac{160}{4} = 40 \text{ mm}$$

2. *Thiessen's polygon method* Join the locations of the adjacent rain gauge stations by straight lines as, AB, AH, BH, BC, BD, CD, DE, DF, EF, FG, GH and HF. Draw perpendicular bisectors of these lines to meet

- Inside the catchment area at M, N, P, Q, R, S
- The catchment area boundary at a, b, c, d, e, f, g, h, i, j, k, l, m, n, y

Each rain gauge will be surrounded by a polygon inside the catchment area and its area will be as follows:

Serial no.	Station	Polygon	Area (km^2)
1	A	k S n m l	3.5
2	B	k S N M j	11.6
3	C	j M h	1.5
4	D	h M N P Q g	11.9
5	E	g Q d	3.2
6	F	d Q P R c	9.7
7	G	c R a	2.0
8	H	a R P N S n	10.6
			Total 54.0

Therefore, the average daily precipitation over the catchment area on 2 August 2005 will be

$$= \frac{(3.5\times40) + (11.6\times30) + (1.5\times25) + (11.9\times35) + (3.2\times45) + (9.7\times50) + (2.0\times55) + (10.6\times45)}{3.5 + 11.6 + 1.5 + 11.9 + 3.2 + 9.7 + 2.0 + 10.6}$$

$$= \frac{2158}{54} = 39.96 \text{ mm}$$

3. *Isohyetal method* Taking into consideration the location of the rain gauge stations inside (B, D, F, H) as well as outside the catchment area (A, C, E, G) and the precipitation recorded by them, isohyets of 30, 35, 40, 45, 50 mm are drawn and area between them was measured accurately (Fig. 4.15).

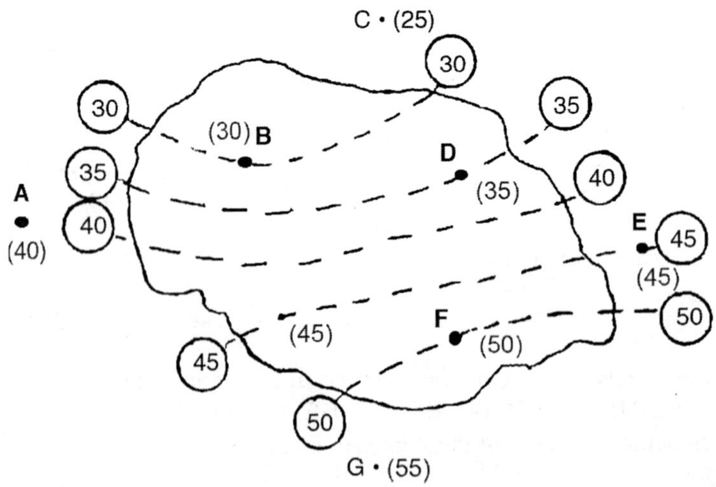

Fig. 4.15 Isohyetal method

Serial no.	Isohyets	Average	Area (km^2)
1	30 and less	30.0	10.0
2	30 and 35	32.5	11.0
3	35 and 40	37.5	11.3
4	40 and 45	42.5	9.3
5	45 and 50	47.5	7.6
6	50 and above	50.0	4.8
			Total 54.0

Therefore, the average daily precipitation over the catchment area on 2 August 2005

$$= \frac{30\times10 + 32.5\times11 + 37.5\times11.3 + 42.5\times9.3 + 47.5\times7.6 + 50\times4.8}{54}$$

$$= \frac{300.0 + 357.5 + 423.75 + 395.25 + 361.0 + 240}{54}$$

$$= 38.47 \text{ mm}$$

4.10 RAIN GAUGE DENSITY

Rainfall is the most fundamental data used in the hydrological studies and hence a well distributed network of rain gauge stations is essential. The average area of influence of the rain gauge stations is indicated as *rain gauge density or network density*. The density of rain gauge stations in an area may be decided taking into consideration the following points:

- Variation in the rainfall in the area. If the area is plain and if there is not much variation in rainfall, the number of rain gauge stations may be small.
- The nature of study for which rainfall data is required
- Cost involved in establishing and maintaining the rain gauge stations

4.10.1 Minimum Density of Rain Gauge Stations

The BIS has recommended the following criteria:

- One rain gauge per 520 km^2 in plain areas. (For area in the path of low pressure systems, denser network is necessary.)
- One rain gauge per 260–390 km^2 for area where average elevation of the area above mean sea level is 1000 m and above.
- One rain gauge per 130 km^2 in hilly areas with heavy rainfall. (Higher density recommended wherever necessary.)

The BIS has recommended the following procedure to calculate the optimum number of rain gauges in a catchment area.

n = Number of rain gauge stations existing in an area

$P_1, P_2, P_3, \ldots, P_n$ = Average rainfall of the 'n' rain gauge stations

Now,

$$P_{av} = \frac{P_1 + P_2 + P_3 + \ldots + P_n}{n}$$

Standard deviation $= sx^2 = \dfrac{(P_i - P_{av})^2}{n-1}$ $\quad (i = 1, 2, 3, \ldots n)$

$$C_v = \text{Coeficient of variance} = \frac{sx \times 100}{P_{av}}$$

N = Number of optimum rain gauge stations $= (C_v/x)^2$

x = Percentage permissible error in the estimation of average rainfall.

Additional rain gauge stations $= N - n$

It may be noted that both C_v and x be mentioned as percentages.

The additional rain gauge stations may be located in addition to the existing rain gauge stations, so that they all are evenly distributed over the entire catchment area.

Example 4.10

In a catchment area covering 100 km^2, the average annual precipitation observed at five rain gauge stations is as under.

Station	1	2	3	4	5
Precipitation (mm)	750	1000	900	650	500

Find the number of additional rain gauge stations and also the rain gauge density if the permissible error is 10%.

Solution:

$$\text{Average precipitation} = \frac{750 + 1000 + 900 + 650 + 500}{5} = 760 \text{ mm}$$

$$sx^2 = \frac{(750 - 760)^2 + (1000 - 760)^2 + (900 - 760)^2 + (650 - 760)^2 + (500 - 760)^2}{5 - 1}$$

$$= \frac{157000}{4} = 39250$$

Therefore, $sx = \sqrt{39250} = 198.1$

$$C_v = \frac{198.1}{760} \times 100 = 26.06\%$$

Since permissible error is 10%,

Number of rain gauge stations required $= (C_v/10)^2 = (26.06/10)^2$

$= 6.79$

Additional rain gauge stations required $= 6.79 - 5 = 1.79 \approx 2$

Rain gauge density $= 100/7 = 14.30$ km^2/rain gauge.

Example 4.11

In a catchment area covering 5600 km^2, the zone-wise existing rain gauge stations were as follows:

Zone	A	B	C	D	E	Total
Catchment area (km^2)	800	500	1600	2500	200	5600
Number of existing stations	1	1	2	3	0	7

Additional nine rain gauge stations are to be installed. Indicate the zone-wise distribution of these nine additional rain gauge stations.

Solution:

The total number rain gauge stations in the catchment N will be 7 + 9 = 16.

The additional nine rain gauge stations will be distributed proportional to the area of the zones and also taking into consideration the number of existing rain gauge stations, as follows.

Zone	A	B	C	D	E	Total
Area (km^2)	800	500	1600	2500	200	5600
P = Area in decimal of total area	0.143	0.089	0.286	0.447	0.035	1.000
$N \times P$	2.288	1.424	4.576	7.152	0.560	16
$N \times P$ *rounded*	2	1	5	7	1	16
Number of existing rain gauge stations	1	1	2	3	0	7
Additional rain gauge	1	0	3	4	1	9

4.10.2 Moving Average Curve

The hyetograph plotted in a usual way will not indicate any trend or cyclic pattern. A moving average curve proposed will smoothen the variables and indicate the trend or the cyclic pattern, if any.

It is also known as *moving mean curve*.

The moving average period 'm' is generally odd, for example 3 or 5, depending upon n where n is the number of observations.

If $X_1, X_2, X_3, \ldots, X_n$ is the sequence of rainfall of n values, then the moving average curve coordinates are as follows (m = 3).

$$Y_2 = \frac{X_1 + X_2 + X_3}{3}$$

$$Y_3 = \frac{X_2 + X_3 + X_4}{3}$$

$$Y_{n-1} = \frac{X_{n-2} + X_{n-1} + X_n}{3}$$

when $X = n$, number of Y terms $= n + 1 - m$,
when $m = n$, it will be a simple arithmetic average
and when $m = 1$, it will be a normal hyetograph.

The 3-year moving average for a specific study is shown in Fig. 4.16.

Example 4.12

The average annual precipitation in millimetres observed over a catchment area from 1980 to 1995 is as follows:

1149, 1260, 1425, 1680, 1200, 1400, 1645, 1500,
1155, 945, 875, 1080, 1480, 1625, 1500, 1470

Construct a 3-year moving average curve and plot it along with the original data.

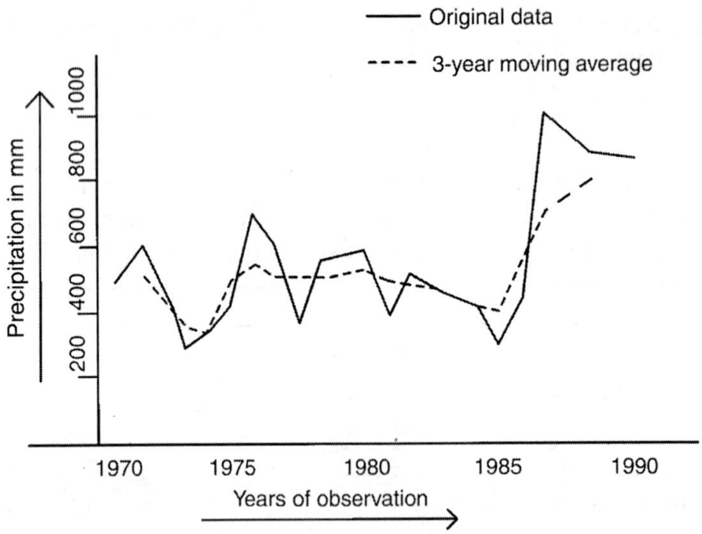

Fig. 4.16 Three-year moving average

Solution:

The first, 3-year moving average curve coordinate $= \dfrac{1149 + 1260 + 1425}{3} = 1278$

The second, 3-year moving average curve coordinate $= \dfrac{1260 + 1425 + 1680}{3} = 1455$

Thus, the 3-year moving average curve coordinates will be as under.

1278, 1455, 1435, 1426, 1415, 1515, 1433,
1200, 991, 966, 1145, 1395, 1535, 1531

The number of coordinates in the 3-year moving average curve = 16 − 3 + 1 = 14. The 3-year moving curve and the original curve will be as shown in Fig. 4.17.

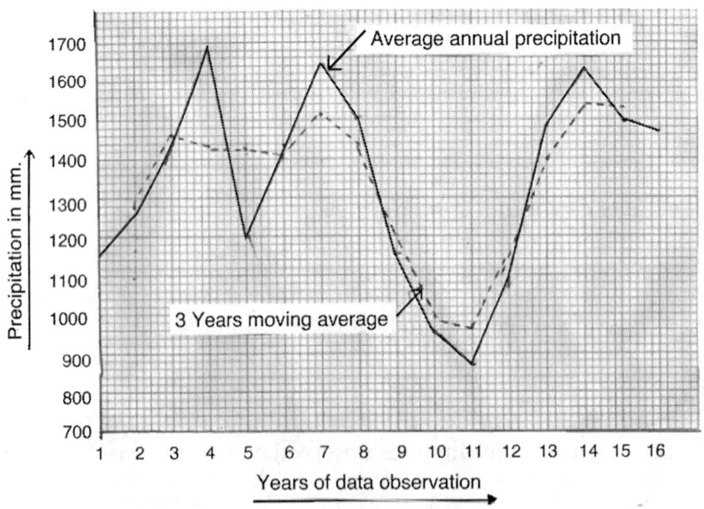

Fig. 4.17 Three-year moving curve and the original curve

4.11 PROBABLE MAXIMUM PRECIPITATION

For an area, the maximum depth of precipitation that may occur for a specific duration is known as *possible maximum precipitation or probable maximum precipitation.* The usual short form used is PMP. This data is required for estimating maximum possible flood from a catchment area. The PMP for a known duration can be correlated as follows:

$$\text{PMP} = P_a + K \times sx$$

where, P_a = Average precipitation over the area
sx = Standard deviation of rainfall series
K = A constant for the area and is in the neighbourhood of 15.

The data of world's greatest rainfall was analysed and it was observed that it follows a curve $\text{PMP} = 42.16 \times D^{0.475}$

where, PMP = Precipitation in centimetres
D = Duration in hours

4.11.1 Relation Between the Extremes and the Average Precipitation

It is observed that normally over a period

The maximum annual rainfall = 1.51 × average annual rainfall
The minimum annual rainfall = 0.60 × average annual rainfall.

4.11.2 Station Year Method

In case of a specific study, the data of rainfall may be required for a long period. This data may not be available in some cases. It is then assumed that the rainfall data of n rain gauge stations for 1 year is equivalent to the data of one rain gauge station for n years and the study is continued. This approach is known as *station year method.*

4.11.3 Recurrence Interval of a Storm

The number of years within which a given storm may equal or even exceed is known as *recurrence interval or return period* and normally is denoted by T_r.

Suppose the record at a station or over an area is available for n years, then the precipitation data may be arranged in the descending order.

The serial number of a specific value of precipitation in the descending order is known as *ranking of the storm*, e.g. m. Then the return period of that specific value of precipitation, $T_r = n/m$. This means that this precipitation value or more than this occurs m times in n years.

Probability

Probability generally denoted by p is reciprocal of the return period, i.e. $p = m/n = 1/T_r$.

Frequency

Probability expressed in terms of percentage is frequency, i.e. frequency $= p \times 100 = m/n \times 100$

Example 4.13

The precipitation in millimetres observed at a rain gauge station for the last 32 years is as follows:

988, 966, 935, 1007, 992, 1050, 975, 920, 1035, 990, 1095, 1015, 986, 927, 1003, 1055, 1135, 955, 1001, 1045, 1090, 997, 1040, 1100, 948, 972, 1012, 950, 1070, 982, 929, 960

Find 1. The return period and frequency of the precipitation of 997 mm
2. The precipitation of return period of 1.33 and its frequency

Solution:

The precipitation figures in millimetre were arranged in a descending order as follows:

Serial no.	1	2	3	4	5	6	7	8	9	10	
	1135	1100	1095	1090	1070	1055	1050	1045	1040	1035	
Serial no.	11	12	13	14	15	16	17	18	19	20	21
	1015	1012	1007	1003	1001	997	992	990	988	986	982
Serial no.	22	23	24	25	26	27	28	29	30	31	32
	975	972	966	960	955	950	948	935	929	927	920

1. The serial order of 997 mm is 16
   Therefore, its return period = 32/16 = 2 and its frequency = 1/2 × 100 = 50%.
2. Return period = 1.33 = n/m = $32/m$
   Therefore, m = 24. The precipitation figure at serial no. 24 is 966 mm and its frequency will be = 24/32 × 100 = 75%.

4.12 INTENSITY DURATION ANALYSIS

It is observed that most intense storms last for a short time. As the intensity reduces, the duration of the storm increases and vice versa. The study of intensity and its duration is known as *intensity duration analysis.*

4.12.1 Intensity–Duration Curve

For an area, the available data of the duration and its intensity is analysed and a graph of duration versus intensity is plotted. This graph is known as *intensity–duration curve or intensity–duration graph.*

It is observed that this graph normally follows the following equation:

$$I = \frac{C}{(t+a)^b}$$

where, I = Intensity in mm/h
t = duration in minutes
C, a, b = constants for the specific area.

The intensity duration graph prepared for a specific study is shown in Fig. 4.18.

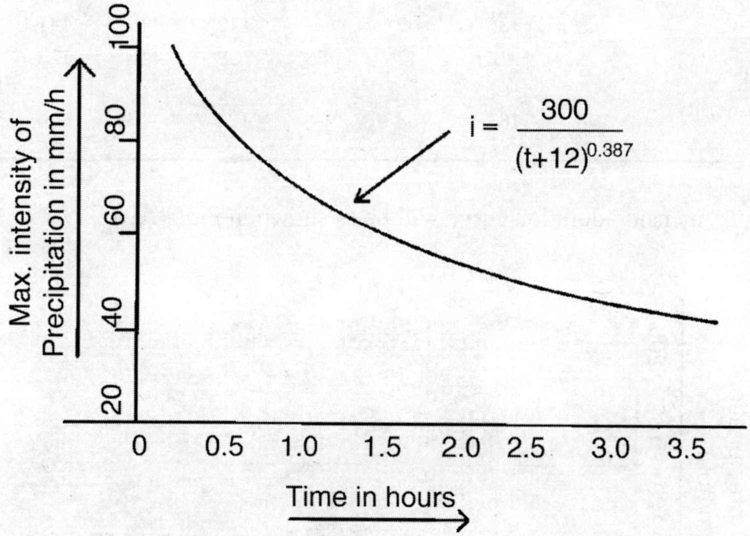

Fig. 4.18 Intensity–duration curve for a specific study

Example 4.14

A storm occurred over a catchment area as under:

Time (min)	0	10	20	30	40	50	60	70	80	90
Precipitation (mm)	0	19	22	7	20	23	33	28	8	6

Plot the maximum intensity–duration curve.

Solution:

Time (min)	Cumulative precipitation in mm at time interval								
	10	20	30	40	50	60	70	80	90
0	0								
10	19								
20	22	41							
30	7	29	48						
40	20	27	49	68					
50	23	43	50	72	91				
60	**33**	56	76	83	105	124			
70	28	**61**	**84**	**104**	111	**133**	**152**		
80	8	36	69	92	**112**	119	141	**160**	
90	6	14	42	75	98	118	125	147	**166**
Maximum intensity (mm/h)	198	183	168	156	134	133	130	120	110

The maximum intensity–duration curve will be as shown in Fig. 4.19.

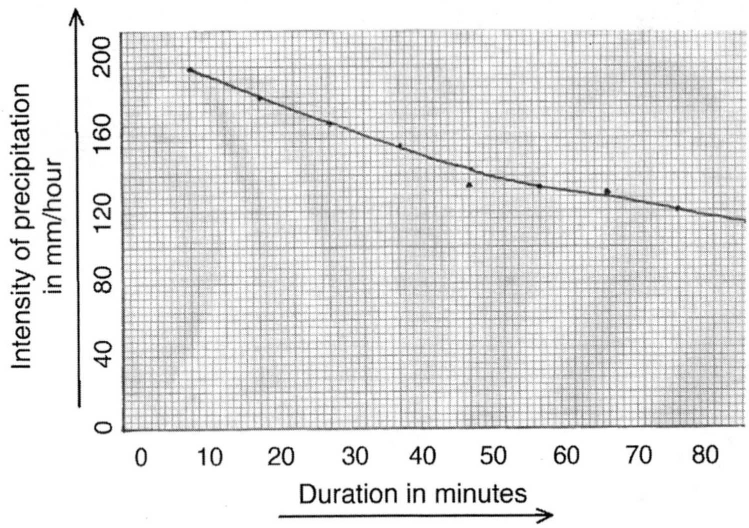

Fig. 4.19 Maximum intensity–duration curve

4.12.2 Intensity–Frequency–Duration Analysis

If for a catchment area sufficient data, for example for more than 50 years, is available, this data is analysed for each storm, for its intensity, frequency as well as its duration. This analysis may be presented as shown in Fig. 4.20.

These graphs are for a specific catchment area and may change for different catchments depending upon their hydrologic character.

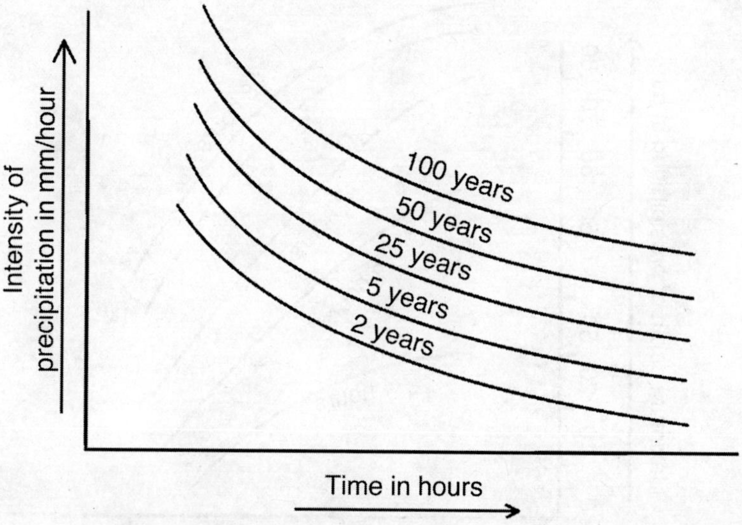

Fig. 4.20 Intensity–frequency–duration curve

4.12.3 Isopluvial Map

The intensity–frequency–duration, curves are prepared for various adjoining areas. A combined map for the large area can be prepared for a maximum rainfall depth for various combinations of a return period and duration.

Such maps for a region for various rainfall depths, return periods and duration are called *isopluvial maps*. An isopluvial map is shown in Fig. 4.21.

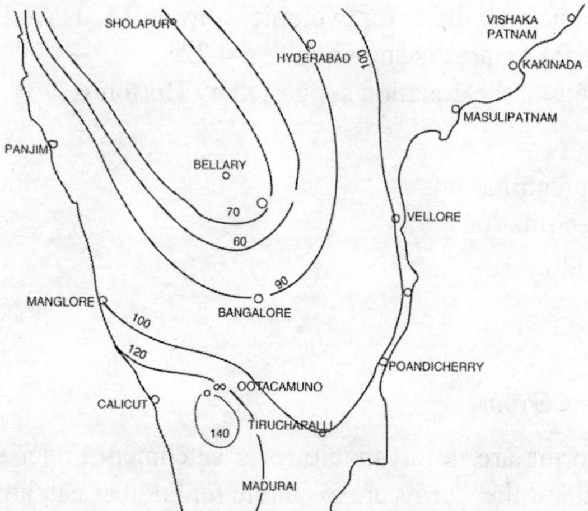

Fig. 4.21 Isopluvial map of South India

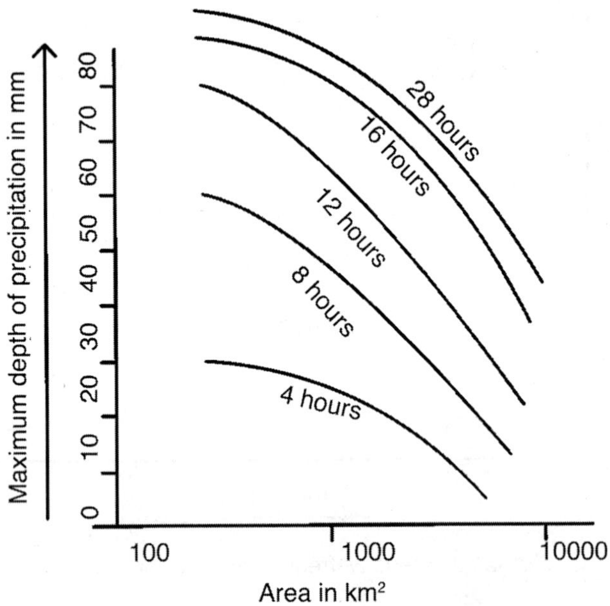

Fig. 4.22 Depth–area–duration curve

4.12.4 Depth–Area–Duration Analysis

In some cases, the average depth of storm and its duration is required for a specific area. Such study is called *depth–area–duration analysis*. The normal short form is DAD study. Every storm has a centre having maximum precipitation, e.g. P_o, and the average precipitation over a specific area, e.g. P_a. Naturally $(P_o - P_a)$ is always positive, and its value increases with bigger catchments and decreases with smaller catchments.

For a catchment, this value may differ for cyclonic, convective, as well as orographic precipitation. A specific study done for an area is shown in Fig. 4.22.

Normally, for each duration, the equation suggested by Horton is

$$P_a = P_o^{-(KA)^n}$$

where, P_o = Maximum precipitation
P_a = Average precipitation
A = Area in sq. km
K, n = Constants

4.12.5 Transposition of a Storm

Sometimes, data about storms are not available for a catchment under study, e.g. catchment A. However, sufficient data about the storms are available for another catchment B, which are meteorologically homogeneous.

A storm occurring over B is considered to be applicable and occurring over A and further studies are continued for A. This procedure is called *transposition of storm*. However, this may not be applicable for hilly catchments.

4.13 PRECIPITATION OVER INDIA*

Rainfall over India is very erratic in terms of both time and space. The coefficient of variation of the annual rainfall varies between 15 and 36, and hence it is said that India's prosperity is a gamble in the monsoon's rains.

The average annual precipitation over India is 1140 mm. Both the maximum and the minimum precipitations in the world are observed in India. The important features affecting precipitation over India are the following:

1. The orographic features
2. The wind currents

The seasons in the Indian subcontinent can be divided into four major seasons.

1. South-west monsoon (June to September)
2. Post-monsoon (October to December)
3. Winter season (January to February)
4. Pre-Monsoon (March to May)

The chief characteristics of these seasons are as follows.

4.13.1 South-West Monsoon

The south-west monsoon is the principal rainy season of India. Over 70% of the annual precipitation occurs over the Indian subcontinent during this season.

Except for the south-eastern part of the peninsula and Jammu and Kashmir, the rest of the country experiences heavy rains during this season. The monsoon enters into the Indian subcontinent through the southern part of Kerala towards the end of May or beginning of June with a very good degree of regularity in the statistical sense.

Under normal monsoon conditions, the distribution clearly shows the marked influence of the effect of orography of the Western Ghats, Khasi–Jantia hills, the Vindhyas, and the Himalayas. These are the regions with heavy precipitation. There is an increase in precipitation from the west coast up to the Western Ghats and then it decreases rapidly. For example, Goa gets over 2000 mm of precipitation, while Pune gets less than 600 mm. Immediately, south of Vindhyas again is the region of heavy precipitation decreasing southwards. There is a constant decrease of precipitation over the Gangetic plane from east to west decreasing from over 1000 mm over West Bengal to less than 100 mm over west Rajasthan. There is another region of heavy precipitation over the foothills of Himalayas reducing sharply towards the west of north-west, with almost negligible in the Ladakh valley that is on the leeward side.

4.13.2 Post-Monsoon

As the southwest monsoon retreats, low-pressure areas develop in the Bay of Bengal and a north-easterly flow of air that picks up moisture in the Bay of Bengal is formed. This air mass strikes the east coast of the southern peninsula, particularly Tamil Nadu, and causes precipitation in those areas. In addition, during this period several tropical cyclones form in the Bay of Bengal and to a lesser extent in the Arabian Sea. These strike the coastal areas and cause intense precipitation.

*Courtesy of National Institute of Hydrology

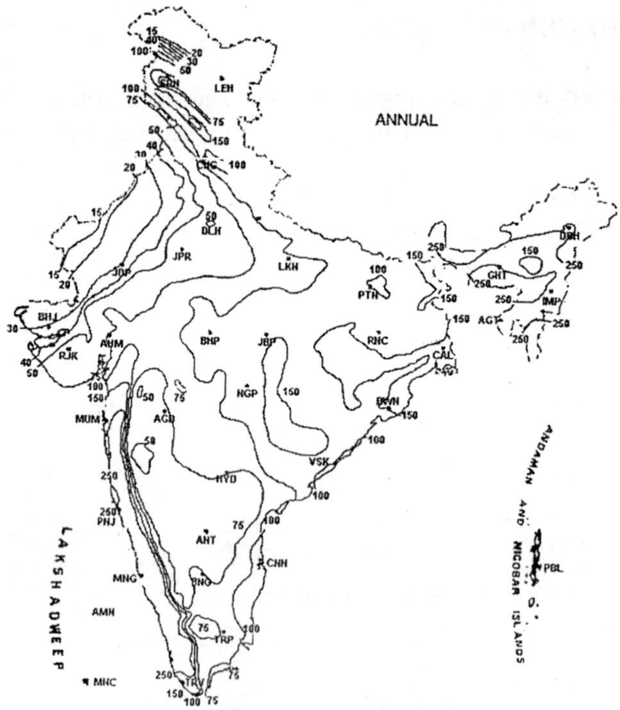

Fig. 4.23 Isohyetal map of India

Courtesy: India Meteorological Department

4.13.3 Winter Rains

Near about the end of December, disturbances originating in the mid-east travel eastwards across Afghanistan and Pakistan, known as western disturbances. They cause moderate to heavy precipitation and snowfall in the Himalayas and Jammu and Kashmir. Some light precipitation occurs in the northern planes. Some precipitation is also experienced in the Tamil Nadu region due to low-pressure areas formed in the Bay of Bengal.

4.13.4 Pre-Monsoon

There is hardly any precipitation during this season. Convective cells cause some thunderstorms, mainly in Kerala, West Bengal and Assam. Some cyclone activity, primarily in the eastern coast, also occurs. Figure 4.23 shows the average annual precipitation over India.

REVIEW QUESTIONS

1. Define precipitation. Explain its importance in the study of hydrology.
2. What are the different forms of precipitation?
3. Discuss the factors affecting precipitation.
4. Explain the different types of precipitation.
5. Explain artificial precipitation. Why is it not followed on a large scale in India?
6. Discuss the considerations for selecting the site for a rain gauge.
7. Explain the different types of rain gauges. Discuss their merits and demerits.

8. Explain with the help of a neat sketch, the working of a non-recording rain gauge. Discuss its merits and demerits.
9. Explain with the help of a neat sketch, the working of a siphon bucket-type rain gauge.
10. Explain with the help of a neat sketch, the working of a tipping bucket-type rain gauge.
11. Explain with a neat sketch, the working of a weighing bucket-type rain gauge.
12. Discuss the difficulties experienced during the measurement of precipitation.
13. Why supplementing of rainfall data is required? State the different methods in use.
14. Explain verification of consistency of a rain gauge.
15. Explain the double mass curve method.
16. How the average precipitation over a catchment is calculated? Discuss the different methods with their merits and demerits.
17. Explain the ISI standards for rain gauge density in a catchment. Also, discuss the procedure to calculate the optimum number of rain gauge stations in a catchment.
18. What is a moving average curve? What are its advantages? Explain the procedure to draw it.
19. Explain the depth–area–duration analysis.
20. What is an intensity–duration curve?
21. Write a detailed note on precipitation over India.
22. Write a note on frequency of a storm.
23. Write short notes on the following:
    a. Terminal velocity
    b. Orographic precipitation
    c. Cloud seeding
    d. History of measurement of precipitation
    e. Use of radar for measurement of precipitation
    f. Index of wetness
    g. Classification of the intensity of rainfall
    h. Rain gauge density
    i. Probable maximum precipitation
    j. Station year method
    k. Recurrence interval of a storm
    l. Isopluvial map
    m. Transposition of a storm
24. Differentiate between
    a. Record graph sheets of siphon bucket-type rain gauge and a weighing bucket-type rain gauge.
    b. Forms of precipitation and types of precipitation.
    c. Cyclonic precipitation and orographic precipitation.
    d. Non-recording-type rain gauge and recording-type rain gauge.
    e. Siphon-type rain gauge and weighing-type rain gauge.

NUMERICAL QUESTIONS

1. During a storm, the hourly precipitation data observed was as follows:

Time (h)	0	1	2	3	4	5	6	7	8	9
Precipitation (mm)	0	25	20	45	30	10	0	15	10	0

Plot the following: (i) Hyetograph
(ii) Ordinate graph
(iii) Mass curve

Ans: The hourly mass curve coordinates are as follows:

Time (h)	0	1	2	3	4	5	6	7	8	9
Precipitation (mm)	0	25	45	90	120	130	130	145	155	155

2. The precipitation coordinates in mm observed on an automatic float-type rain gauge are as follows:

Time (h)	8	9	10	11	12	13	14	15	16	17	18	19	20
Precipitation (mm)	0	3	11	21	35	40	40	50	65	75	84	88	92
Time (h)	21	22	23	24	1	2	3	4	5	6	7		
Precipitation (mm)	100	4	8	11	21	21	21	28	32	40	42		

Find: (i) Daily precipitation
(ii) Time when siphon operated
(iii) Time of no precipitation
(iv) Maximum intensity of precipitation

Ans: (i) Daily precipitation 5 142 mm
(ii) Time when siphon operated 5 At 21 h
(iii) Time of no precipitation 5 between 13 and 14 h and 2, 3 and 4 h
(iv) Maximum intensity of precipitation = 15 mm/h between 15 and 16 h

3. The precipitation coordinates observed on a weighing-type automatic rain gauge were as follows:

Time (h)	8	9	10	11	12	13	14	15	16	17	18	19	20
Precipitation (mm)	0	5	16	28	46	56	58	58	68	77	81	84	88
Time (h)	21	22	23	24	1	2	3	4	5	6	7		
Precipitation (mm)	100	97	85	80	75	75	75	72	68	65	62		

Find: (i) Daily precipitation
(ii) Time when siphon operated
(iii) Time of no precipitation
(iv) Maximum intensity of precipitation

Ans: (i) Daily precipitation = 138 mm
(ii) Time when siphon operated = At 21 h
(iii) Time of no precipitation = Between 14 and 15 h and 1, 2 and 3 h
(iv) Maximum intensity of precipitation = 18 mm/h between 11 and 12 h

4. Five rain gauge stations located in a catchment indicate the average annual precipitation as follows:

Station	A	B	C	D	E
Precipitation (mm)	2456	2341	2472	2300	2491

The precipitation at B was missing for the year 1998 since the rain gauge was out of order. The precipitation observed at other stations for 1998 are as follows:

Station	A	B	C	D	E
Precipitation (mm)	2465	?	2492	2285	2511

Calculate the missing data for B for the year 1998.

Ans: 2348 mm

5. The location coordinates in km of five rain gauges w.r.t. P are as follows:

Station	P	Q	R	S	T
Precipitation (mm)	0,0	25,20	−45,17	−35,215	35,−16

The annual average precipitation at the Station P was missing for the year 2004.

The average annual precipitation at other stations for the year 2004 is as follows:

Station	P	Q	R	S	T
Precipitation (mm)	?	2735	2800	2831	2655

Evaluate the missing precipitation at P for the year 2004.

Ans: 2750 mm

6. The annual precipitation of the rain gauge Station A for the year 1990–1999 was as given below. Similarly, the average annual precipitation of the remaining five rain gauge stations for the same period was as given below. The precipitation recorded at A was doubtful because there were changes at the location. Verify the precipitation at A is consistent and correct it if necessary.

Serial no.	Year	Precipitation at A (mm)	Average precipitation of the five rain gauges (mm)
1	1990	1388	8250
2	1991	1412	8539
3	1992	1398	8157
4	1993	1443	8755
5	1994	1384	8242
6	1995	1456	8852
7	1996	1412	8450
8	1997	1727	9247
9	1998	1710	8848
10	1999	1690	8656

Ans: The precipitation data at Station A is wrong from the year 1996 and the corrected precipitation figures from 1990 in mm are 1388, 1412, 1398, 1443, 1384, 1456, 1412, 1489, 1464, 1427, respectively.

7. A catchment area is in the form of an equilateral triangle ABC of side 10 km. Four rain gauge stations are located at A, B, C and D. Station D is the centroid of the triangle. The average annual precipitation observed at these stations in mm are 1145, 1252, 1184 and 1056, respectively. Find the average annual precipitation of the catchment area by all the three methods.

Ans: 1101 mm

8. The average annual precipitations of the five rain gauge stations in a catchment are as follows:

Station	A	B	C	D	E
Precipitation (mm)	1100	890	1000	1370	1805

If the catchment covers an area of 110 km^2, find the number of additional rain gauge stations required. Acceptable permissible error is 10%.

Ans: 4

9. The average annual precipitation in mm observed over a catchment area for the last 16 years is as follows:

   1154, 1265, 1430, 1683, 1205, 1390, 1650, 1497, 1160, 940, 870, 1080, 1490, 1630, 1505, 1475

   Construct a 3-year moving average curve and compare with the original data.

10. The average annual precipitation in mm observed at a rain gauge station for the last 32 years is as follows:

    966, 992, 920, 1045, 1012, 1070, 960, 935, 1050, 1035, 1015, 948, 972, 950, 929, 1007, 975, 990, 1095, 986, 988, 927, 1003, 1055, 982, 1135, 955, 1001, 1040, 1090, 997, 1100

    Find: (i) The return period and the frequency of the precipitation of 1045 mm
    (ii) The precipitation of the return period of 2 and its frequency

    Ans: (i) 4, 25%
    (ii) 997, 50%

11. The storm occurred over a catchment area is a under:

Time (s)	0	60	120	180	240	300	360	420	480	540	100
Precipitation (mm)	0	20	23	8	21	24	35	27	9	5	0

Plot the Maximum intensity duration curve

MULTIPLE CHOICE QUESTIONS

1. The terminal velocity of a raindrop depends upon
   (a) Diameter of the raindrop (b) Velocity of air
   (c) Temperature of the air (d) Specific gravity of water

2. Cyclone is also known as
   (a) Hurricane (b) Tornado
   (c) Typhoon (d) All the above

3. Raindrop is always
   (a) Spherical (b) Conical
   (c) Elliptical (d) Cubical

4. Hygroscopic nuclei used for artificial rain are of
   (a) Dry ice (b) Silver iodide
   (c) Sodium chloride (d) Portland cement
   (e) Any of the above

5. Artificial rain method is normally not followed because
   (a) It is a costly process.
   (b) It may affect the rainfall in the adjoining area.
   (c) The success rate is not very high.
   (d) All the above.
6. In case of a siphon bucket-type rain gauge, the siphon acts when
   (a) The rain stops
   (b) The rain starts
   (c) Water in the float chamber reaches a specific level
   (d) After a specific time.
7. In case of a mass curve of rainfall, when there is no rainfall, the graph is a
   (a) Vertical line
   (b) Blank
   (c) Horizontal line
   (d) Downward line
8. In case of a tipping bucket-type rain gauge, the bucket tips when
   (a) The rain starts
   (b) The bucket receives a rainfall of 25 mm
   (c) The rain stops
   (d) After a specific time
9. In case of a weighing-type rain gauge, when the water level reaches a specific level in the chamber
   (a) The pointer moves down
   (b) The pointer reverses its movement
   (c) The pointer stops
   (d) The drum stops its rotational movement
10. Cloud seeding is accomplished by the following methods
   (a) Emitting the hygroscopic nuclei with the help of a special gun
   (b) Using the aeroplanes for spreading the nuclei
   (c) Using rockets by suitable delivery systems
   (d) Any of the above
11. In India, the precipitation received is mostly
   (a) Cyclonic precipitation
   (b) Convective precipitation
   (c) Orographic precipitation
   (d) None of the above
12. Clear distance between the rain gauge and the obstruction should be equivalent to at least
   (a) Height of the obstruction
   (b) Twice the height of the obstruction
   (c) Three times the height of the obstruction
   (d) Four times the height of the obstruction
13. A plot of time versus rainfall is called a
   (a) Hydrograph
   (b) Hyetograph
   (c) Isohyet
   (d) None of the above
14. An isohyet is a line joining
   (a) Equal precipitation intensity
   (b) Equal precipitation depth
   (c) Equal storm duration
   (d) Equal height of stations
15. The characteristics of convective precipitation are
   (a) High intensity and long duration
   (b) High intensity and low duration
   (c) Low intensity and low duration
   (d) Low intensity and long duration

16. The chart fixed to an automatic rain gauge gives
    (a) A rainfall hyetograph
    (b) An intensity–duration curve
    (c) An isohyetal map
    (d) A precipitation mass curve

17. The average annual precipitation over India is
    (a) 1000 mm
    (b) 1500 mm
    (c) 1140 mm
    (d) 1350 mm

18. Orographic precipitation occurs due to lifting of air mass because of
    (a) Presence of mountain barriers
    (b) Extratropical cyclones
    (c) Density difference of air mass
    (d) Difference of air temperature

19. The double mass curve technique is followed to
    (a) Check the consistency of rainfall at a station
    (b) Find average rainfall of a station
    (c) Find number of rain gauge stations required in a catchment
    (d) Estimate the missing data of rainfall of a station

20. The probable maximum precipitation is given by the relation
    (a) $PMP = P_a + K\sigma$
    (b) $PMP = P_a - K\sigma$
    (c) $PMP = P_a e^{K\sigma}$
    (d) $PMP = P_a e^{-K\sigma}$

21. A rain gauge inclined at 10° will catch
    (a) 0.5% less as compared to vertical
    (b) 1.0% less as compared to vertical
    (c) 1.5% less as compared to vertical
    (d) 2.0% less as compared to vertical

22. In olden days, a rain gauge was also known as
    (a) Hyetometer
    (b) Ombrometer
    (c) Pluviometer
    (d) All of the above

23. Radar means
    (a) Radio Detection and Recording
    (b) Radio Detection and Ranging
    (c) Radiation Detection and Recording
    (d) Radiation Detection and Ranging

24. A day is known as rainy day when the rainfall during the day is
    (a) 2.5 mm or more
    (b) 5.0 mm or more
    (c) 7.5 mm or more
    (d) 10.0 mm or more

25. Average precipitation over a catchment can be calculated by
    (a) Arithmetic mean method
    (b) Thiessen's polygon method
    (c) Isohyetal method
    (d) Any of the above methods

26. The density of rain gauge stations in a catchment is decided considering
    (a) Variation of rainfall in that area
    (b) The nature of study for which rainfall data are required
    (c) Cost involved in establishing and maintaining the rain gauge stations
    (d) All the above considerations

27. Over a period, the maximum annual rainfall is equal to
    (a) 1.26 times the average annual rainfall (b) 1.51 times the average annual rainfall
    (c) 1.75 times the average annual rainfall (d) 1.95 times the average annual rainfall

28. A raindrop deforms and splits because of the resistance of air when it is more than
    (a) 4.0 mm (b) 5.5 mm
    (c) 7.5 mm (d) 10.0 mm

29. Over a period the minimum annual rainfall is equal to
    (a) 0.50 times the average annual rainfall (b) 0.60 times the average annual rainfall
    (c) 0.70 times the average annual rainfall (d) 0.80 times the average annual rainfall

30. Sunspots have a cycle of variation of
    (a) 7 years (b) 9 years
    (c) 11 years (d) 13 years

31. The frequency of a storm in years is given by
    (a) $n - m$ (b) $n + m$
    (c) $n \times m$ (d) $\frac{n}{m}$

32. The equation of the intensity–duration curve is
    (a) (b)
    (c) (d)

33. A mass curve is a graphical representation of
    (a) Rainfall intensity versus time in chronological order (b) Accumulated rainfall versus time in chronological order
    (c) Accumulated rainfall intensity versus time in chronological order (d) None of the above

ANSWERS TO MULTIPLE CHOICE QUESTIONS

1. a	2. d	3. a	4. e	5. d	6. c	7. c	8. b
9. b	10. d	11. c	12. b	13. b	14. b	15. b	16. d
17. b	18. a	19. a	20. a	21. c	22. d	23. b	24. a
25. d	26. d	27. b	28. b	29. b	30. c	31. d	32. a
33. b							

*** ——— ***

5 Infiltration

Chapter Outline

5.1 DEFINITION

Infiltration may be defined as *entry and movement of water through the land surface into the sub-strata below.*

5.1.1 Infiltration and Percolation

Infiltration includes entry of water into the soil surface and its movement, while percolation is the movement of water under gravity. These two phenomena are confusing as they are closely related, but technically there is a difference. Percolation starts after infiltration.

5.1.2 Abstractions

When there is precipitation, it may or may not result in overland flow into a stream depending upon its intensity and duration. The part of precipitation that is not available as surface runoff is referred to as *precipitation loss* or *abstraction*. The abstractions include (1) interception, (2) depression storage, (3) evaporation and (4) infiltration.

Generally, interception and depression storages are together termed as '*surface retention*'. Knowledge of these abstractions, their rate, and so on, is necessary for the determination of the surface flow.

5.1.3 Dominant Abstraction

Out of these abstractions, evaporation during precipitation is very negligible. Interception and depression storage are comparatively small as compared to infiltration. Hence, infiltration is called as *dominant abstraction*.

5.1.4 Excess Rainfall

The part of precipitation that is available in the form of surface flow after meeting all the abstractions is known as *excess rainfall*. It is also known as *effective rainfall*.

5.2 PROCESS OF INFILTRATION

When rainwater falls on the ground, there is some resistance offered by the soil surface for the entry of rainwater and also to the flow of water through the soil. There are cracks, vertical as well as lateral in the soil, so also there are some voids between the soil particles, which are ordinarily occupied by air or water. Water flows through these cracks and gaps until it reaches the saturated zone below. Naturally, the passage of water experiences some resistance.

The rate at which water enters the ground surface and then flows downwards is known as *infiltration rate*. This rate is high in the beginning because it has to meet the requirements of the dry soil. However, it attains a steady constant lower value after a passage of time.

The unit for the rate of infiltration is mm/h.

5.3 FACTORS AFFECTING INFILTRATION

The factors affecting infiltration are discussed below:

5.3.1 Rainfall Characteristics

The duration as well as the intensity of rainfall influences infiltration. If the intensity of rainfall is more than the infiltration rate, then only surface runoff is noticed. On the contrary, if the intensity of rainfall is less than the infiltration rate, no surface flow is observed and all the rainfall is abstracted as infiltration.

The rate of infiltration is high in the beginning and goes on reducing and attains a steady state after some time. At earlier stage the intensity of rainfall may be less than the rate of infiltration. Under this condition, the ground will absorb all the rainfall and there will not be any surface flow.

However, when the infiltration rate reduces and becomes less than the intensity of precipitation, surface flow will be noticed, which might increase till the infiltration rate stabilizes. In addition to these considerations, rainfall has some additional effects as follows:

- When rainwater strikes the bare soil, there is mechanical compaction of soil due to the impact, which may reduce the infiltration rate.

- Fine soil particles are carried down due to rain water resulting in choking of the pore spaces in the soil, and consequently resulting in reducing the infiltration rate.

5.3.2 Ground Surface Condition

The land surface that receives rain may be bare, vegetated or covered with mulch or litter. The bare ground receiving rainfall may be subjected to the effects of impact, and so on. If impervious material is exposed at the surface, naturally infiltration is small or negligible.

The vegetated ground reduces the impact effect of the raindrops, avoids dislodging of particles by mechanical binding action of the roots and hence provides a high rate of infiltration by maintaining open soil structure and also slows down the rate of runoff.

Thus, due to vegetated ground the rate of infiltration increases.

Mulch cover has got a similar effect. Because of mulch and litter, the infiltration rate increases. This has been proved by the field experiments. Contour ploughing and terracing in the agricultural land delays the surface flow and increases the infiltration rate.

Small surface slope has little effect on the infiltration rate. If the slope of the ground is steeper than 16°, then the rate of infiltration reduces due to the velocity of water causing more overland flow.

5.3.3 Soil Characteristics

Soil characteristics have definite effect on infiltration.

A uniformly graded material will have more pores and hence the infiltration rate will be more than a well-graded material. In clayey soils, because of the removal of moisture due to evaporation, some shrinkage cracks may be observed. These are termed as *Sun cracks*. Because of these cracks, infiltration rate may increase in the initial stage. Subsequently, when the soil gets wet, the cracks get closed and may not affect the infiltration rate.

Because of the presence of water, some soils have a tendency to form aggregation that reduces the infiltration rate. Also, the organic material present helps to promote soil aggregation. In some cases, puddle is formed at the surface where the soil has a fair proportion of clay, which becomes relatively impervious when it dries.

In the case of urbanization, because of concrete buildings, asphalt pavement, and so on, the infiltration rate is considerably reduced.

5.3.4 Soil Moisture

Even if the soil contains some moisture, there is no effect, practically, on the rate of infiltration except that the rate is reduced at the initial stage, as shown in Fig. 5.1.

5.3.5 Human Activities

Cultivation of land disturbs the soil structure, closes the openings made by burrowing animals and insects as well as decaying roots and thus reduces the rate of infiltration.

5.3.6 Climatic Conditions

The flow of water through the soil is laminar. Change in temperature may cause change in the viscosity of water and consequently may cause change in the velocity of water, and thus may affect the rate of infiltration.

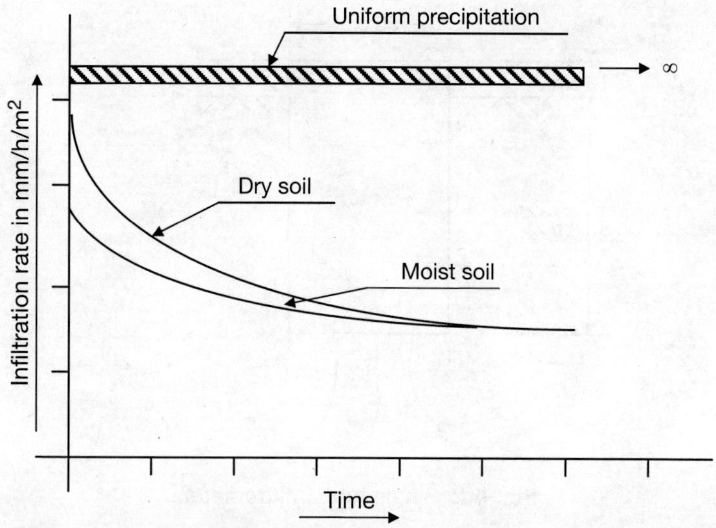

Fig. 5.1 Rate of infiltration w.r.t. time

5.3.7 Entrapped Air

If infiltration process covers a large area, there may not be an exit passage to the entrapped air in the soil. Also, because of the downward passage of water, the air entrapped may get compressed and may offer more resistance to the flow, which may result in reducing the infiltration rate.

5.3.8 Other Minor Factors

Other factors, such as depth of water on the ground, groundwater table, and so on, have practically no effect on the infiltration rate.

5.4 MEASUREMENT OF INFILTRATION

The rate of infiltration is initially high. It goes on reducing with time and after some time it becomes steady. A usual graph of the rate of infiltration is shown in Fig. 5.1. The rate of infiltration for a soil is measured in the field as well as in the laboratory. These are known as *infiltrometers*.

The most common types are the following:

- Flooding-type infiltrometers
- Sprinkling-type infiltrometers or rain simulators

5.4.1 Flooding-Type Infiltrometers

There are two types of flooding-type infiltrometers:

1. Single-tube flooding infiltrometer
2. Double-tube flooding infiltrometer

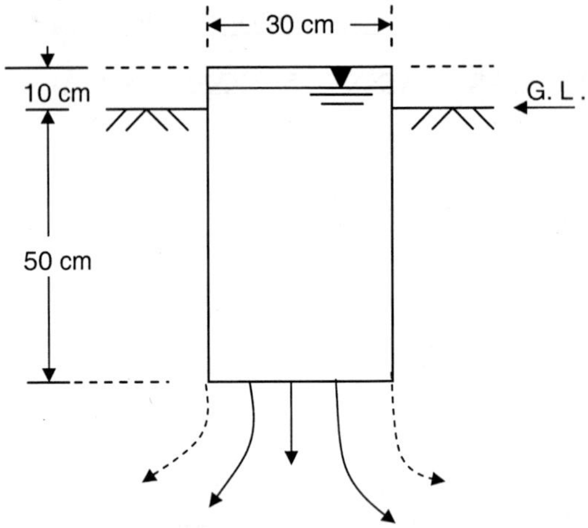

Fig. 5.2 Single-tube infiltrometer

5.4.1.1 *Single-tube flooding infiltrometer*

The single-tube flooding infiltrometer consists of a metal tube 250–300 mm in diameter with both ends open. This tube is driven in a vertical position into an open level ground surface up to a depth of 500 mm, leaving about 100 mm above the ground. The tube is so driven into the ground that the soil is disturbed to a minimum.

Water is then added to this tube to maintain a constant level, sufficiently deep to submerge the plant or the grass crowns. If the soil is bare, i.e., there is no vegetation or grass, the soil is protected by a perforated metal disc to avoid turbidity.

A pointer gauge is used to measure the water level accurately. As the infiltration starts, the water level may drop down but it is maintained at a constant level by adding a measured quantity of water at successive time intervals, till a constant rate of infiltration is achieved.

Knowing the quantity of water added, the time interval and the area of tube, the rate of infiltration can be worked out. The infiltration rate thus worked out is more than the actual value.

Figure 5.2. shows a single-tube infiltrometer. It is also known as *single-ring infiltrometer*.

5.4.1.2 *Double-tube flooding infiltrometer*

In the case of a single-tube infiltrometer, water may flow sideways. Such loss is controlled to some extent by using one more ring outside the test ring. Instead of one tube, two concentric circular tubes are used. These may be of size 300 and 600 mm in diameter. These are driven 150 mm in the ground in a vertical position leaving 100 mm above the ground level. The outside metal ring is used to avoid the side and border effects.

The procedure followed for single tube is followed in this double-tube case also. Water is added into both the tubes and the same level of water is maintained in both tubes by adding the measured quantity of water at successive time intervals.

The observations for the inner tube are used for working out the infiltration rate. Knowing the water added, the time interval and the area of cross section of the inner tube, the infiltration rate with respect to time can be calculated.

Figure 5.3. shows a double-ring infiltrometer. It is also known as *double-ring infiltrometer*.

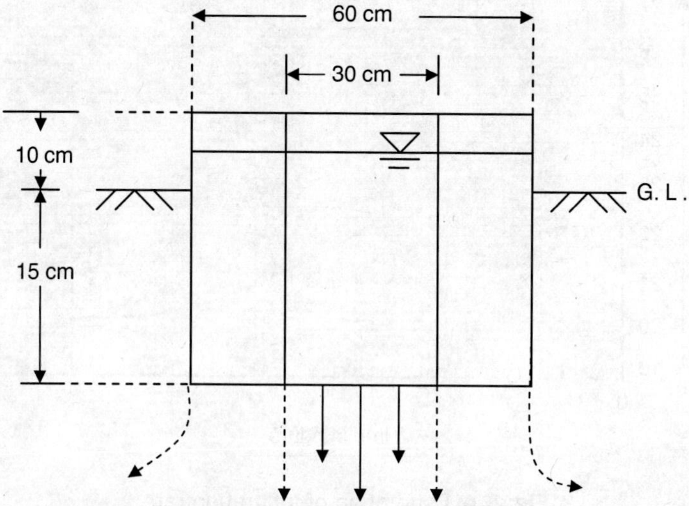

Fig. 5.3 Double-tube infiltrometer

5.4.1.3 Drawbacks in tube infiltrometers

The main drawbacks in the tube infiltrometers are as follows:

- The soil is disturbed to some extent when the tubes are driven into the soil.
- There may be lateral flow of infiltrated water. In double-tube infiltrometers, efforts are made to reduce it, but some lateral flow still occurs.
- Similarly, air entrapped in the soil may escape laterally. In a double-tube infiltrometer, this effect persists.
- Effect of the raindrop impact on soil is not accounted.
- Effect of the slope of ground is not accounted.
- Experiments cannot be conducted on soil with boulders etc.
- The infiltration is affected because of the ring size. Smaller the diameter of the ring, more will be the rate of infiltration.

Example 5.1

The quantity of water added to a double-ring infiltrometer of 1.00-m diameter at 30 min interval to keep the water level constant is as follows:

Time (min)	0	30	60	90	120	150	180
Quantity of water added (lit)	0	10	9.2	8.6	8.2	8.0	8.0

Find: 1. Rate of infiltration for every 30 min and plot the graph
2. Average rate of infiltration

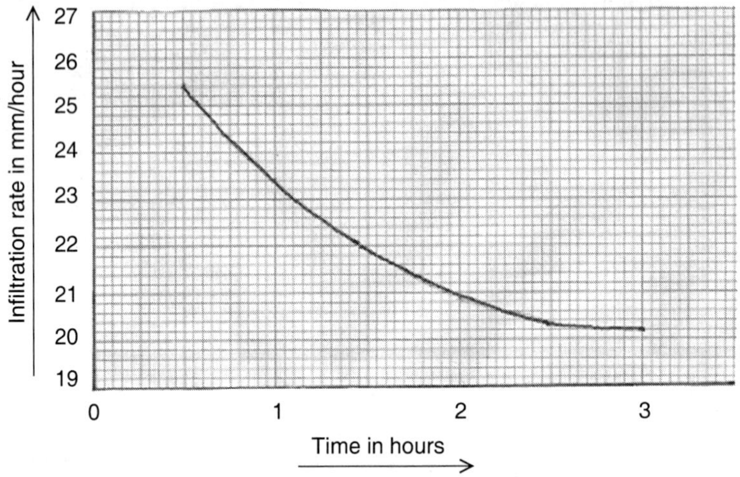

Fig. 5.4 Estimation of infiltration rate

Solution:

$$\text{Area of ring} = \frac{\pi}{4} \times 1^2 = 0.786 \text{ m}^2$$

$$\text{Infiltration rate for 1st 30 min} = \frac{10 \times 10^{-3}}{0.786} \times 10^3 \times \frac{60}{30} = 25.44 \text{ mm/h}$$

$$\text{Infiltration rate for next 30 min} = \frac{9.2 \times 10^{-3}}{0.786} \times 10^3 \times \frac{60}{30} = 23.40 \text{ mm/h}$$

$$\text{Infiltration rate for next 30 min} = \frac{8.6 \times 10^{-3}}{0.786} \times 10^3 \times \frac{60}{30} = 21.88 \text{ mm/h}$$

$$\text{Infiltration rate for next 30 min} = \frac{8.2 \times 10^{-3}}{0.786} \times 10^3 \times \frac{60}{30} = 20.86 \text{ mm/h}$$

$$\text{Infiltration rate for next 30 min} = \frac{8 \times 10^{-3}}{0.786} \times 10^3 \times \frac{60}{30} = 20.35 \text{ mm/h}$$

The graph of time in hours versus infiltration rate is as shown in Fig. 5.4.
Total quantity of water added in 150 min till a steady state was achieved = 10.0 + 9.2 + 8.6 + 8.2 + 8.0 = 44.0 lit

$$\text{Therefore, average rate of infiltration} = \frac{44 \times 10^{-3} \times 10^3 \times 60}{150 \times 0.786} = 22.39 \text{ mm/h/m}^2$$

5.4.2 Sprinkling-Type Infiltrometer

This experiment is conducted in the laboratory under controlled conditions.

For this purpose, a very small catchment area, about 5 m^2, is selected. Water is sprinkled over the catchment area to represent rainfall at a uniform rate of about 50 mm/h. Sprinkling of water may be from a height of 5 m. The trial starts with the sprinkling of water. It is, therefore, called *sprinkling-type infiltrometer*. Sprinkling of water represents rain, and hence it is also called *rain simulator*.

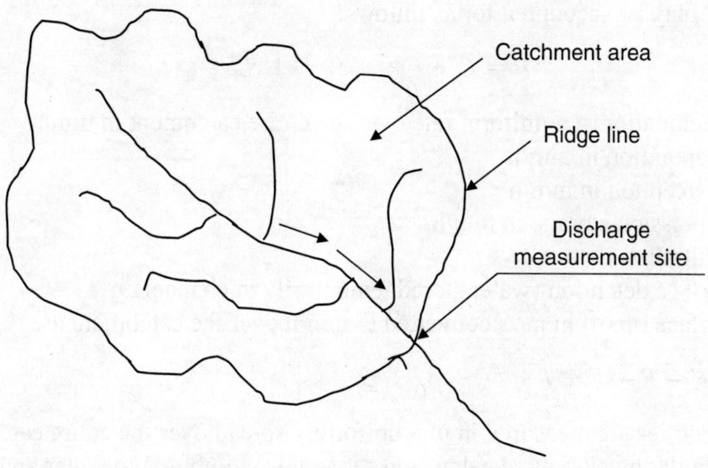

Fig. 5.5 Selected catchment area

The surface runoff is measured very accurately at the catchment outlet and thus knowing all the abstractions such as 'evaporation, interception, depression storage, surface detention' and surface runoff, the infiltration rate during the time of experiment can be worked out.

Consider the following particulars of a specific experiment.

A small catchment was selected for the experiment as shown in Fig. 5.5.

The experiment started at 8:00 a.m. by sprinkling water over the entire catchment area at a uniform rate, which was stopped at 12:00 noon. There was no surface runoff noticed up to 8:30 a.m. The discharge was noticed at 8:30 a.m. and it went on increasing up to 10:00 a.m. and remained constant up to 12:00 noon. When the sprinkling of water, i.e. rainfall, was stopped, the discharge went on reducing and finally stopped at 2:00 p.m. as shown in Fig. 5.6.

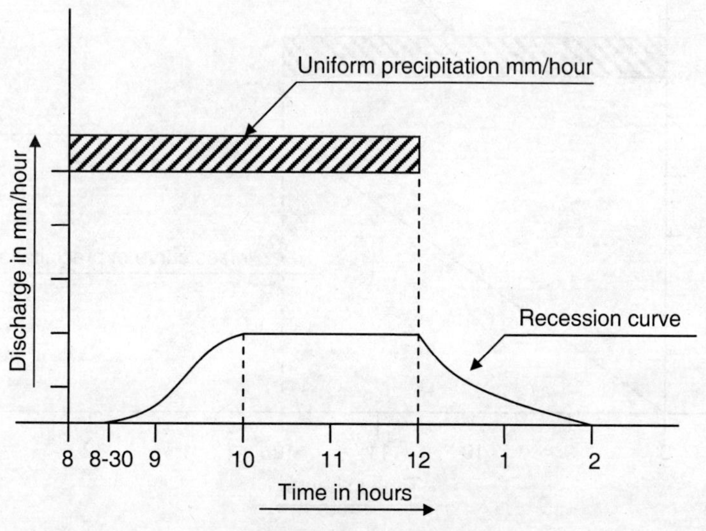

Fig. 5.6 Observed discharge

The rainfall P may be accounted for as follows:

$$P = E + I + D_S + F + S_D + Q$$

where, P = Precipitation at a uniform rate over the entire catchment in mm/h
E = Evaporation in mm/h
I = Interception in mm/h
D_S = Depression storage in mm/h
F = Infiltration in mm/h
S_D = Surface detention (water stored temporarily in channel)
Q = Surface runoff in m^3/s converted to mm/h over the catchment area

Thus, $F = P - (E + I + D_S + S_D + Q)$

(All these quantities are measured in mm uniformly spread over the entire catchment area.) The parameters in the bracket were calculated separately, combined together and then subtracted from the precipitation and thus the infiltration rate was calculated. Infiltration started immediately at 8:00 a.m. The following assumptions were made:

1. Evaporation being negligible during the experiment was neglected.
2. As usual, interception and depression storage were combined together and treated as S_R *surface retention*. This abstraction started immediately at 8:00 a.m.

The precipitation was started at 8:00 a.m. and stopped at 12:00 noon. The precipitation graph with respect to time and its mass curve is shown in Fig. 5.7.

A judgement was taken as to how much water was lost as surface retention, and its requirement would be first met with completely from 8:00 a.m. to 8.30 a.m. and there would be no addition

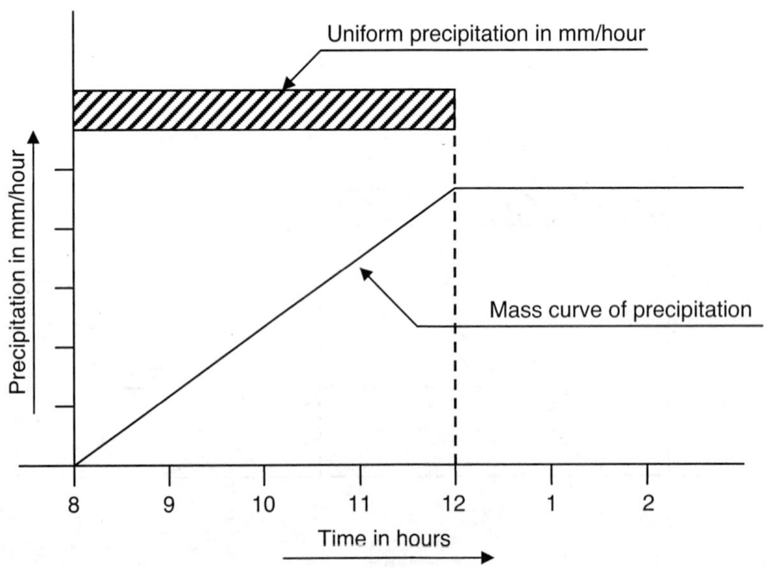

Fig. 5.7 Mass curve of precipitation

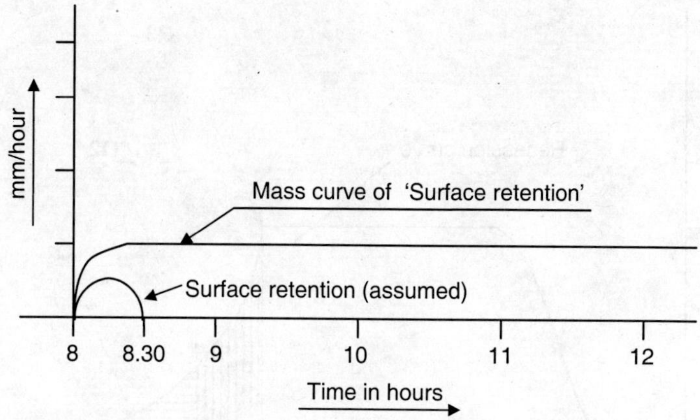

Fig. 5.8 Surface retention

in this abstraction after 8:30 a.m. Surface runoff started only after the requirement of the surface retention was met with. The surface retention and its mass curve is shown in Fig. 5.8.

The discharge was observed from 8:30 a.m. onwards. It went on increasing up to 10:00 a.m. It was constant from 10:00 a.m. to 12:00 noon and then reduced from 12:00 noon to 2:00 p.m. The discharge as well as its mass curve is shown in Fig. 5.9.

The S_D is in mm/s, i.e. water stored in the channel is a function of depth in a channel. More the discharge in the channel, more will be the depth in the channel and more will be the storage in the channel. Thus, indirectly S_D will be a function of discharge. Here the discharge was started at 8:30 a.m. Similarly, S_D also started at 8:30 a.m. It went on increasing up to 10:00 a.m. It remained constant from 10:00 a.m. to 12:00 noon and then was reduced from 12:00 noon to 2:00 p.m.

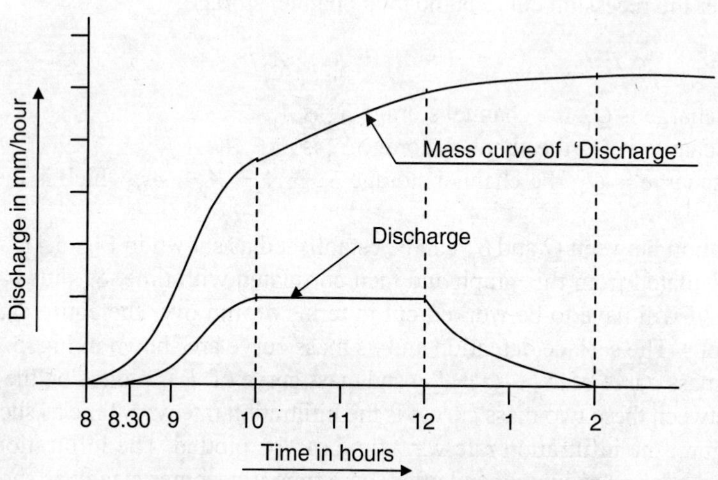

Fig. 5.9 Mass curve of discharge

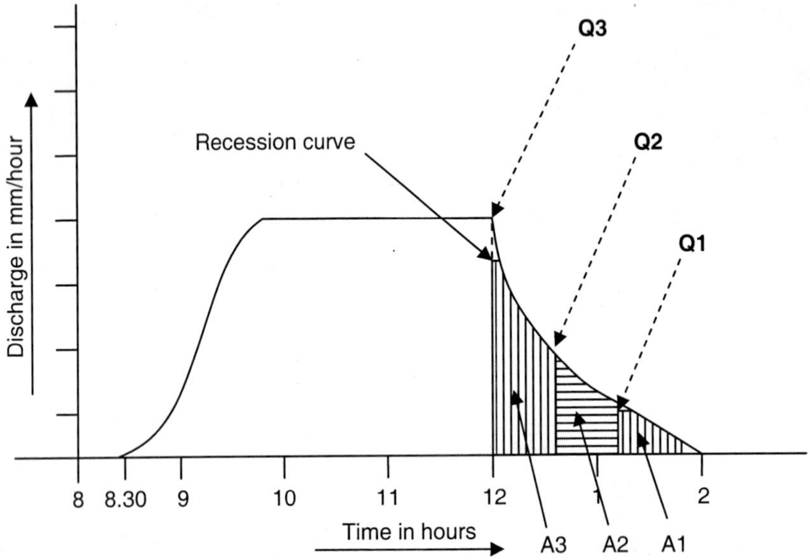

Fig. 5.10 Estimation of surface detention

Since initially and also after the experiment the channel carried no water, the water stored in the channel from 8:30 a.m. to 12:00 a.m. was drained completely from 12:00 noon to 2:00 p.m.

When the precipitation stopped at 12:00 a.m., there was still flow in the stream. This was due to draining the water stored in the channel, i.e. S_D. Immediately after the precipitation stopped, the discharge was maximum. S_D was also maximum. The total volume of water stored in the channel was the area under the recession curve. As the discharge reduced, so also did the storage in the channel. The storage in the channel will be the area under the recession curve. The recession curve is shown in Fig. 5.10.

The area under the recession curve is the total channel storage.

$$A = A_1 + A_2 + A_3$$

When the discharge is Q_1, the channel storage S_D is A_1.

When the discharge is Q_2, the channel storage S_D is $A_1 + A_2$.

When the discharge is Q_3, the channel storage S_D is $A_1 + A_2 + A_3$, which is the area under the recession curve.

Thus, the relation between Q and S_D can be established as shown in Fig. 5.11.

S_D can be calculated from this graph and then correlated with time. S_D thus calculated will be in terms of m^3 and will have to be worked out in terms of mm over the entire catchment area for further calculations. The surface detention and its mass curve are shown in Fig. 5.12.

A combined mass curve of ($S_R + Q + S_D$) and mass curve of P is plotted on the graph paper and the difference between these two mass curves is the infiltration rate w.r.t. time as shown in Fig. 5.13.

From this figure, the infiltration rate w.r.t. time can be plotted. The infiltration rate is more at the beginning and goes on reducing, and after some time it becomes steady as shown in Fig. 5.14.

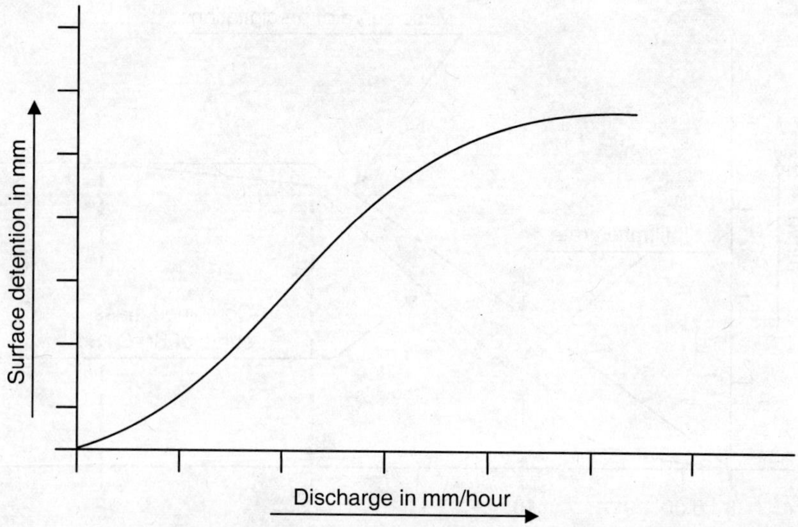

Fig. 5.11 Surface detention w.r.t. discharge

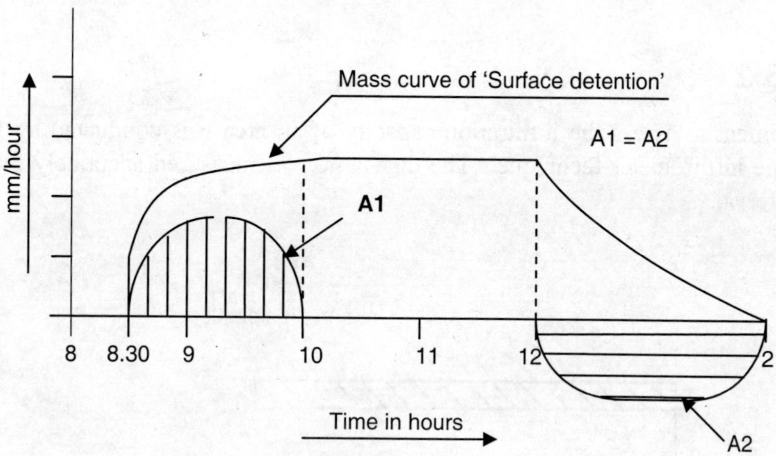

Fig. 5.12 Mass curve of surface detention

5.4.2.1 Drawbacks in sprinkling-type infiltrometer

The main drawbacks in the sprinkling-type infiltrometer are as follows:

- It is very difficult to have a uniform sprinkling of water of a uniform intensity over the entire catchment area for a sufficient duration.
- The discharge has to be measured very accurately.
- The assumption of S_R requires good judgement.
- The area under the experiment being small, lateral flow may be significant.

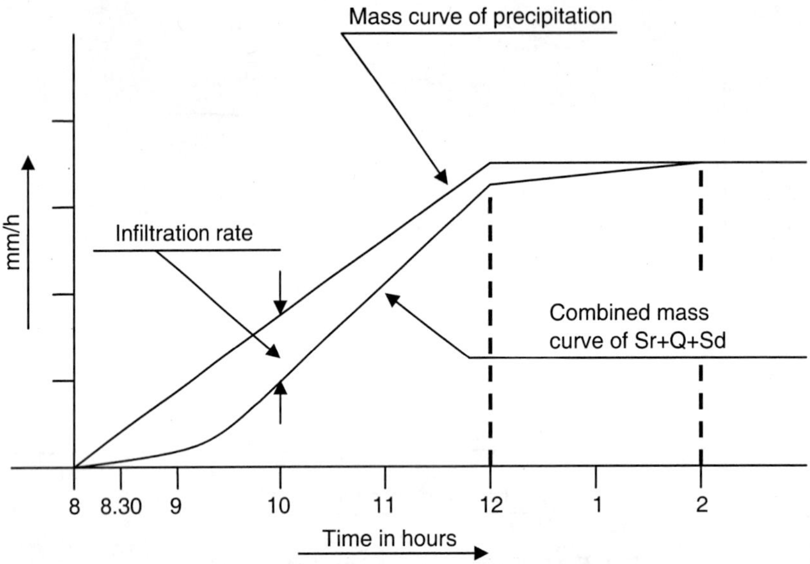

Fig. 5.13 Estimation of infiltration by mass curves

Example 5.2

A field experiment to assess the infiltration capacity of an area was conducted by following the Sprinkling type infiltrometer technique'. The discharge was measured accurately over a notch at 10 minutes interval.

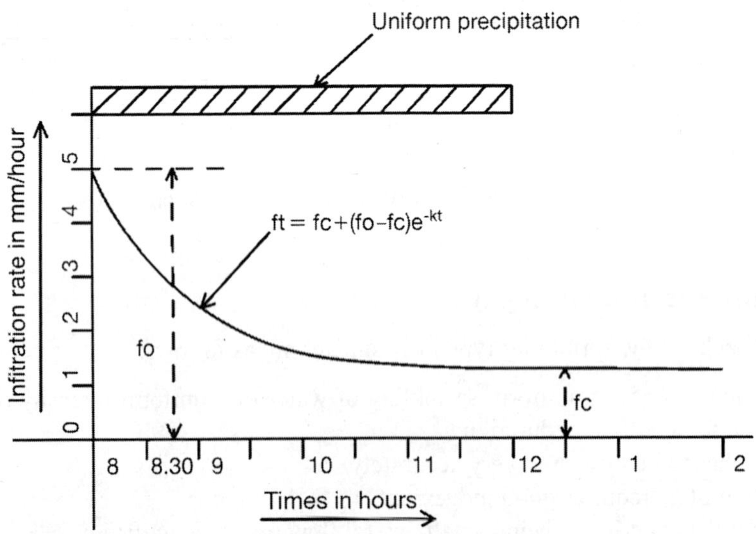

Fig. 5.14 Horton's equation of infiltration rate

The details are as follows.

1. Catchment area 4.5 m^2
2. Experiment started at 8:00 a.m.
3. Experiment concluded at 12:50 a.m.
4. Uniform rainfall intensity used 4 cm/h

The discharge measured is as follows.

1 Sr. no.	2 Time	3 Discharge in c. c. /s	4 Discharge in cm / h
1	8–00	0.00	0.00
2	–10	0.00	0.00
3	–20	1.88	0.15
4	–30	5.00	0.40
5	–40	7.38	0.59
6	–50	-10.00	0.80
7	9–00	12.88	1.03
8	–10	16.00	1.28
9	–20	19.25	1.54
10	–30	22.50	1.80
11	40	25.00	2.00
12	–50	27.25	2.18
13	10–00	28.75	2.30
14	–10	29.75	2.38
15	–20	30.75	2.46
16	–30	31.25	2.50
17	–40	31.75	2.50
18	–50	31.75	2.50
19	11–00	17.00	1.36
20	–10	12.25	0.98
21	–20	9.50	0.76
22	–30	7.88	0.63
23	–40	6.63	0.53
24	–50	5.75	0.46
25	12–00	4.63	0.37
26	–10	4.25	0.34
27	–20	3.88	0.31
28	–30	2.50	0.20
29	–40	1.88	0.15
30	–50	0.00	0.00

The discharge measured on the notch in c. c. /s, was converted for simplicity, into depth in cm/h uniformly distributed over the entire catchment area of 4.5 m^2, for these calculations. 12.5 c.c./s is equivalent to 1.00 cm /h as follows

$$\frac{1 \times 4.5 \times 10^6}{100 \times 3600} = 12.5 \text{ c.c./s}$$

The discharge measured in c.c./s converted in cm / h is entered in col. 4 in the statement. The calculations done to evaluate the infiltration rate are tabulated in the following statement.

1	2	3	4	5	6	7	8=5+6+7	9=3−8	10
Sr. no.	Time	Mass curve rainfall cm	Q cm	Mass curve Q cm	Mass* curve S_D cm	Mass** curve S_R cm	Q mass + S_R mass + S_D mass cm	Mass curve F cm	Infiltration rate cm
1	8–00	4	0.00	0.00	0.00	0.00	0	4.00	4.00
2	–10	8	0.00	0.00	0.00	1.15	1.15	6.85	2.85
3	–20	12	0.15	0.15	0.15	2.35	2.65	9.35	2.50
4	–30	16	0.40	0.55	1.50	2.35	4.40	11.60	2.25
5	–40	20	0.59	1.14	1.83	2.35	6.32	13.68	2.08
6	–50	24	0.80	1.94	4.08	2.35	8.37	15.63	1.95
7	9–00	28	1.03	2.97	5.20	2.35	10.52	17.48	1.85
8	–10	32	1.28	4.25	6.14	2.35	12.74	19.26	1.78
9	–20	36	1.54	5.79	6.90	2.35	15.04	20.96	1.70
10	–30	40	1.80	7.59	7.47	2.35	17.41	22.59	1.63
11	–40	44	2.00	9.59	7.88	2.35	19.82	24.18	1.59
12	–50	48	2.18	11.77	8.14	2.35	22.26	25.74	1.56
13	10–00	52	2.30	14.07	8.30	2.35	24.72	27.28	1.54
14	–10	56	2.38	16.45	8.40	2.35	27.20	28.80	1.52
15	–20	60	2.46	18.91	8.43	2.35	29.69	30.31	1.51
16	–30	64	2.50	21.41	8.43	2.35	32.19	31.81	1.50
17	–40	68	2.50	23.91	8.43	2.35	34.69	33.31	1.50
18	–50	72	2.50	26.41	8.43	2.35	37.19	34.81	1.50
19	11–00	72	1.36	27.77	6.37	2.35	36.49	35.51	0.70
20	–10	72	0.98	28.75	4.94	2.35	36.04	35.96	0.45
21	–20	72	0.76	29.51	3.88	2.35	35.74	36.26	0.30
22	–30	72	0.64	30.15	3.04	2.35	35.54	36.46	0.20
23	–40	72	0.53	30.68	2.36	2.35	35.39	36.61	0.15
24	–50	72	0.46	31.14	1.80	2.35	35.29	36.71	0.10
25	12–00	72	0.37	31.51	1.38	2.35	35.24	36.76	0.05
26	–10	72	0.34	31.85	1.04	2.35	35.24	36.76	0.00
27	–20	72	0.31	32.16	0.73	2.35	35.24	36.76	0.00
28	–30	72	0.20	32.36	0.53	2.35	35.24	36.76	0.00
29	–40	72	0.15	32.51	0.15	2.35	35.24	36.74	0.00
30	–50	72	0.00	32.51	0.00	2.35	35.24	36.74	0.00

* Read from the graph in Fig. 5.16.

** Assumed

The experiment was started at 8:00 a.m. by starting sprinkling water at the rate of 4.00 cm / h over the entire catchment at a uniform rate. There was no flow of water upto 8:10 a.m. since the requirement of S_R [Interception and depression storage] was to be met with.

It was assumed plot S_R is equivalent to 2.35 cm. The discharge started flowing at 8:20 a.m. but was very low because the requirement of S_R was not fully met with. It was assumed that the requirement of S_R of 2.35 cm. was completely met with upto 8:20 a.m.

The discharge as it started at 8:20 went on increasing upto 10:30 a.m. and reached 2.50 cm and remained constant from 10:30 a.m. upto 10:50 a.m.. This indicated that the infiltration rate has reached a steady state.

From 10:50 a.m., sprinkling of water was stopped.

Naturally the discharge started reducing from 10:50 a.m. and was completely stopped at 12:50 p.m. Fig. 5.15. shows the discharge observed.

The storage capacity S_D of the channel w.r.t. discharge flowing in the channel is calculated from the falling graph of discharge from 10:50 a.m. onwards and is shown in Fig. 5.16.

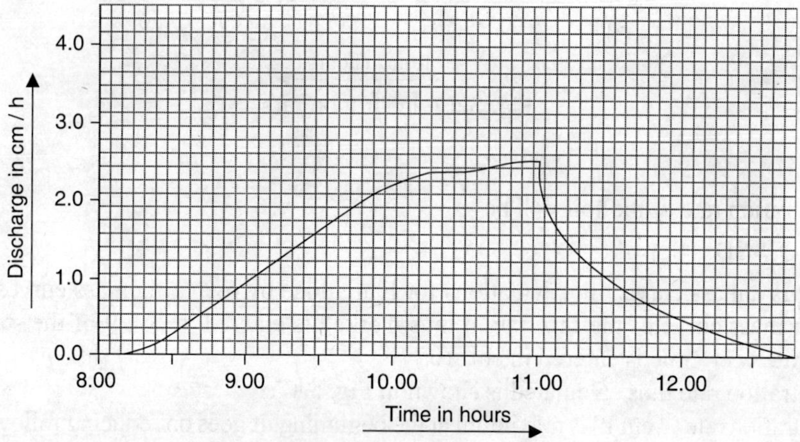

Fig. 5.15 Discharge

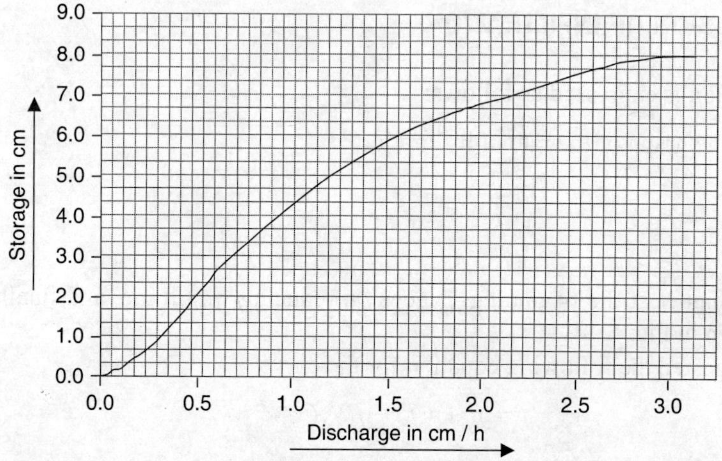

Fig. 5.16 Discharge-Storage curve

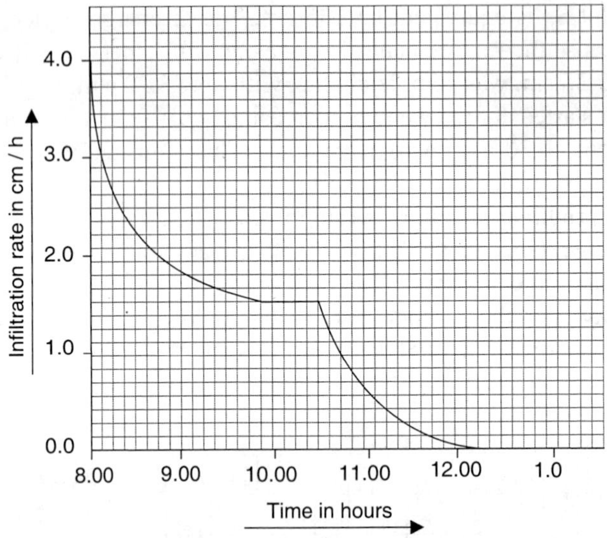

Fig. 5.17 Infiltration rate

The infiltration rate of the soil will be as

$F = P - [Q + SD + S_R]$

Since S_D is a mass curve, all calculations for F are done by preparing mass curves of P and S_D Q + S_R. The mass curve of the infiltration thus worked is entered in col. 9 of the statement. The infiltration rate w.r.t. time is entered in col. 10.

The infiltration rate thus calculated is shown in Fig. 5.17.

The infiltration rate '4 cm/h' is maximum at the beginning. It goes on reducing following a curved path with concavity upwards. It is constant '1.5 cm/h' for some time indicating a steady state. Then it goes on reducing. This curve is also having concavity upwards and finally it reduces to zero.

5.5 EXPRESSION OF INFILTRATION

The rate of infiltration is expressed as follows:

- An equation for the graph of the rate of infiltration
- Infiltration index

5.5.1 Equation

Infiltration rate is higher at the beginning. It goes on reducing w.r.t. time, and finally it attains a steady rate as shown in Fig. 5.14.

Horton has suggested the following equation for the curve

$$f = f_c + (f_o - f_c) \times e^{-kt}$$

where, f = Infiltration rate at time t

f_o = Initial infiltration rate

f_c = Final infiltration rate
k = A constant
t = Time in hours

when $t = 0$, $e^{-kt} = 1$ then $f = f_o$
when $t = \infty$, $e^{-kt} = 0$ then $f = f_c$
Value of k is normally between 2 to 5.

Example 5.3

The value of k in the Horton's equation for infiltration is 2. The maximum and the minimum rates of infiltration are 2 cm/h and 0.5 cm/h. Plot the infiltration rate curve.

Solution:

The Horton's equation will be:

$f = 0.5 + (2.0 - 0.5)\, e^{-2t} = 0.5 + 1.5^{-2t}$ and the hourly rate of infiltration will be as under:

Serial no.	Time (h) = t	$2t$	e^{2t}	e^{-2t}	$1.5\, e^{-2t}$	f (cm/h)
1	0	0	—	—	0	2.000
2	0.5	1.0	2.718	0.368	0.551	1.051
3	1.0	2.0	7.389	0.135	0.203	0.703
4	1.5	3.0	20.085	0.049	0.074	0.574
5	2.0	4.0	54.598	0.018	0.027	0.527
6	2.5	5.0	148.413	0.007	0.010	0.510

The infiltration rate curve will be as shown in Fig. 5.18.

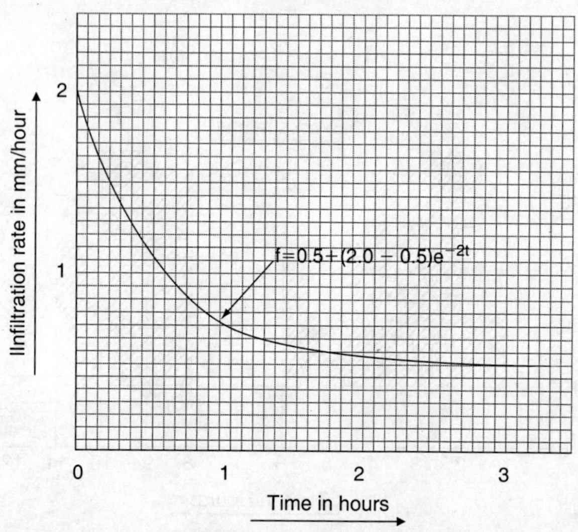

Fig. 5.18 Estimation of infiltration rate

5.5.2 Infiltration Indices

For simplicity in calculations, the infiltration rate is expressed as a uniform average rate. This is expressed in the following ways:

- ϕ index
- W index
- W_{min} index
- f_{AV}

5.5.2.1 ϕ index

The most common way of expressing infiltration rate is ϕ index.

Now, $P = E + S_R + S_D + Q + I$

or, $I = P - (E + S_R + S_D) - Q$

where, P = Precipitation
E = Evaporation
S_R = Surface retention (interception + depression storage)
S_D = Surface detention (finally it flows as surface runoff)
Q = Surface runoff
I = Infiltration rate

Neglecting E, S_R and S_D, one can say that

$$\phi \text{ index} = P - Q$$

Indirectly, ϕ index includes all abstractions, e.g. evaporation, surface retention, and so on. Figure 5.19 shows definition sketch of ϕ index.

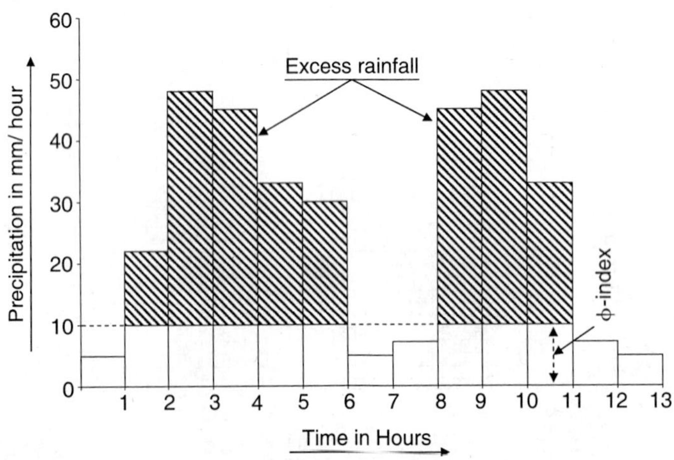

Fig. 5.19 Definition sketch of ϕ index

5.5.2.2 W index

W index is the rate of infiltration when all other abstractions are accounted, viz,

$$W \text{ index} = P - (E + S_R + S_D + Q)$$

Therefore, *W* index < ϕ index

5.5.2.3 W_{min} index

It is the minimum rate of infiltration when a uniform stage after stabilization is attained.

W_{min} = ϕ index after it is stabilized

= $k\phi$ index, where k will be always less than one

5.5.2.4 f_{AV}

This approach is a slight modification of ϕ index. In a storm, when there is no precipitation or when the precipitation is very low, infiltration is still going on because of the previous high precipitation. A provision is, thus, made for this infiltration when precipitation is very low from the previous precipitation when it is high. Consider a storm as shown in Fig. 5.20.

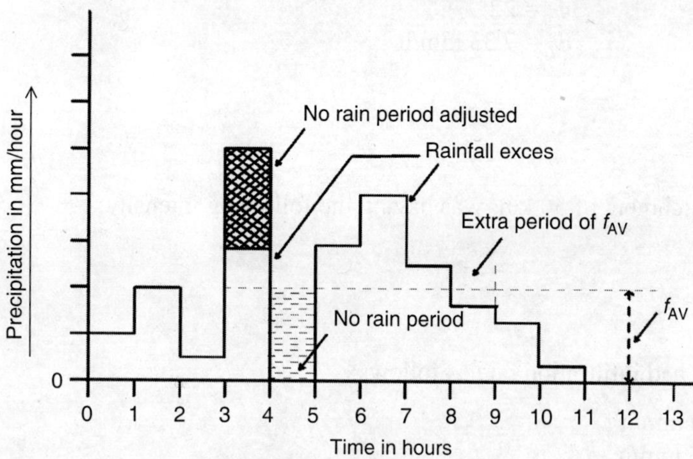

Fig. 5.20 Definition diagram of f_{AV}

A provision for infiltration when there is no precipitation, i.e. between 4 to 5 h, is made from the storm precipitation between 3 to 4 h and hence is deducted from the excess rainfall from 3 to 4 h. f_{AV} is calculated and then the excess rainfall is worked out. It is slightly less than ϕ index.

Example 5.4

The average precipitation during a storm over a catchment area of 10 km^2 is as follow:

1. 40 mm/h for 1 h
2. 60 mm/h for 1 h
3. 30 mm/h for 1 h

The resulting hydrograph was plotted on a graph paper with the following scale.

1 cm = 1 h on x-axis
1 cm = 10 m^3/s on y-axis

If the area of the hydrograph was measured as 30 cm^2, find the ϕ index of infiltration.

Solution:

$$\text{Total runoff observed} = 30 \times 10 \times 3600 = 1.08 \times 10^6 \text{ m}^3$$

Runoff from the storm assuming rate of infiltration to be ϕ

$$= \left(\frac{(40-\phi)\times 1}{1000} + \frac{(60-\phi)\times 1}{1000} + \frac{(30-\phi)\times 1}{1000}\right) \times 10 \times 10^6$$

$$= (130 - 3\phi) \times 10^4 \text{ m}^3$$

Therefore, $(130 - 3\phi) \times 10^4 = 1.08 \times 10^6$

$$3\phi = 22$$

$$\phi = 7.33 \text{ mm/h}$$

Example 5.5

The storm over a catchment of 50 km^2 was having the following intensity:

1. 40 mm/h for 1 h
2. 70 mm/h for 2 h
3. 30 mm/h for 1 h

The catchment area had infiltration rate as follows:

a. 20% area ϕ =10 mm/h
b. 60% area ϕ =15 mm/h
c. Balance impervious

Find the runoff due to the storm.

Solution:

A) The runoff from the area having ϕ=10 mm/h will be

$$= \left(\frac{(40-10)\times 1}{1000} + \frac{(70-10)\times 2}{1000} + \frac{(30-10)\times 1}{1000}\right) \times 0.2 \times 50 \times 10^6$$

$$= \frac{30 + 120 + 20}{1000} \times 0.2 \times 50 \times 10^6 = 1.7 \times 10^6 \text{ m}^3$$

B) The runoff from the area having ϕ=15 mm/h will be

$$= \left(\frac{(40-15)\times 1}{1000} + \frac{(70-15)\times 2}{1000} + \frac{(30-15)\times 1}{1000}\right)\times 0.6\times 50\times 10^6$$

$$= \frac{25+110+15}{1000}\times 30\times 10^6 = 4.5\times 10^6\ \text{m}^3$$

C) The runoff from impermeable area will be

$$= \left(\frac{40\times 1}{1000} + \frac{70\times 2}{1000} + \frac{30\times 1}{1000}\right)\times 0.2\times 50\times 10^6$$

$$= \frac{210}{1000}\times 10\times 10^6 = 2.1\times 10^6\ \text{m}^3$$

Therefore, total runoff $= 1.7\times 10^6 + 4.5\times 10^6 + 2.1\times 10^6$
$= 8.3\times 10^6\ \text{m}^3$

Example 5.6

A catchment area having an average rate of infiltration of 15 mm/h experienced the storm of the following intensities.

1. 50 mm/h for 2 h
2. 30 mm/h for $\frac{1}{2}$ h

The resulting runoff was $10\times 10^6\ \text{m}^3$.
Find the catchment area.

Solution:

$$\left[\frac{(50-15)\times 2}{1000} + \frac{(30-15)\times 0.5}{1000}\right]\times A\times 10^6 = 10\times 10^6$$

Therefore, $\left(\frac{70+7.5}{1000}\right)A = 10$

$$A = \frac{10\times 1000}{77.5} = 129.03\ \text{km}^2$$

Example 5.7

For a catchment area of 12 km², a 7-h storm was as follows:

Time (h)	1	2	3	4	5	6	7
Precipitation (mm)	20	40	0	30	50	40	5

The discharge observed at the gauging site was as follows:

Time (h)	0	1	2	3	4	5	6	7	8	9	10	11	12	13	14	15
Discharge (m^3/s)	0	7	17	32	66	55	45	36	28	21	15	10	6	3	1	0

Assume evaporation loss to be 0.6 mm/h/$m^2$and seepage loss equal to 50% of the evaporation loss. Find the ϕ index f_{AV} index and W index.

Solution:

Total precipitation = (20 + 40 + 0 + 30 + 50 + 40 +5) = 185 mm
∑ Discharge observed =
(0+ 7+ 17 + 32 + 66 + 55 + 45 + 36 + 28 + 21 + 15 + 10 + 6 + 3 + 1 + 0) = 342 m^3/s

ϕ index

Assume ϕ index = x mm/h

Assuming that the value of ϕ index will be more than 5 mm/s, the last rainfall slab of 5 mm/s is neglected.

$$\left\{\frac{(20-x)\times 1+(40-x)\times 1+(30-x)\times 1+(50-x)\times 1+(40-x)\times 1}{1000}\right\}\times 12\times 10^6$$

$$= 342\times 60\times 60$$

$$\therefore \qquad 180-5x=\frac{342\times 3.6}{12}$$

$$5x = 180-\frac{342\times 3.6}{12}=180-102.6$$

$$\text{Therefore, } x=\frac{180-102.6}{5}=15.48\text{ mm/h}$$

f_{AV}

Assume f_{AV} = y mm/h

Even though there is no precipitation between 2 and 3 h, there will be infiltration. This is taken into consideration. So also the precipitation between 6 and 7 h is far less than the f_{AV} (expected). Infiltration during this period is also taken into consideration.

$$\frac{(185-7y)\times 12\times 10^6}{1000}=342\times 60\times 60$$

$$\text{Therefore,}\quad f_{AV}=y=11.77\text{ mm/h}$$

W index

Assume W index = z mm/h

Evaporation loss + seepage loss = 1.5 × 5 × 0.6 = 4.5 mm

$$\frac{(180-4.5-5z)}{1000}\times 12\times 10^6=342\times 60\times 60$$

Therefore, W index $= z = 14.58$ mm/h

Figure 5.21 shows the three indices thus worked.

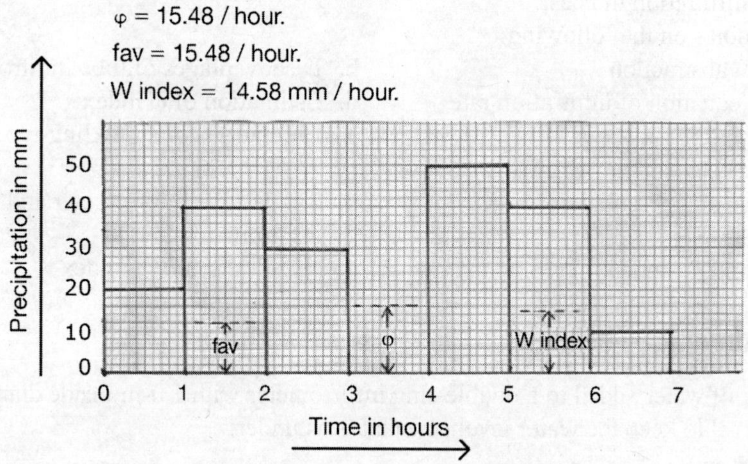

Fig. 5.21 Estimation of indices

Example 5.8

The rate of precipitation in mm/h observed over a catchment of 30 km^2 for successive 30 min is as follows:

16, 20, 24, 36, 28, 12, 4

If the value of $\phi = 22$ mm/h, find

(1) Total precipitation and (2) Runoff in ha. m

Solution:

Since the precipitation observed is for a duration of 1/2 hour, the total precipitation will be as follows:

$$\frac{1}{2}(16 + 20 + 24 + 36 + 28 + 12 + 4) = \frac{140}{2} = 70 \text{ mm}$$

Since the rate of infiltration is higher than the first, second, sixth and seventh slab, all the precipitation will be lost in the infiltration and the runoff will be only from third, fourth and fifth slab. It will be as follows:

$$\text{Runoff} = \frac{1}{2}[(24 - 22)+(36 - 22)+(28 - 22)]$$

$$= \frac{1}{2}(2 + 14 + 6) = 22 = 11 \text{ mm}$$

$$= \frac{11 \times 30 \times 10^6}{1000} = 33 \times 10^4 \text{ m}^3 = 33 \text{ ha m}$$

REVIEW QUESTIONS

1. Define infiltration. Explain the process of infiltration.
2. Discuss the factors affecting infiltration.

3. Explain with the help of neat sketches the flooding-type infiltrometers. What are their advantages and disadvantages?
4. Explain sprinkling-type infiltrometer.
5. Discuss the infiltration indices.
6. Write short notes on the following:
   a. Dominant abstraction
   b. Disadvantages of tube infiltrometers
   c. Horton's equation of infiltration rate
   d. Estimation of ϕ index
   e. Excess rainfall
   f. Limitations of sprinkling-type infiltrometer
7. Differentiate between the following:
   a. Infiltration and percolation
   b. Single-tube infiltrometer and double-tube infiltrometer
   c. ϕ index and f_{AV} index
   d. W index and W_{min} index

NUMERICAL QUESTIONS

1. The quantity of water added to a double-ring infiltrometer with 1.0-m inside diameter at 30-min interval to keep the water level constant is as under:

Time (min)	0	30	60	90	120	150	180
Quantity of water added (l)	0	9.5	9.0	8.4	8.0	7.9	7.9

Find, (i) Rate of infiltration for every 30 min and plot the graph
(ii) Average rate of infiltration

Ans: (ii) 11.49 mm/h/m^2

2. The value of K in the Horton's equation for infiltration is 2.2. And the maximum and minimum rates of infiltration are 2.1 cm/h and 0.5 cm/h. Plot the infiltration rate curve.

3. A catchment area of 13 km^2 experiences a storm of average precipitation as: (1) 45 mm/h for 1 h, (2) 50 mm/h for 1.5 h, (3) 25 mm/h for 1 h and (4) 8 mm/h for 1 h.
The resulting hydrograph was plotted on a graph with the following scales:

1 cm = 1 h on x-axis
1 cm = 10 m^3/s on y-axis

The area under the hydrograph thus plotted was measured and was found to be 40 cm^2. Find the ϕ index of infiltration.

Ans: 10 mm/h

4. The storm over a catchment of 60 km^2 was having the intensity as: (1) 45 mm/h for 1 h, (2) 60 mm/h for 2 h and (3) 35 mm/h for 1 h.
The catchment area had infiltration rates as under:
(a) ϕ for 25% area = 10 mm/h
(b) ϕ for 50% area = 15 mm/h
(c) Balance impervious

Find the runoff due to this storm.

Ans: 8.53 × 10^6 m^3

5. For a catchment area of 12 km^2, a storm of 6-h duration was observed as under:

Time (h)	1	2	3	4	5	6
Precipitation (mm)	22	41	0	32	55	6

The discharge observed at the gauging site was measured as follows:

Time (h)	0	1	2	3	4	5	6	7	8	9	10	11	12	13	14
Discharge (m^3/s)	0	8	19	34	68	58	48	40	25	19	15	11	6	3	0

Assume the evaporation loss to be 1 mm/day/m^2 and the seepage loss equal to 50% of the evaporation loss. Find, ϕ index, f_{AV} and W index.

Ans: ϕ index = 10.95 mm/h
f_{AV} index = 7.30 mm/h
W index = 9.45 mm/h

MULTIPLE CHOICE QUESTIONS

1. Because of mulch and litter, the infiltration rate
   (a) Reduces
   (b) Increases
   (c) Has no effect

2. When the slope of the ground is more than 16°, it has an effect in
   (a) Increasing the rate of infiltration
   (b) Reducing the rate of infiltration
   (c) Has no effect on the rate of infiltration

3. The rate of infiltration observed on a single-tube infiltrometer is
   (a) Less than that observed on a double-ring infiltrometer.
   (b) More than that observed on a double-ring infiltrometer.
   (c) Equal to that observed on a double-ring infiltrometer.

4. The curve of the infiltration rate is
   (a) Concave upwards
   (b) Concave downwards
   (c) A straight line

5. W_{min} index will always be
   (a) Less than ϕ index
   (b) More than ϕ index
   (c) Equal to ϕ index
   (d) None of the above

6. Which of the following will have the maximum rate of infiltration?
   (a) Forest area
   (b) Grazed land
   (c) Rock outcrops
   (d) Concrete Pavement

7. Horton's equation of infiltration rate is given by

(a) $f = f_c + (f_o - f_c)\, e^{-kt}$ (b) $f = f_c + (f_o - f_c)\, e^{-kt}$
(c) $f = f_c - (f_o - f_c)\, e^{-kt}$ (d) $f = f_c + (f_o + f_c)\, e^{-kt}$

ANSWERS TO MULTIPLE CHOICE QUESTIONS

1. b 2. b 3. b 4. a 5. a 6. a 7. a

*** ——— ***

6 Hydrograph

Chapter Outline

6.1 DEFINITION

A hydrograph may be defined as *a graphical representation of time versus discharge* (Fig. 6.1).

6.1.1 Units

The unit for time on the x-axis may be:

1. Hours
2. Days
3. Months

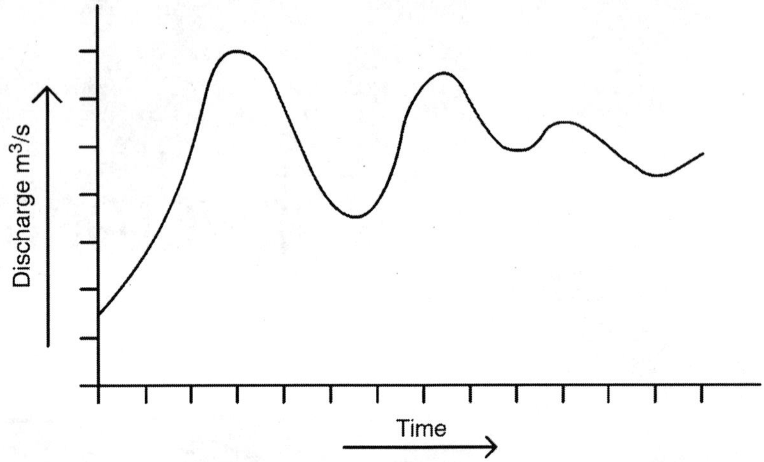

Fig. 6.1 Hydrograph

This unit is selected depending upon the purpose and the nature of the study. If the study is floods or flood routing, then the scale may be in hours. If the study is for estimation of runoff for a year, then it may be in days or in months.

On the y-axis, the unit for discharge is m^3/s. Sometimes, the discharge is expressed in cm/s, i.e. the depth of water per unit area of the catchment area per second expressed in $cm/s/m^2$.

6.1.2 Area of Hydrograph

The area under the hydrograph is the runoff, i.e. the volume of water (Fig. 6.2).

Example 6.1

The scales for the hydrograph are

1 cm = 1 h on x-axis
1 cm = 10 m^3/s on y-axis

If the area under the hydrograph is 45 cm^2, (vide Fig. 6.2) find the runoff.

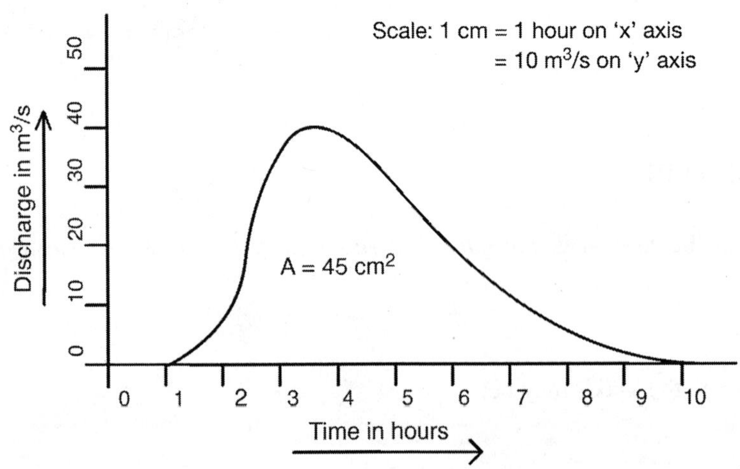

Fig. 6.2 Resultant hydrograph

Solution:

$$\text{Volume of water} = 45 \times (1 \times 3600) \times (1 \times 10)\ \text{m}^3$$
$$= 1620 \times 10^3\ \text{m}^3 = 1.62 \times 10^6\ \text{m}^3 = 1.62\ \text{million m}^3 = 162\ \text{ha m}$$

6.1.3 Isolated Storm

When subsequent storm does not occur before the runoff of the previous storm ceases, that storm is called isolated storm (Fig. 6.3).

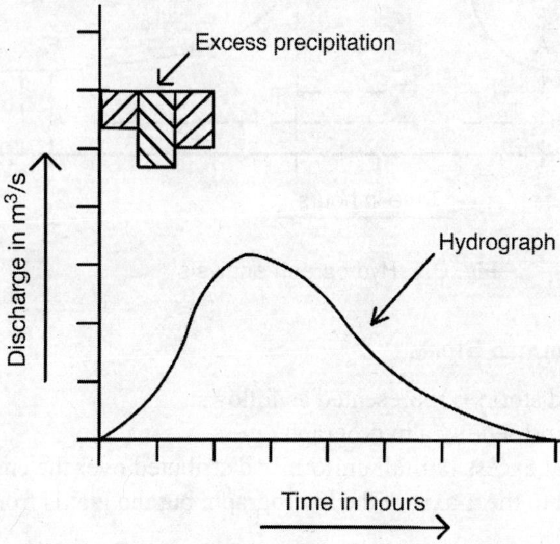

Fig. 6.3 Hydrograph due to an isolated storm

6.1.4 Complex Storm

When subsequent storm occurs before the runoff of the previous storm ceases, such combined storms are called complex storm (Fig. 6.4).

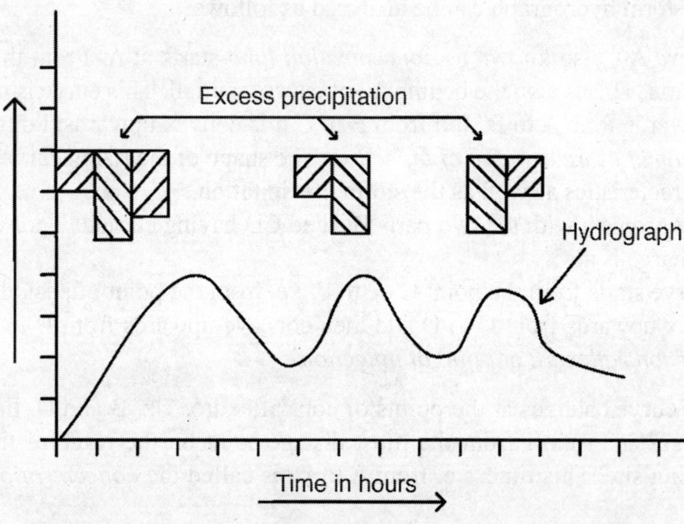

Fig. 6.4 Hydrograph due to a complex storm

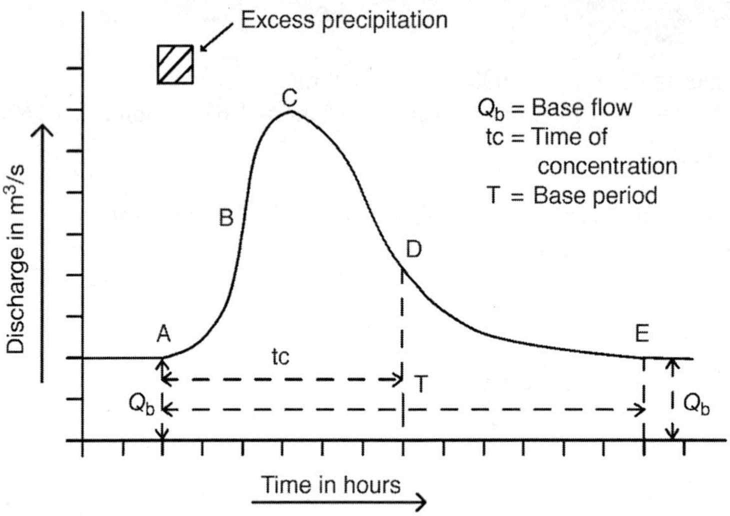

Fig. 6.5 Hydrograph analysis

6.1.5 Hydrograph of an Isolated Storm

The hydrograph of an isolated storm is represented as follows:

This is the normal way of presenting a hydrograph.

The hyetograph or the total excess rainfall uniformly distributed over the entire catchment area is shown with an axis parallel to the x-axis of the hydrograph, but the y-axis from a distance and in the reverse direction (Fig. 6.5).

The discharge Q_b is the discharge in the river before the storm. This does not include the storm runoff. This discharge is known as *base flow* or *basic flow* or *sustained flow* or *fair weather runoff.* Thus, a hydrograph covers the base flow as well as storm runoff.

6.1.6 Analysis of a Hydrograph Due to an Isolated Storm

The shape of the storm hydrograph can be analysed as follows:

1. *The rising curve* AC also known as *concentration limb* starts at A. From this point, the flood hydrograph starts. This is also the beginning of excess rainfall. This curve is normally steep and is concave upwards from A to B, and from B to C it is convex upwards. Point B is, thus, called the *point of contraflexure or point of inflection.* The shape of the rising curve depends upon the catchment characteristics as well as the storm precipitation.
2. BD is the *crest segment* with the two parts BC and CD having curvature convex upwards.
3. The peak discharge is at C.
4. The falling curve starts from the point C up to E, i.e. from the point of peak discharge to E. The curve is convex upwards from C to D and then concave upwards from D to E. Point D is also called *point of contraflexure* or *point of inflection.*

The nature of curve changes at the points of contraflexure, viz. B and D. Point D is the point up to which the surface runoff from the most distant point on the ridge of the catchment area reaches the gauging site. This time, i.e. from A to D, is called the *concentration time* or *the time of concentration.*

From point D onwards, there is no addition to the surface runoff. However, when the river flows, there is some water stored in the flowing river channel. The discharge at the gauging site is due to draining of this water stored in the river channel, i.e. draining of surface retention, and hence the nature of the recession curve is always the same for a specific catchment. It depends only on the basin characteristics of the catchment area and is, therefore, independent of the precipitation.

The curve from D to E is called *recession curve*.

An equation in the following form is derived for the recession curve:

$$Q_{t+1} = Q_t K^{(t-t_0)}$$

where, Q_t = Discharge in m³/s at time t

Q_{t+1} = Discharge in m³/s at time t_0

K = A constant known as recession constant

By observing recession curves of a number of storm runoffs of a specific catchment, a *master recession curve* may be prepared with the help of a tracing paper. This is also known as *composite recession curve, master depletion curve* or *type curve*. Point E denotes the end of surface runoff due to the storm precipitation. The discharge in the stream after E is again the base flow. It is rather difficult to locate the point E. (Theoretically, the recession curve is tangential to the base flow at E.)

Thus, the storm hydrograph is from A to E and the time interval from A to E is known as *base period*. It may be noted that the period for the rising curve is much less than that of the falling curve.

6.2 SEPARATION OF BASE FLOW

The flood hydrograph observed during the storm is a combination of the flood runoff and the base flow. It is necessary to study the excess rainfall and its corresponding runoff from a catchment. Hence, it is necessary to separate the base flow from the combined hydrograph to arrive at flood runoff.

When the flood occurs, the water level in the river rises substantially. The groundwater inflow, if any, depends mainly on the difference between the level of groundwater table and water surface level in the channel. It is a major portion of the base flow. When the water level in the channel rises, this inflow may reduce during the floods and may be less than what it was at the beginning.

A typical hydrograph with base flow is shown in Fig. 6.6.

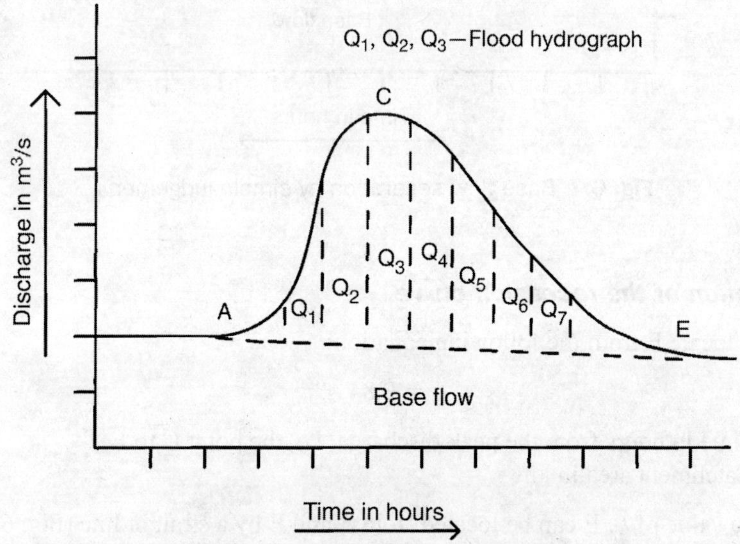

Fig. 6.6 Hydrograph and base flow

6.2.1 Procedure to Separate Base Flow

All the methods for the separation of base flow follow a step-by-step procedure as indicated below:

1. Locate the point A
2. Locate the point E
3. Join A and E by a suitable curve

The discharge below the line AE is the base flow and above it is the flood runoff.

Normally, locating A is not difficult because there is a sudden rise in the hydrograph from this point.

The different methods to separate base flow are as follows:

6.2.1.1 Simple judgement

Make a guess. Locate A and E by judgement and join these two points by a straight line (Fig. 6.7).

The discharge below the line AE is the base flow and above it is the flood runoff.

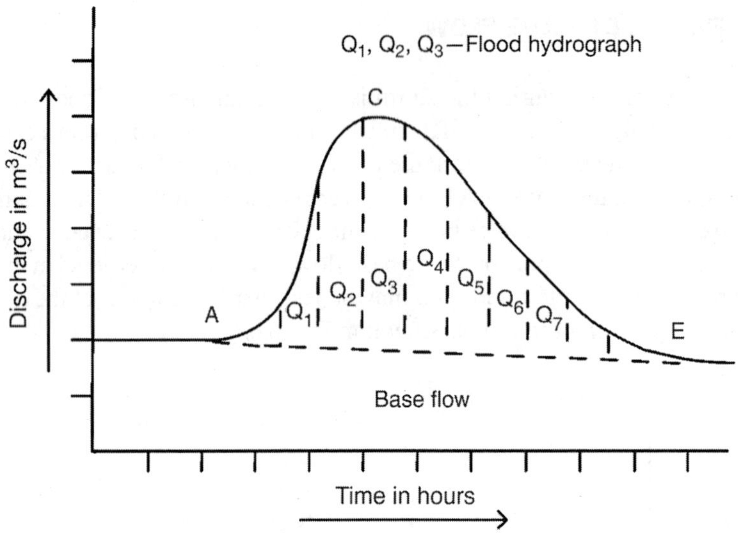

Fig. 6.7 Base flow separation by simple judgement

6.2.1.2 Equation of the recession curve

Locate A. Then locate E from the following equation:

$$t_0 = 0.84\, A^{0.2}$$

where, t_0 = Time in hours from the peak discharge, i.e. the point C to E
A = Catchment area in km²

Knowing the value of t_0, E can be located. Join A and E by a straight line (Fig. 6.8).

The discharge below AE is the base flow and above AE is the flood runoff.

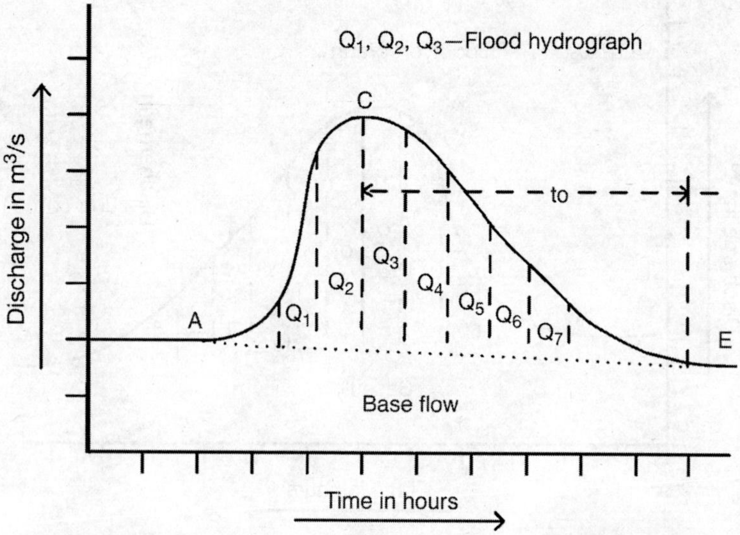

Fig. 6.8 Base flow separation by the equation of recession curve

6.2.1.3 Master depletion curve

Trace the master depletion curve of the catchment area on a paper and try to match it with the hydrograph in question and then locate E. Join A and E by a straight line (Fig. 6.9).

The discharge below AE is the base flow and above AE is the storm runoff.

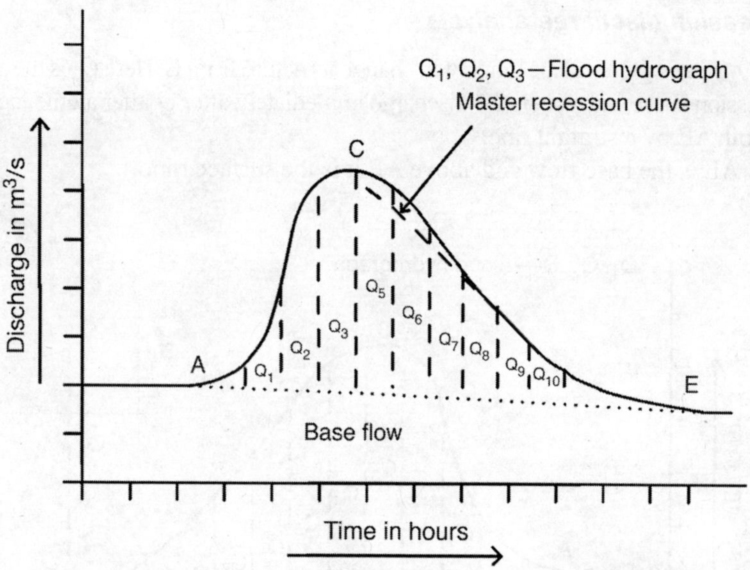

Fig. 6.9 Base flow separation by master recession curve

6.2.1.4 Semi-log analysis

Plot the recession curve including some part of the base flow on a semi-log paper (Fig. 6.10).

Since the recession curve is plotted on a semi-log paper, it will be represented by two straight lines in the region close to E. There will be a sudden change in slope at a point. This will be

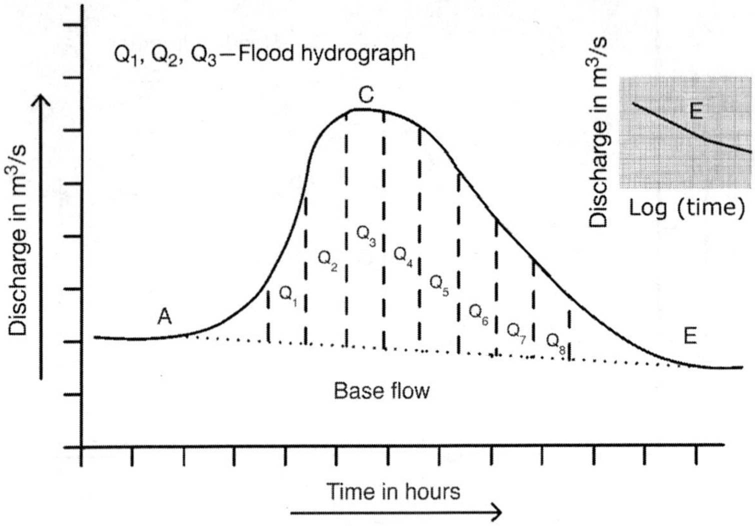

Fig. 6.10 Base flow separation by semi-log analysis

E since the recession curve and the base flow curve have different equations and hence will have different slopes.

Join AE by a straight line.

The flow below AE is the base flow and above AE it is the flood runoff.

6.2.1.5 Recession discharge analysis

Plot a graph of Q_n versus Q_{n+1} and the graph will change its nature from E. Here Q_n is the discharge at any time on the recession curve and Q_{n+1} is the discharge immediately after Q_n after a time lag (Fig. 6.11).

Locate E. Join AE by a straight line.

Flow below AE is the base flow and above AE it is the surface runoff.

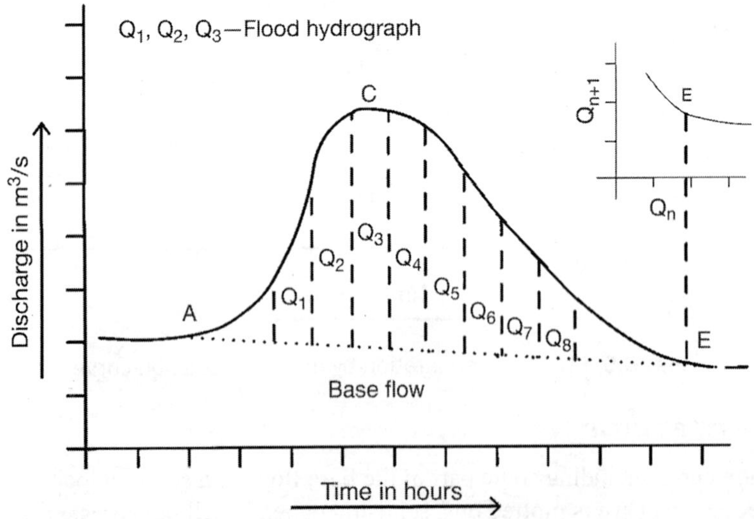

Fig. 6.11 Base flow separation by recession discharge analysis

6.2.1.6 *Extension of base flow graph*

Extend forward the base flow graph, prior to storm, from A up to the peak discharge by a straight line. Similarly, extend backward the base flow up to the peak point by a straight line. Join these two straight lines smoothly by making a good judgement (Fig. 6.12).

The flow below this line AE is the base flow and above it is flood discharge.

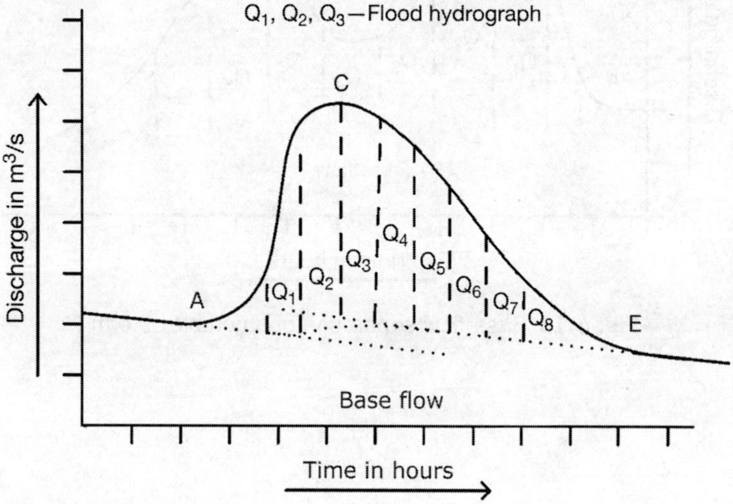

Fig. 6.12 Base flow separation by extension of base flow graph

6.2.2 Review of the Methods

Firstly, all these methods are arbitrary. Not all may be applicable for a specific hydrograph. Proper judgement will have to be made before applying any method.

Secondly, if there is some error in assessing the base flow, the result will be as follows:

1. The base flow time T will slightly change.
2. The runoff volume also will slightly change.

However, it will not make much difference in the storm runoff.

6.2.3 Base Flow Separation of a Complex Storm

Separating the base flow from flood hydrograph resulting from a complex storm is a difficult job. The basic principles are the same and it is a matter of experience.

Figure 6.13 shows the base flow separation for a complex storm.

Example 6.2

Analyse the flood hydrograph (Fig. 6.14).

Solution:

The hydrograph is during an isolated storm. The base flow is separated from the flood hydrograph. The base flow goes on reducing and is negative for some period. It then increases and becomes positive and reaches the point E.

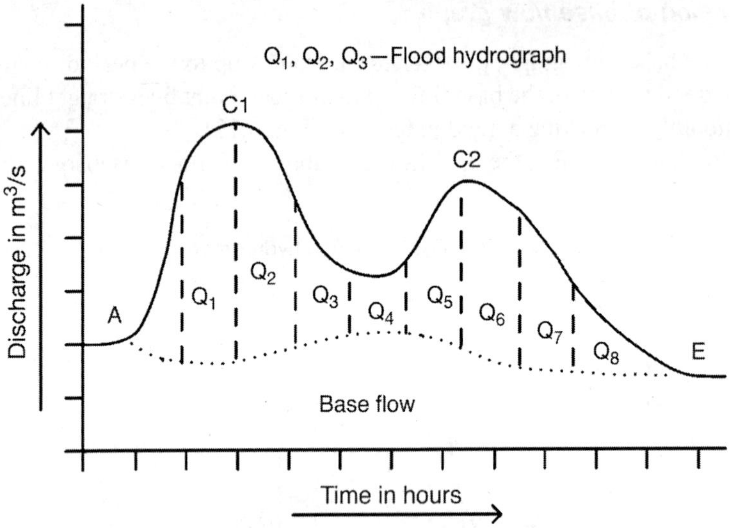

Fig. 6.13 Base flow separation of a complex storm

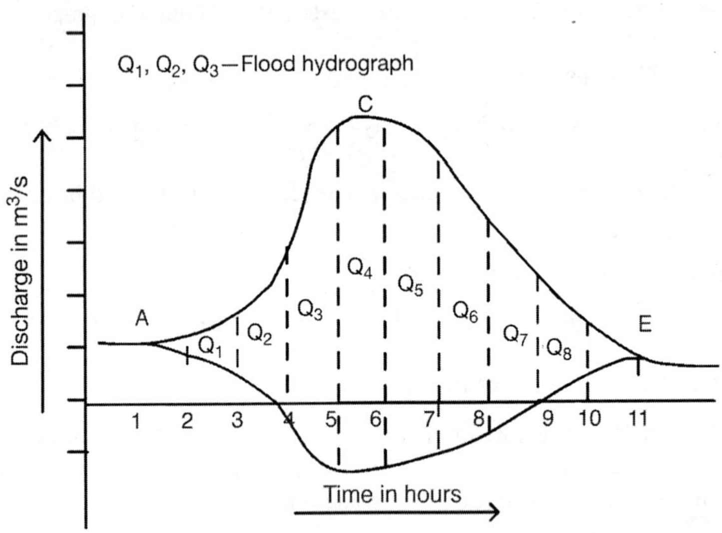

a	12	14	32	57	65	58	40	25	18	12	11
b	12	10	7	0	−9	−13	−12	−11	0	9	11
c	0	4	25	57	74	71	52	36	18	3	0

a–observed flood hydrograph in m^3/s

b–base flow in m^3/s

c–surface flow in m^3/s

Fig. 6.14 Combined flood hydrograph

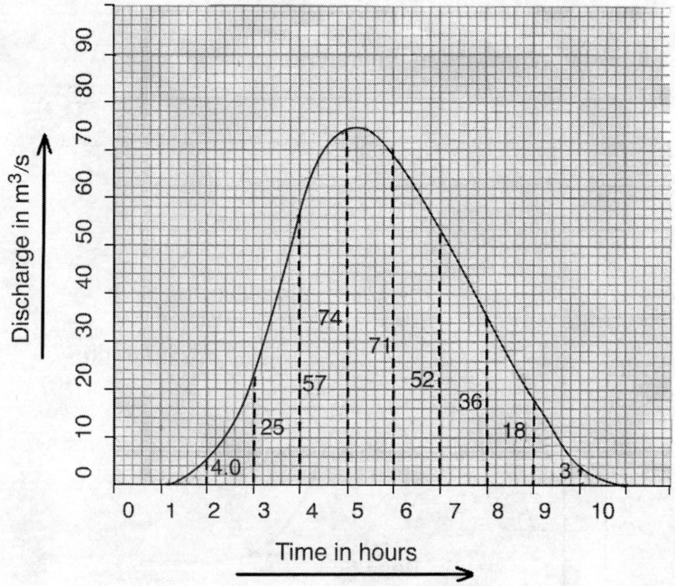

Fig. 6.15 Derived flood hydrograph

The negative flow indicates that prior to the storm there is flow from groundwater to the river channel. However, during floods the water level in the river channel is raised abruptly, and hence water flows from river channel to groundwater, i.e. the base flow is negative.

The flood hydrograph will be as shown in Fig. 6.15.

6.3 EXCESS RAINFALL

When the base flow is separated from the combined hydrograph, the storm runoff will be arrived at. The excess rainfall causing this storm runoff can be calculated as follows:

The hydrograph of the storm runoff is plotted and the volume of the storm runoff is calculated by measuring the area under the storm hydrograph and also taking into consideration the scales adopted for plotting the hydrograph for time and discharge on x- and y-axes as discussed in Subsection 6.1.2.

Assuming a value for ϕ index, the storm runoff is also calculated from the storm precipitation in terms of ϕ and equating it to the storm runoff already calculated. The value of ϕ can thus be calculated.

Example 6.3

Calculate the excess rainfall when

$$P_1, P_2, P_3 = \text{The storm rainfall in cm/h}$$
$$= 3 \text{ cm/h}; 5 \text{ cm/h}; 4 \text{ cm/h}$$

$$A = \text{Catchment area} = 50 \text{ km}^2$$
$$A_H = \text{Area of storm hydrograph} = 100 \text{ cm}^2$$

S_1 = Scale of time adopted to plot flood hydrograph: 1 cm = 1 h

S_2 = Scale of Q adopted to plot flood hydrograph: 1 cm = 10 m^3/s

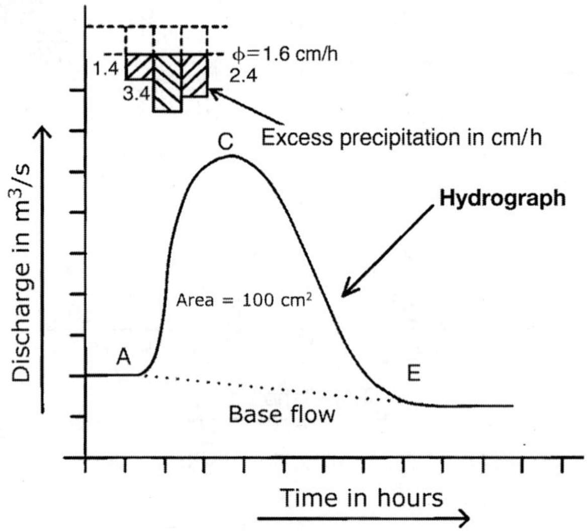

Fig. 6.16 Flood hydrograph and excess rainfall

Solution:

Assume ϕ index of infiltration = x cm/h.
Thus,

$$\frac{[(3-x)\times 1 + (5-x)\times 1 + (4-x)\times 1]\times 50\times 10^6}{100} = 100\times 3600\times 10$$

$$12 - 3x = \frac{100\times 3600\times 100\times 10}{50\times 10^6} = \frac{360}{50} = 7.2$$

Therefore, $3x = 12 - 7.2 = 4.8$

Therefore, $x = 1.6$ cm/h

And excess rainfall will be 1.4 cm/h; 3.4 cm/h; 2.4 cm/h.

It may be noted that the excess rainfall and the flood hydrograph start at the same time (Fig 6.16).

6.4 UNIT HYDROGRAPH

For identical precipitation, hydrographs observed for different catchments will have different shapes. This is because the shape of the hydrograph of every catchment depends on the characteristics of the catchment.

An attempt is made to correlate the hydrograph of each catchment with the precipitation. This is done by the *unit hydrograph theory*. The theory of unit hydrograph was first presented by L. K. Sherman in *Engineering News Record* in April 1932. Originally, this theory was known

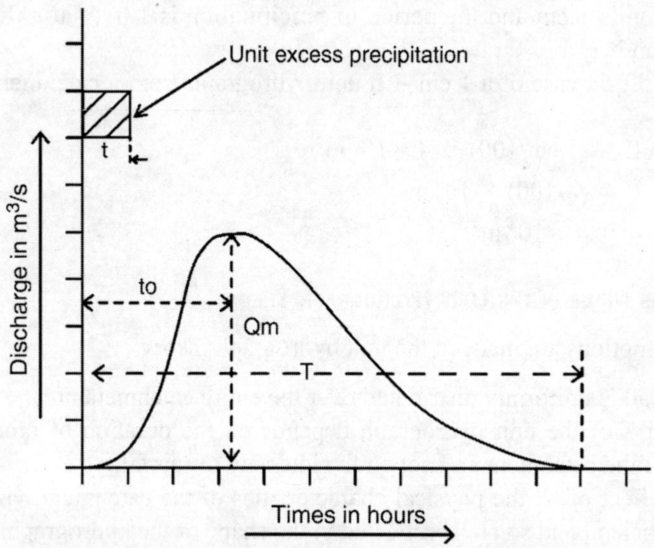

Fig. 6.17 Definition sketch of unit hydrograph

as unit graph. However, this title unit graph was misinterpreted, and hence it was modified as *unit hydrograph*.

Subsequent to the introduction of this theory by Sherman, it underwent a number of modifications, but the basic principle as presented by Sherman remained the same.

6.4.1 Definition of Unit Hydrograph

When a unit excess rainfall occurs uniformly distributed over a catchment area, then the resultant hydrograph is known as unit hydrograph (Fig. 6.17).

6.4.2 Definitions Involved in Unit Hydrograph Theory

The theory of unit hydrograph involves the following definitions.

1. The excess rainfall means the precipitation after all the abstractions, such as interception, evaporation, depression storage, infiltration, etc., are met with.
2. The excess rainfall is 1 unit. This unit may be 1 cm or 5 cm. If the unit is 1 cm, then the excess rainfall may be 1 cm/h for 1 h, or 1/2 cm/h for 2 h, or 2 cm/h for 1/2 h. So that the total excess rainfall will be 1 cm, i.e. 1 unit.
3. The duration of the excess rainfall should be sufficiently less than the time of concentration. Preferably, $t = \frac{t_0}{4}$.
4. The base period T is the total time of the flood hydrograph at the gauging site.

6.4.3 Specifications of Unit Hydrograph

The unit of precipitation as well as intensity of the excess precipitation are the controlling parameters. The unit hydrograph is, therefore, specified as '1 cm–1 h unit hydrograph'. Here the unit precipitation is 1 cm and the period of precipitation is 1 h. (Naturally, the intensity of precipitation is 1 cm/h.)

The surface runoff, in case of a 1 cm–1 h unit hydrograph from a catchment area A km^2, will be as below:

$$\text{The surface runoff} = (1\text{ cm}/100) \times A \times 10^6 \text{ m}^3$$
$$= (A/100) \times 10^6 \text{ m}^3$$
$$= A \times 10^4 \text{ m}^3.$$

6.4.4 Assumptions Made in the Unit Hydrograph Theory

The following assumptions are made in the unit hydrograph theory:

1. The excess rainfall is uniformly distributed over the entire catchment area.
2. The base period T of the unit hydrograph depends on the duration of rainfall and the basic characteristics of the catchment and not on the intensity of rainfall.
3. The combined effect of all the physical characteristics of the catchment, viz. its slope, shape, Manning's coefficient, and so on, is reflected in the shape of the hydrograph.
4. The unit hydrograph coordinates are *time invariant*. This means that the coordinates do not change with rerpect to period. This means that the hydrograph coordinates will remain unchanged for any season, any month, any day or even any year.
5. The coordinates of the resulting surface runoff hydrograph are directly proportional to the intensity of rainfall. This is known as *principle of linearity* or *principle of superposition*.
   This will be clear from the following example (6.4).

Example 6.4

The unit hydrograph coordinates of a 1 cm–1 h unit hydrograph are as follows:

Time (h)	0	1	2	3	4	5	6	7	8	9	10
Discharge (m^3/s)	0	6	13	22	16	11	7	4	2	1	0

Find (A) Flood hydrograph for a storm of 2 cm/h for 1 h.
(B) The catchment.

Solution:

(A) For the required flood hydrograph, the time of precipitation is the same as that of unit hydrograph, but the intensity of precipitation is double that of the unit hydrograph.

The hourly unit hydrograph coordinates are for a precipitation of 1 cm/h for 1 h. The flood hydrograph coordinates for a precipitation of 2 cm/h for 1 h will be double the coordinates of unit hydrographs, since the intensity of precipitation of the flood hydrograph required is double the intensity of precipitation of the unit hydrograph.

The flood hydrograph coordinates will be as follows:

Time (h)	0	1	2	3	4	5	6	7	8	9	10
Discharge(m^3/s)	0	12	26	44	32	22	14	8	4	2	0

The unit hydrograph and the flood hydrograph will be as shown in Fig. 6.18.

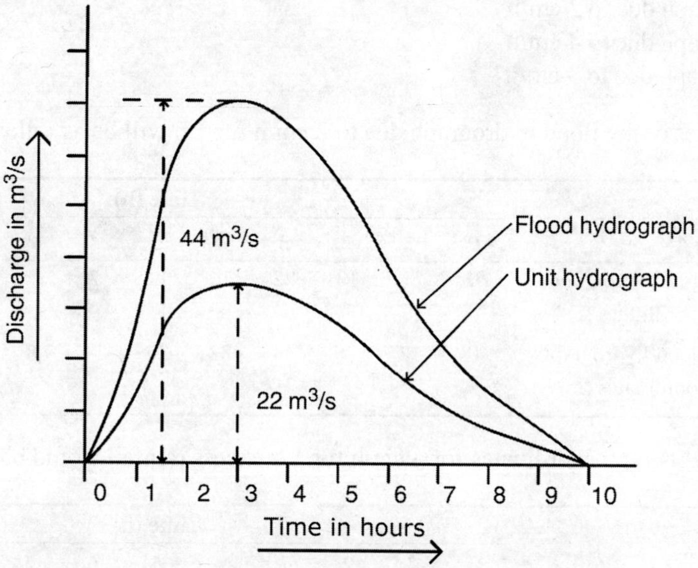

Fig. 6.18 Derived hydrograph

(B) By definition of the unit hydrograph, the precipitation over the entire catchment is 1.0 cm. Therefore, the total expected runoff from catchment A would be $= A \times 10^6 \times 1/100\ m^3$.

$$= A \times 10^4\ m^3$$

The runoff calculated from the unit hydrograph would be $= \Sigma$(hourly ordinates) $\times$ 3600
These two figures should be equal
Therefore, $A \times 10^4 = 82 \times 3600$

$$A = 29.52\ km^2.$$

6.4.5 Use of Unit Hydrograph to Derive Flood Hydrograph

The flood hydrograph can be derived from a unit hydrograph if the excess rainfall storm is known. For this purpose, the principle of linearity is used.

Example 6.5

Find the flood hydrograph from a catchment for excess rainfall of 2 cm/h for 1 h followed by 4 cm/h for 1 h followed by 3 cm/h for 1 h.

Given 1 h–1 cm unit hydrograph coordinates

Time (h)	0	1	2	3	4	5	6	7	8	9	10
Discharge (m^3/s)	0	6	13	22	16	11	7	4	2	1	0

Solution:

The flood hydrograph of the excess precipitation mentioned will thus be a summation of three flood hydrographs, viz.

i. Hydrograph due to 2 cm/h
ii. Hydrograph due to 4 cm/h
iii. Hydrograph due to 3 cm/h

i. The coordinates of the flood hydrograph due to 2 cm/h for 1 h will be as follows:

Serial no.		Time (h) 0	1	2	3	4	5	6	7	8	9	10
1	Unit hydrograph coordinates (m^3/s)	0	6	13	22	16	11	7	4	2	1	0
2	Flood hydrograph coordinates (m^3/s)	0	12	26	44	32	22	14	8	4	2	0

ii. The flood hydrograph coordinates for 4 cm/h for 1 h excess rainfall would be as follows:

Serial no.		Time (h) 0	1	2	3	4	5	6	7	8	9	10
1	Unit hydrograph coordinates (m^3/s)	0	6	13	22	16	11	7	4	2	1	0
2	Flood hydrograph coordinates (m^3/s)	0	24	52	88	64	44	28	16	8	4	0

This hydrograph will start 1 h after the hydrograph due to 2 cm/h.

iii. The flood hydrograph coordinates due to 3 cm/h for 1 h excess rainfall would be as follows:

Serial no.		Time (h) 0	1	2	3	4	5	6	7	8	9	10
1	Unit hydrograph coordinates (m^3/s)	0	6	13	22	16	11	7	4	2	1	0
2	Flood hydrograph coordinates (m^3/s)	0	18	39	66	48	33	21	12	6	3	0

This hydrograph will start 2 h later than the hydrograph due to 2 cm/h and 1 h later than the hydrograph due to 4 cm/h.

Multiplications of unit hydrograph coordinates by the intensity of respective precipitation are because of the assumed principle of linearity.

The combined hydrograph can be calculated by simple addition of these three hydrographs, considering the proper time lag, as follows:

Serial no.	Hydrograph coordinates (m^3/s)	Time (h) 0	1	2	3	4	5	6	7	8	9	10	11	12
1	Due to 2 cm/h	0	12	26	44	32	22	14	8	4	2	0		
2	Due to 4 cm/h		0	24	52	88	64	44	28	16	8	4	0	
3	Due to 3 cm/h			0	18	39	66	48	33	21	12	6	3	0
4	Combined	0	12	50	114	159	152	106	69	41	22	10	3	0

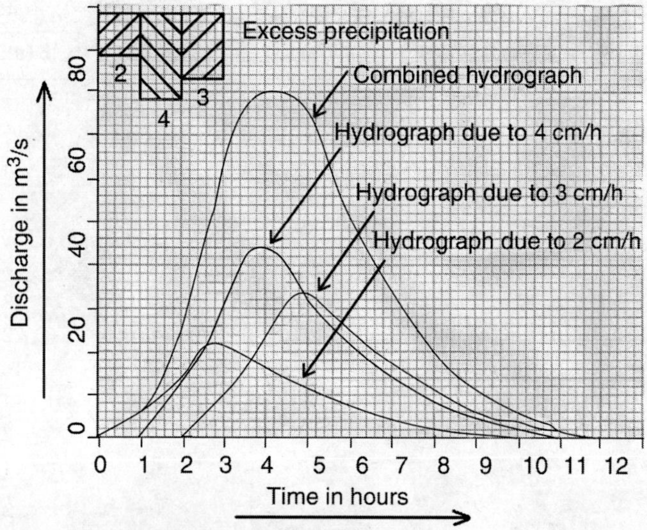

Fig. 6.19 Derivation of flood hydrograph

Fig. 6.19 shows the combined flood hydrograph.

It may be noted that the base period of the resultant flood hydrograph (Fig. 6.19) is equal to

$$T + x - t = 10 + 3 - 1 = 12 \text{ h}$$

where, T = Base period of unit hydrograph

x = Total time of excess rainfall storm

t = Time of excess rainfall in unit hydrograph

Example 6.6

Derive a flood hydrograph from a 1 cm–1 h unit hydrograph for the storm of excess rainfall of 2 cm/h for 1 h followed by 4 cm/h for 2 h then there is a gap of 1 h then 1 cm/h for 1 h.

The 1 cm–1 h unit hydrograph coordinates are as follows:

Time (h)	0	1	2	3	4	5	6	7	8	9	10
Discharge (m^3/s)	0	6	13	22	16	11	7	4	2	1	0

Solution:

The calculations for the flood hydrograph are as per the table below. It may be noted that the precipitation of 4 cm/h for 2 h is split into two parts each of 4 cm/h for 1 h and their hydrographs worked separately. The hydrograph due to 1 cm/h precipitation is taken after a gap of 2 h since there is a gap of 1 h after the 4-cm/h precipitation.

1	2	3	4	5	6	7
Time (h)	A (m^3/s)	B (m^3/s)	C (m^3/s)	D (m^3/s)	E (m^3/s)	F (m^3/s)
0	0	0	–	–	–	0
1	6	12	0	–	–	12
2	13	26	24	0	–	50
3	22	44	52	24	–	120
4	16	32	88	52	0	172
5	11	22	64	88	6	180
6	7	14	44	64	13	135
7	4	8	28	44	22	102
8	2	4	16	28	16	64
9	1	2	8	16	11	37
10	0	0	4	8	7	19
11	–	–	0	4	4	8
12	–	–	–	0	2	2
13	–	–	–	–	1	1
14	–	–	–	–	0	0

Figure 6.20 shows the computed flood hydrograph.

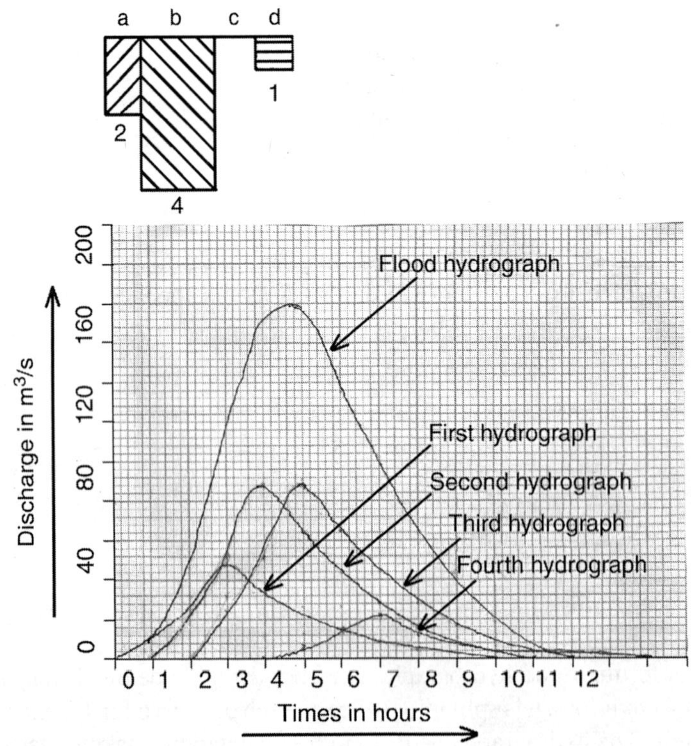

Fig. 6.20 Computation of flood hydrograph

Here, A = 1 cm–1 h unit hydrograph coordinates

B = Hydrograph due to 2-cm/h precipitation

C = Hydrograph due to 4-cm/h precipitation

D = Hydrograph due to 4-cm/h precipitation

E = Hydrograph due to 1-cm/h precipitation

F = Resulting flood hydrograph

The flood hydrograph is as per column F with maximum discharge of 180 m^3/s.

It may be noted that the base period of the flood hydrograph is 14 h, which is equal to 10 + 5 − 1 = 14. (10 is the base period of the unit hydrograph, 5 is the period of storm.)

6.4.6 S Hydrograph

Consider a 1 cm–1 h unit hydrograph as follows: (Fig. 6.21).

Time (h)	0	1	2	3	4	5	6	7	8	9	10
Discharge (m^3/s)	0	6	13	22	16	11	7	4	2	1	0

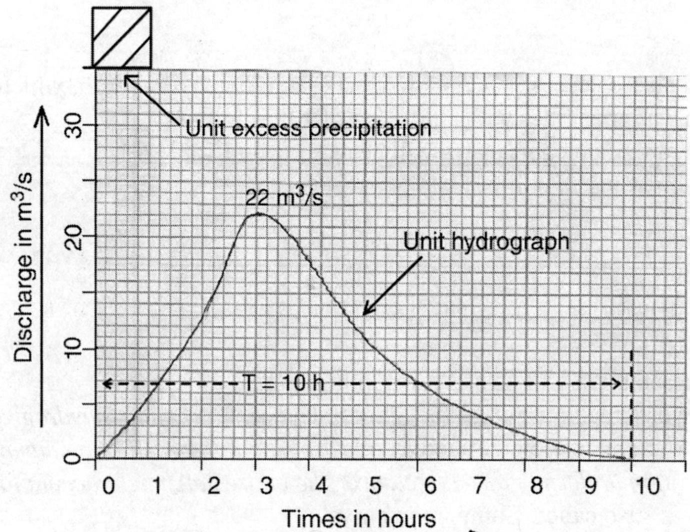

Fig. 6.21 Unit hydrograph

Now imagine that the unit excess rainfall of the same intensity occurs for a period that tends to infinity means the excess rainfall of that intensity occurs continuously up to infinite time.

This excess precipitation tending to infinity can be divided into time intervals equal to unit hydrograph excess precipitation of 1 cm/h for 1 h one after another, tending to infinity.

1	2	3	4	5	6	
1 cm/h	1 cm/h	1 cm/h	1 cm/h	1 cm/h	1 cm/h	→ ∞

Table 6.1 Resulting 'S' hydrograph

Time (h)	Discharge (m^3/s) due to each slab											Total resulting hydrograph (m^3/s)
	1	2	3	4	5	6	7	8	9	10	11	
0	0											0
1	6	0										6
2	13	6	0									19
3	22	13	6	0								41
4	16	22	13	6	0							57
5	11	16	22	13	6	0						68
6	7	11	16	22	13	6	0					75
7	4	7	11	16	22	13	6	0				79
8	2	4	7	11	16	22	13	6	0			81
9	1	2	4	7	11	16	22	13	6	0		82
10	0	1	2	4	7	11	16	22	13	6	0	82
11		0	1	2	4	7	11	16	22	13	6	82
12			0	1	2	4	7	11	16	22	13	82
13				0	1	2	4	7	11	16	22	82
14					0	1	2	4	7	11	16	82
						↓	↓	↓	↓	↓	↓	↓ 8

Naturally, each slab of excess precipitation will have one hydrograph, which is equivalent to unit hydrograph one after another with a time lag of 1 h as follows:

The resulting hydrograph coordinates will be as shown in Table 6.1.

The S hydrograph coordinates will be as follows:

Time (h)	0	1	2	3	4	5	6	7	8	9	10	11	12	Constant value of
Q (m^3/s)	0	6	19	41	57	68	75	79	81	82	82	82	82	82 m^3/s tending to infinity

The graphical representation will be as shown in Fig. 6.22.

The resulting hydrograph resembles the letter S. Hence, it is called *S hydrograph*.

Thus, an S hydrograph may be defined as *a hydrograph observed at a catchment outlet, when the excess precipitation uniformly distributed over the entire catchment occurs for a period which tends to infinity*. It is also called 'Summation curve'.

Note the following:

1. The S hydrograph is a mass curve.
2. The S hydrograph discharge is constant (82) after the base period T (10 h).
3. The constant value of discharge is the summation of all the unit hydrograph coordinates (82 m^3/s).
4. Being a mass curve, it will not have negative slope at any point. The slope is horizontal after the base period T.

It may also be noted here that, in practice, there are so many errors committed in observing the discharge flowing in a river, and as such the S curve plotted from the field observations is normally not a smooth curve as shown in Fig. 6.23.

In such case, a smooth curve is drawn and used for further calculations.

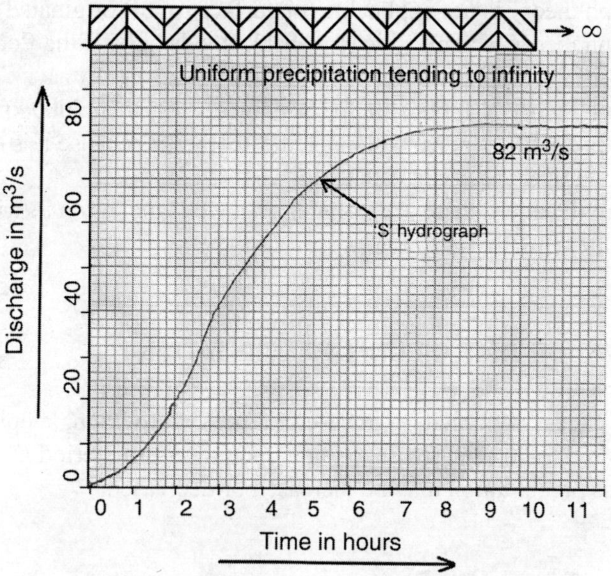

Fig. 6.22 Definition sketch of S hydrograph

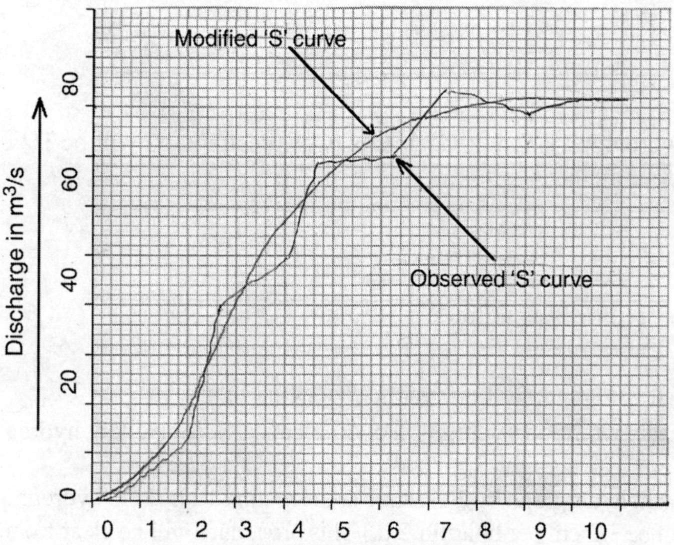

Fig. 6.23 Modification of S hydrograph

6.4.7 Limitations of Unit Hydrograph Theory

The limitations of the unit hydrograph theory are as follows:

1. The basic assumption that the excess rainfall is uniformly distributed over the entire area may not be practicable. It is very difficult to have this condition fulfilled in practice.
2. The principle of linearly assumed is not correct.

3. The unit hydrograph theory is not applicable for surface runoff originated from snow and ice.
4. The unit hydrograph theory is applicable to in-bank floods only. If the flood water overtops the bank, this theory will not be applicable.
5. The unit hydrograph theory is applicable for catchments less then 5000 km^2.
6. The theory is not applicable to narrow elongated catchments because it is not possible to have a uniform precipitation over the entire catchment.
7. The theory is not applicable if there are storages on the channel or on its tributaries in the catchment upstream of the gauging station.

6.5 CHANGING OF TIME OF UNIT PRECIPITATION

At times, the time of excess precipitation in case of a unit hydrograph, keeping the unit precipitation the same, is required to be changed depending upon the storm period *t*.

This time of excess precipitation may be increased or decreased.

6.5.1 Increase in the Time of Unit Precipitation

Suppose the unit hydrograph of 1 h–1 cm precipitation is available as follows:

Time (h)	0	1	2	3	4	5	6	7	8	9	10
Discharge (m^3/s)	0	6	13	22	16	11	7	4	2	1	0

A 2 h–1 cm unit hydrograph is required. (Intensity of precipitation will be 1/2 cm/h for 2 h.) This can be derived by four methods as follows:

6.5.1.1 Method 1

The step-by-step procedure is as follows:

1. Plot the given 1 h–1 cm unit hydrograph.
2. Plot the same 1 h–1 cm unit hydrograph by a time lag of 1 h.
3. Add together these two flood hydrograph coordinates. Thus, resultant hydrograph after addition will be a 2 h–2 cm hydrograph (Fig. 6.24).
4. Divide the resultant hydrograph by 2 to arrive at the required 2 h–1 cm hydrograph (The intensity of precipitation will be 1/2 cm per hour for 2 h.) This procedure will be clear from Table 6.2.

The same procedure can also be followed graphically.

This new flood hydrograph will be 2 h–1 cm unit hydrograph with an intensity of 1/2 cm/h for 2 h, resulting into unit precipitation of 1 cm.

It should be noted that,

1. Maximum discharge has reduced.
2. The base period has increased by 1 h (2 − 1 = 1 h).
3. The total of the new unit hydrograph coordinates (2 h–1 cm) remains the same (82 m^3/s).

Table 6.2 Increasing time of unit precipitation

1	2	3	4	5 = 3 + 4	6
Serial no.	Time (h)	Unit Hydrograph coordinates (m^3/s)	Unit Hydrograph coordinates after a lag of 1 h (m^3/s)	Total (m^3/s)	Required unit hydrograph (m^3/s)
1	0	0	0	0	0
2	1	6	0	6	3
3	2	13	6	19	9.5
4	3	22	13	35	17.5
5	4	16	22	38	19.0
6	5	11	16	27	13.5
7	6	7	11	18	9.0
8	7	4	7	11	5.5
9	8	2	4	6	3.0
10	9	1	2	3	1.5
11	10	0	1	1	0.5
12	11	0	0	0	0
	Total	82 m^3/s	82 m^3/s	164 m^3/s	82 m^3/s

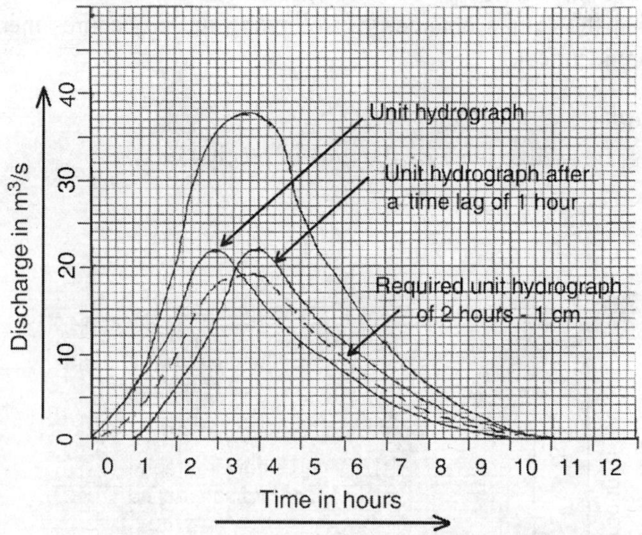

Fig. 6.24 Method 1 for increasing the time of unit precipitation

6.5.1.2 Method 2

The same procedure can be followed by a different step-by-step procedure as follows:

1. Plot the given 1 h–1 cm unit hydrograph.
2. Divide these unit hydrograph coordinates by 2, so that a 1 h–1/2 cm hydrograph is obtained.
3. Plot this 1 h–1/2 cm hydrograph by a time lag of 1 h.
4. Add these two 1 h–1/2 cm hydrographs to arrive at the required 2 h–1 cm unit hydrograph.

Table 6.3 Increasing time of unit precipitation

1		2	3	4	5 = 3 + 4
Serial no.	Time (h)	Unit hydrograph coordinates (m^3/s)	Unit hydrograph coordinates divided by 2, (m^3/s)	Unit hydrograph coordinates divided by 2, with a time lag of 1 h (m^3/s)	Total required unit hydrograph (m^3/s)
1	0	0	0	0	0
2	1	6	3	0	3
3	2	13	6.5	3	9.5
4	3	22	11.0	6.5	17.5
5	4	16	8.0	11.0	19.0
6	5	11	5.5	8.0	13.5
7	6	7	3.5	5.5	9.0
8	7	4	2.0	3.5	5.5
9	8	2	1.0	2.0	3.0
10	9	1	0.5	1.0	1.5
11	10	0	0	0.5	0.5
12	11	0	0	0	0
	Total	82	41	41	82

This procedure will be clear from Table 6.3.

This procedure can also be followed graphically as shown in Fig. 6.25.

It can be seen that the results obtained by following the procedures mentioned in Sections 6.5.1.1 and 6.5.1.2 are identical.

Use of S hydrograph

S hydrograph can also be used to increase the time of precipitation.

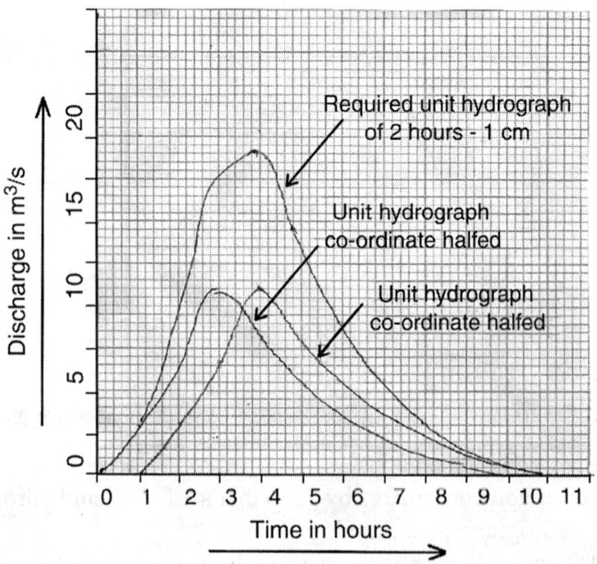

Fig. 6.25 Method 2 for increasing the time of unit precipitation

6.5.1.3 Method 3

The step-by-step procedure will be as follows:

1. Prepare an S hydrograph of 1 h–1 cm unit hydrograph and plot the S hydrograph.
2. Plot the same S hydrograph by a time lag of 2 h, say S^1.
3. Deduct the second, i.e. S^1 hydrograph from the first one S.
4. After deducting S^1 hydrograph from S hydrograph, divide the resultant hydrograph by 2 to arrive at the required 2 h–1/2 cm unit hydrograph. The intensity of precipitation will be $\frac{1}{2}$ cm/h.

The procedure will be clear from Table 6.4.
The same procedure can also be followed graphically as shown in Fig. 6.26.

Table 6.4 Increasing time of unit precipitation by S curve

1	2	3	4	5	6 = (4 − 5)/2
Serial no.	Time (h)	Unit hydrograph coordinates (m^3/s)	S hydrograph coordinates (m^3/s)	S^1 hydrograph coordinates after 2-h time lag(m^3/s)	Required unit hydrograph coordinates (m^3/s)
1	0	0	0	0	0
2	1	6	6	0	3
3	2	13	19	0	9.5
4	3	22	41	6	17.5
5	4	16	57	19	19.0
6	5	11	68	41	13.5
7	6	7	75	57	9.0
8	7	4	79	68	5.5
9	8	2	81	75	3.0
10	9	1	82	79	1.5
11	10	0	82	81	0.5
12	11	0	82	82	0
	Total	82			82

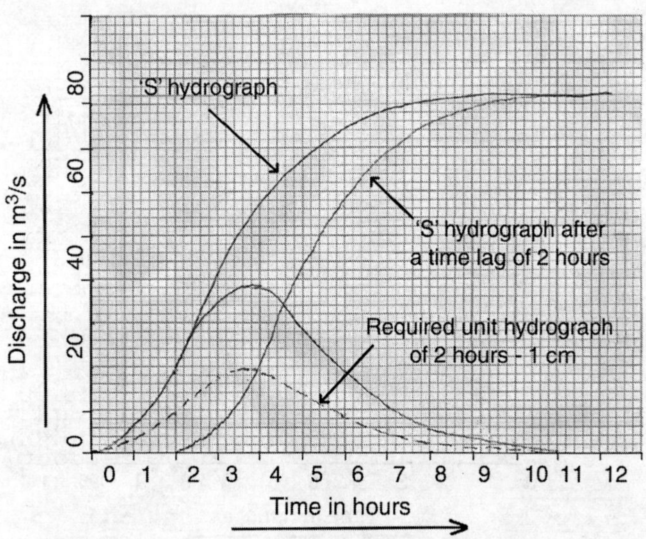

Fig. 6.26 Method 3 for increasing the time of unit precipitation

6.5.1.4 Method 4

The same procedure as followed in Section 6.5.1.3. in a different step-by-step procedure can be followed as follows:

1. Prepare the S hydrograph of the given 1 h–1 cm unit hydrograph and plot it.
2. Divide these S hydrograph coordinates by 2 so that it will be a 1 h–1/2 cm S^1 hydrograph. Plot this S^1 hydrograph with a time lag of 2 h.
3. Deduct this S^1 hydrograph from the original S hydrograph to arrive at the required 2 h–1/2 cm unit hydrograph. (The intensity of precipitation will be $\frac{1}{2}$ cm for 2 h.)

The procedure will be clear from Table 6.5.
The same procedure can be followed graphically as shown in Fig. 6.27.

Table 6.5 Increasing time of unit precipitation by S curve

1	2	3	4	5	6 = 4 – 5
Serial no.	Time (h) coordinates	Unit hydrograph coordinates (m^3/s)	S hydrograph coordinates (m^3/s) lag (m^3/s)	S1 hydrograph coordinates after 2-h time	Required unit hydrograph coordinates (m^3/s)
0	0	0	0	0	0
1	6	6	3	0	3
2	13	19	9.5	0	9.5
3	22	41	20.5	3	17.5
4	16	57	28.5	9.5	19.0
5	11	68	34.0	20.5	13.5
6	7	75	37.5	28.5	9.0
7	4	79	39.5	34.0	5.5
8	2	81	40.5	37.5	3.0
9	1	82	41.0	39.5	1.5
10	0	82	41.0	40.5	0.5
11	0	82	41.0	41.0	0
Total	82				82

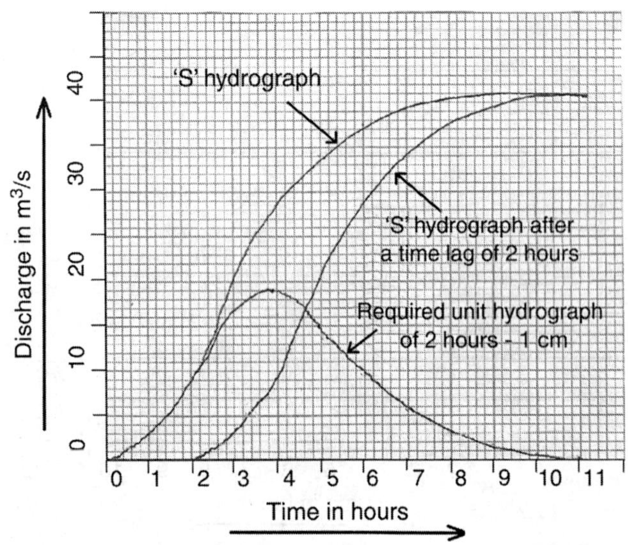

Fig. 6.27 Method 4 for increasing the time of unit precipitation

Example 6.7

Derive a 1 cm–5 h unit hydrograph from a 1 cm–3 h unit hydrograph.
The 1 cm–3 h unit hydrograph coordinates are as under:

Time (h)	0	1	2	3	4	5	6	7	8	9	10	11	12
Discharge (m^3/s)	0	2.00	6.34	13.66	17.00	16.32	11.34	7.34	4.34	2.32	1.00	0.33	0

Solution:

The given unit hydrograph period 3 h and the required unit hydrograph period 5 h are not exact multiples of each other. Therefore, the required unit hydrograph can be derived only from the S curve. The step-by-step solution will be as follows: vide Table 6.6

1. Write down the given unit hydrograph coordinates. (Since the given unit hydrograph is 1 cm–3 h, the intensity of the precipitation is $\frac{1}{3}$ cm/h.) (column A)
2. Repeat these coordinates each after 3 h to obtain the S hydrograph. (Even though the given unit hydrograph is of 3 h, the unit hydrograph coordinates are at 1-h interval.) (columns B, C, D, E, F)

Table 6.6 Increasing time of unit precipitation

1	2	3	4	5	6	7	8	9	10	11
Time (h)	A (m^3/s)	B (m^3/s)	C (m^3/s)	D (m^3/s)	E (m^3/s)	F (m^3/s)	G (m^3/s)	H (m^3/s)	I (m^3/s)	J (m^3/s)
0	0	–	–	–	–	–	0	–	0	0
1	2.00	–	–	–	–	–	2.00	–	2.00	1.20
2	6.34	–	–	–	–	–	6.34	–	6.34	3.80
3	13.66	0	–	–	–	–	13.66	–	13.66	8.20
4	17.00	2.00	–	–	–	–	19.00	–	19.00	11.40
5	16.32	6.34	–	–	–	–	22.66	0	22.66	13.60
6	11.34	13.66	0	–	–	–	25.00	2.00	23.00	13.80
7	7.34	17.00	2.00	–	–	–	26.34	6.34	20.00	12.00
8	4.34	16.32	6.34	–	–	–	27.00	13.66	13.34	8.00
9	2.32	11.34	13.66	0	–	–	27.32	19.00	8.32	5.00
10	1.00	7.34	17.00	2.00	–	–	27.34	22.66	4.68	2.80
11	0.33	4.34	16.32	6.34	–	–	27.33	25.00	2.33	1.40
12	0	2.32	11.34	13.66	0	–	27.32	26.34	1.00	0.60
13	–	1.00	7.34	17.00	2.00	–	27.34	27.00	0.34	0.20
14	–	0.33	4.34	16.32	6.34	–	27.33	27.33	0	0
15	–	0	2.32	11.34	13.66	0	27.32	27.33	0	0
16	–	–	1.00	7.34	17.00	2.00	27.34	27.34	0	0
–	–	–	–	–	–	–	27.33	27.33	–	–
–	–	–	–	–	–	–	27.32	27.32	–	–
–	–	–	–	–	–	–	27.34	27.34	–	–
Total	82.00	82.00	–	–	–	–	–	–	136.67	82.00

3. Add the repeated coordinates (column A + column B + column C + column D + column E + column F) so that a $\frac{1}{3}$-cm/h S hydrograph will be obtained. (column G)
4. Write down the S hydrograph coordinates after 5 h (column H) and deduct them from the S hydrograph in column G to arrive at a hydrograph of $\frac{5}{3}$-cm/h precipitation for 3 h (column I) (Actually, this is a 5 h–$\frac{5}{3}$ cm unit hydrograph.)
5. Multiply the coordinates of the 5 h–5/3 cm unit hydrograph in column I by 3/5 to arrive at the required 5 h–1 cm unit hydrograph (column J). Here the intensity will be 1/5 cm/h.

The required 1 cm–5 h unit hydrograph is as per column J.

6.5.2 Reduction in the Time of Unit Precipitation

Sometimes the time of unit hydrograph is required to be reduced to suit the field conditions.

Consider a 2 h–1 cm unit hydrograph as follows:

Time (h)	0	1	2	3	4	5	6	7	8	9	10	11
Discharge (m^3/s)	0	3.0	9.5	17.5	19.0	13.5	9.0	5.5	3.0	1.5	0.5	0

A 1 h–1 cm unit hydrograph is to be derived from this 2 h–1 cm unit hydrograph.

If a similar procedure, as discussed in Sections 6.5.1.1–6.5.1.2, is attempted to reduce the time of precipitation of a unit hydrograph, then at some step, negative values are arrived at. Thus, the procedures followed in Sections 6.7.1.1–6.7.1.2 cannot be applied here.

Hence, the only method to reduce the time of precipitation of a unit hydrograph is to use the S hydrograph approach, which can be done by two methods.

6.5.2.1 Method 1

The step-by-step procedure will be as follows:

1. Prepare an S hydrograph from the given 2 h–1 cm unit hydrograph and plot it.
2. Plot this S hydrograph with a time lag of 1 h, say S^1 hydrograph and plot it.
3. Deduct this S^1 hydrograph from S hydrograph. After deduction, multiply it by 2 to arrive at the required 1 h–1 cm unit hydrograph (Fig. 6.28).

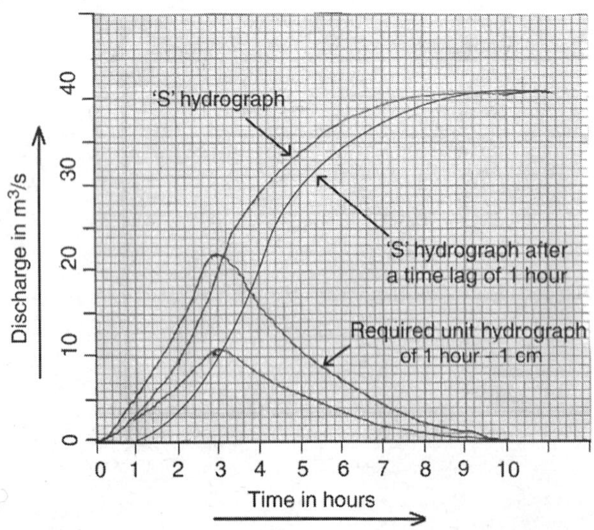

Fig. 6.28 Method 1 for reducing the time of unit precipitation

1	2	3	4	5 = 2 (3 − 4)
coordinates Time (h)	Unit hydrograph coordinates (m³/s)	S hydrograph coordinares (m³/s)	S¹ hydrograph coordinate after 1-h time lag (m³/s)	Required unit hydrograph (m³/s)
0	0	0	0	0
1	3.0	3	0	6
2	9.5	9.5	3	13
3	17.5	20.5	9.5	22
4	19.0	28.5	20.5	16
5	13.5	34.0	28.5	11
6	9.0	37.5	34.0	7
7	5.5	39.5	37.5	4
8	3.0	40.5	39.5	2
9	1.5	41	40.5	1
10	0.5	41	41	0
Total	82			82

Table 6.7 Reducing time of unit precipitation by S hydrograph

The procedure will be clear from Table 6.7.

The same procedure can be followed graphically as under.

6.5.2.2 Method 2

The same procedure can be followed in a different step-by-step procedure as below:

1. Prepare the S hydrograph from the given 2 h–1 cm unit hydrograph and plot it.
2. Multiply the coordinates of this S hydrograph by 2 so that a 2 h–2 cm S¹ hydrograph is obtained.
3. Enter this S¹ hydrograph with a time lag of 1 h.
4. Deduct the S¹ hydrograph from the S hydrograph to arrive at the required 1 h–1 cm unit hydrograph (Fig. 6.29).

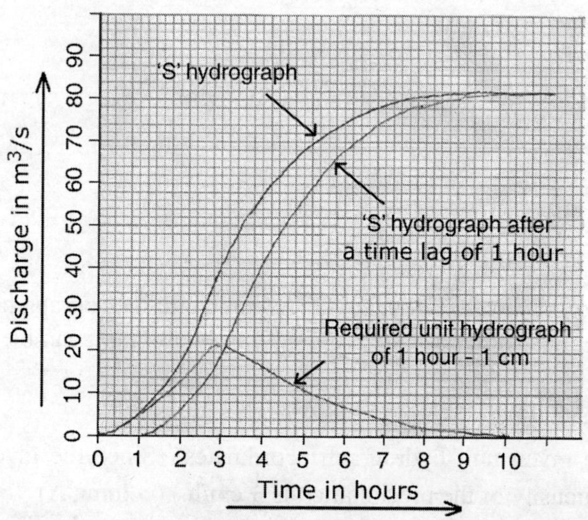

Fig. 6.29 Method 2 for reducing the time of unit precipitation

The above procedure will be clear from Table 6.8.
The same procedure can be followed as under.

Table 6.8 Reducing time of unit precipitation by S hydrograph

1	2	3	4	5	6 = 4 − 5
Time (h)	Unit hydrograph coordinates (m^3/s)	S hydrograph coordinates (m^3/s)	S^1 hydrograph coordinates (m^3/s)	S hydrograph coordinates after 1-hour time lag (m^3/s)	Required unit hydrograph (m^3/s)
0	0	0	0	0	0
1	3	3	6	0	6
2	9.5	9.5	19	6	13
3	17.5	20.5	41	19	22
4	19.0	28.5	57	41	16
5	13.5	34.0	68	57	11
6	9.0	37.5	75	68	7
7	5.5	39.5	79	75	4
8	3.0	40.5	81	79	2
9	1.5	41	82	81	1
10	0.5	41	82	82	0
11	0	41	82	82	0

Example 6.8

Derive a 1 cm–3 h unit hydrograph from a 1 cm–4 h unit hydrograph.
The 1 cm–4 h unit hydrograph coordinates are as under:

Time (h)	0	1	2	3	4	5	6	7	8	9	10	11	12	13
Discharge (m^3/s)	0	1.50	4.75	10.25	14.25	15.50	14.00	9.50	6.00	3.50	1.75	0.75	0.25	0

Solution:

The given unit hydrograph period 4 h and the required unit hydrograph period 3 h are not exact multiples of each other. Therefore, the required unit hydrograph can be derived only from the S curve.

The step-by-step solution will be as follows: vide Table 6.9

1. Write down the given unit hydrograph coordinates. (Since the given unit hydrograph is 1 cm–4 h, the intensity of the precipitation is $\frac{1}{4}$ cm/h.) (column A)
2. Repeat these coordinates each after 4 h to obtain the S hydrograph. (Even though the given unit hydrograph is of 4 h, the unit hydrograph coordinates are at 1-h interval.) (columns B, C, D)

Table 6.9 Reducing time of unit precipitation

1	2	3	4	5	6	7	8	9
Time (h)	A (m^3/s)	B (m^3/s)	C (m^3/s)	D (m^3/s)	E (m^3/s)	F (m^3/s)	G (m^3/s)	H(m^3/s)
0	0	–	–	–	0	–	0	0
1	1.50	–	–	–	1.50	–	1.50	2.00
2	4.75	–	–	–	4.75	–	4.75	6.33
3	10.25	–	–	–	10.25	0	10.25	13.67
4	14.25	0	–	–	14.25	1.50	12.75	17.00
5	15.50	1.50	–	–	17.00	4.75	12.25	16.33
6	14.00	4.75	–	–	18.75	10.25	8.50	11.33
7	9.50	10.25	–	–	19.75	14.25	5.50	7.33
8	6.00	14.25	0	–	20.25	17.00	3.25	4.33
9	3.50	15.50	1.50	–	20.50	18.75	1.75	2.33
10	1.75	14.00	4.75	–	20.50	19.75	0.75	1.00
11	0.75	9.50	10.25	–	20.50	20.25	0.25	0.33
12	0.25	6.00	14.25	0	20.50	20.50	0	0
13	0	3.50	15.50	1.50	20.50	20.50	0	0
14	0	1.75	14.00	4.75	20.50	20.50	0	0
15	0	0.75	9.50	10.25	20.50	20.50	0	0
16	0	0.25	6.00	14.25	20.50	20.50	0	0
Total	82.00	82.00	82.00	82.00	–	–	61.50	81.98

3. Add the repeated coordinates (column A + column B + column C + column D) so that a 1/4 cm/h S hydrograph will be obtained. (column E)
4. Write down the S hydrograph coordinates after 3 h (column F) and deduct them from the S hydrograph in column E to arrive at a hydrograph of $\frac{1}{4}$-cm/h precipitation for 3 h. (column G) (Actually, this is a 3 h $\frac{3}{4}$ cm unit hydrograph.)
5. Multiply the coordinates of the 3 h–$\frac{3}{4}$ cm unit hydrograph in column G by $\frac{4}{3}$ to arrive at the required 3 h–1 cm unit hydrograph (column H). Here the intensity will be $\frac{1}{3}$ cm/h.

The required 1 cm–3 h unit hydrograph is as per column H.

6.6 DERIVATION OF UNIT HYDROGRAPH

Unit hydrograph may be derived from the observed rainfall and its resultant hydrograph. For this, the available data will have to be scanned before it is used.

The step-by-step procedure is as follows:

1. From the observed data, select an isolated storm and its resultant hydrograph.
2. Check the precipitation data for the accuracy of any rain gauge station. If required, evaluate the missing data. Find the average hourly precipitation over the catchment and plot the hyetograph.

   If the unit hydrograph required is for t hours, then the hyetograph may be divided into slabs of t hours. Assuming a ϕ index for infiltration, evaluate the average hourly excess rainfall hyetograph.
3. Check the resultant hydrograph for its accuracy. Separate the base flow. Then the correct observed flood hydrograph will be arrived at.

   Then the observed corrected data of excess precipitation and its resultant storm hydrograph can be used for further analysis.

Consider a specific case

After scanning the data, the excess rainfall hyetograph and the resultant storm hydrograph are as follows:

Excess precipitation: 2 cm/h for 1 h followed by 4 cm/h for 1 h followed by 3 cm/h for 1 h. The resultant flood hydrograph is as below:

Time (h)	0	1	2	3	4	5	6	7	8	9	10	11	12
Discharge (m^3/s)	0	12	50	114	159	152	106	69	41	22	10	3	0

This scanned data can be represented graphically as shown in Fig. 6.30.

A 1 h–1 cm unit hydrograph is to be derived. After examining the observed data, Following observations can be made:

1. The storm is of 3 h.
2. The base period of the combined flood hydrograph is 12 h.
3. For the 1 h–1 cm unit hydrograph, the base period T will be $12 - 3 + 1 = 10$ h.
4. The total resultant flood hydrograph is the summation of three hydrographs, viz.

   (a) Flood hydrograph due to 2-cm/h precipitation
   (b) Flood hydrograph due to 4-cm/h precipitation with a time lag of 1 h from the beginning
   (c) Flood hydrograph due to 3-cm/h precipitation with a time lag of 2 h from the beginning

Since the unit hydrograph base period is 10 h, assume the unit hydrograph coordinates for 10 h to be U_0, U_1, U_2, U_3, U_4, U_5, U_6, U_7, U_8, U_9, U_{10}. out of which U_0 and U_{10} are equal to zero.

The unit hydrograph can be derived by the following methods.

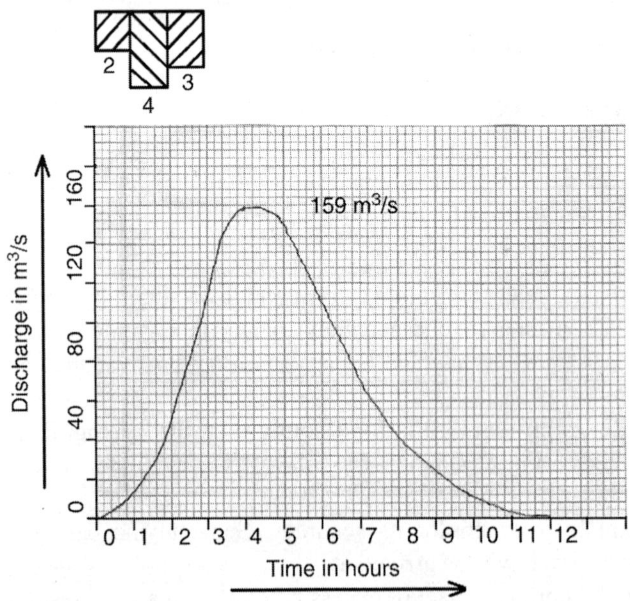

Fig. 6.30 Scanned flood hydrograph

6.6.1 Ven Te Chow's Forward Substitution Method

The total resultant flood hydrograph is the summation of the three hydrographs. Thus, the total flood hydrograph can be expressed in terms of the assumed hourly unit hydrograph coordinates as follows:

$$
\begin{aligned}
Q_0 &= && = 0\\
Q_1 &= 2U_1 && = 12\\
Q_2 &= 2U_2 + 4U_1 && = 50\\
Q_3 &= 2U_3 + 4U_2 + 3U_1 && = 114\\
Q_4 &= 2U_4 + 4U_3 + 3U_2 && = 159\\
Q_5 &= 2U_5 + 4U_4 + 3U_3 && = 152\\
Q_6 &= 2U_6 + 4U_5 + 3U_4 && = 106\\
Q_7 &= 2U_7 + 4U_6 + 3U_5 && = 69\\
Q_8 &= 2U_8 + 4U_7 + 3U_6 && = 41\\
Q_9 &= 2U_9 + 4U_8 + 3U_7 && = 22\\
Q_{10} &= 2U_{10} + 4U_9 + 3U_8 && = 10\\
Q_{11} &= 2U_{11} + 4U_{10} + 3U_9 && = 3\\
Q_{12} &= 2U_{12} + 4U_{11} + 3U_{10} && = 0
\end{aligned}
$$

From the above equations

$$
\begin{array}{lllllll}
Q_0 = & 0 = & U_0 & & & & U_0 = 0\\
Q_1 = & 12 = & 2U_1 & & & & U_1 = 6\\
Q_2 = & 50 = & 2U_2 + 4U_1 & = & 2U_2 + 4\times 6 & & U_2 = 13\\
Q_3 = & 114 = & 2U_3 + 4U_2 + 3U_1 & = & 2U_3 + 4\times 13 + 3\times 6 & & U_3 = 22\\
Q_4 = & 159 = & 2U_4 + 4U_3 + 3U_2 & = & 2U_4 + 4\times 22 + 3\times 13 & & U_4 = 16\\
Q_5 = & 152 = & 2U_5 + 4U_4 + 3U_3 & = & 2U_5 + 4\times 16 + 3\times 22 & & U_5 = 11\\
Q_6 = & 106 = & 2U_6 + 4U_5 + 3U_4 & = & 2U_6 + 4\times 11 + 3\times 16 & & U_6 = 7\\
Q_7 = & 69 = & 2U_7 + 4U_6 + 3U_5 & = & 2U_7 + 4\times 7 + 3\times 11 & & U_7 = 4\\
Q_8 = & 41 = & 2U_8 + 4U_7 + 3U_6 & = & 2U_8 + 4\times 4 + 3\times 7 & & U_8 = 2\\
Q_9 = & 22 = & 2U_9 + 4U_8 + 3U_7 & = & 2U_9 + 4\times 2 + 3\times 4 & & U_9 = 1\\
Q_{10} = & 10 = & 2U_{10} + 4U_9 + 3U_8 & = & 2U_{10} + 4\times 1 + 3\times 2 & & U_{10} = 0\\
Q_{11} = & 3 = & 2U_{11} + 4U_{10} + 3U_9 & = & + 3\times 1 & & U_{11} = 0
\end{array}
$$

From the first equation, we get $Q_0 = 0$. Substitute its value in the next equation and we get $Q_1 = 6$. Thus, by substituting the unit hydrograph coordinate, calculated in the previous step, in the next step, all the unit hydrograph coordinates can be evaluated.

The procedure was first suggested by Prof. Dr Ven Te Chow, hence this method is known as Ven Te Chow's forward substitution method.

6.6.2 Ven Te Chow's Backward Substitution Method

The same analogy as followed for the forward substitution method can be applied here and the unit hydrograph coordinates can be worked out by backward substitution as follows:

Consider the last equation. Solve it. We get $U_9 = 1$. Substitute this value in the previous equation. We get $U_8 = 2$. Thus, by substituting the values in the previous equation, all the coordinates can be evaluated.

$$
\begin{array}{lllll}
Q_{11} = 3 & = 3U_9\ [U_{10}, U_{11}, U_{12}, \ldots \text{ are equal to zero}] & & & U_9 = 1 \\
Q_{10} = 10 & = 3U_8 + 4U_9 & = 3U_8 + 4\times1 & & U_8 = 2 \\
Q_9 = 22 & = 3U_7 + 4U_8 + 2U_9 & = 3U_7 + 4\times2 + 2\times1 & & U_7 = 4 \\
Q_8 = 41 & = 3U_6 + 4U_7 + 2U_8 & = 3U_6 + 4\times4 + 2\times2 & & U_6 = 7 \\
Q_7 = 69 & = 3U_5 + 4U_6 + 2U_7 & = 3U_5 + 4\times7 + 2\times4 & & U_5 = 11 \\
Q_6 = 106 & = 3U_4 + 4U_5 + 2U_6 & = 3U_4 + 4\times11 + 2\times7 & & U_4 = 16 \\
Q_5 = 152 & = 3U_3 + 4U_4 + 2U_5 & = 3U_3 + 4\times16 + 2\times11 & & U_3 = 22 \\
Q_4 = 159 & = 3U_2 + 4U_3 + 2U_4 & = 3U_2 + 4\times22 + 2\times16 & & U_2 = 13 \\
Q_3 = 114 & = 3U_1 + 4U_2 + 2U_3 & = 3U_1 + 4\times13 + 2\times22 & & U_1 = 6 \\
Q_2 = 50 & = \quad + 4U_1 + 2U_2 & & & \\
Q_1 = 12 & = \quad\quad + 2U_1 & & &
\end{array}
$$

This backward substitution method was also suggested by Prof. Dr Ven Te Chow, and hence it is known as Ven Te Chow's backward substitution method.

For both the methods, the unit hydrograph is checked so that the area of the hydrograph is one unit for unit excess precipitation uniformly distributed over the basin. The unit hydrograph thus derived by both the methods is shown in Fig. 6.31.

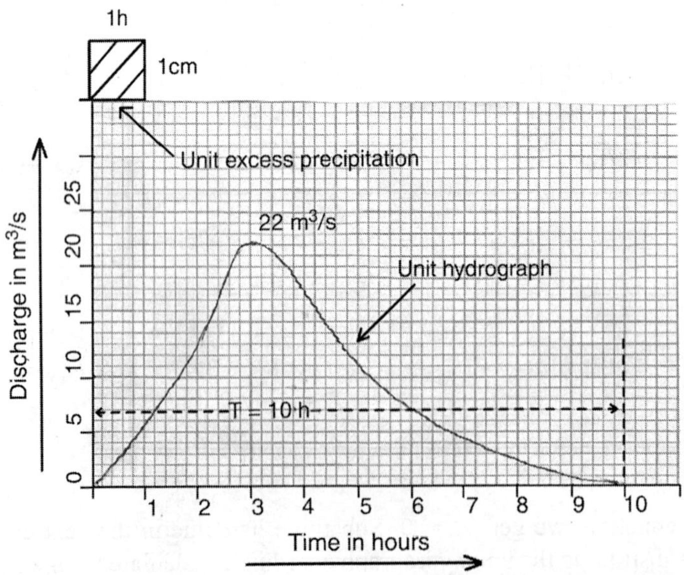

Fig. 6.31 Derived unit hydrograph

6.6.3 Collin's Method

This is a trial and error method, very commonly used for deriving unit hydrograph from a complex storm and resultant hydrograph.

The procedure followed is as follows:

1. The numerical values of the unit hydrograph coordinates are assumed. The flood hydrograph coordinates for the excess rainfall calculated are worked out and then compared with the observed ones. These two coordinates may not tally initially.
2. The assumed unit hydrograph coordinates are modified suitably and the flood hydrographs for these modified coordinates and the excess precipitation are again worked out and compared with the observed ones.
3. Again, the modified coordinates of the unit hydrograph are remodified, and this trial and error is continued till the calculated hydrograph tallies with the observed hydrograph, with an acceptable error.

Then the unit hydrograph coordinates are finalized.

6.7 AVERAGING OF UNIT HYDROGRAPH

There are a number of assumptions made in the unit hydrograph derivation. So also there might be some errors in the observed precipitation and the discharge data, and hence unit hydrographs derived from the different observed data for the same catchment may not tally with each other. Thus, an average of all the unit hydrographs derived is worked out for further studies.

The procedure followed is as stated below:

Plot all the unit hydrographs on a simple graph paper

$$t_p = \frac{t_1 + t_2 + t_3 + \ldots + t_n}{n}$$

where, t_1, t_2, t_3 = Time in hours of the maximum discharge from the beginning of the different unit hydrographs derived

t_p = Time in hours of the maximum discharge from the start of the final average unit hydrograph

$$Q = \frac{Q_1 + Q_2 + Q_3 + \ldots + Q_n}{n}$$

Q_1, Q_2, Q_3 = Maximum flood discharge of the various unit hydrographs in m^3/s

Q = Maximum flood discharge of the final average unit hydrograph

$$T = \frac{T_1 + T_2 + T_3 + \ldots + T_n}{n}$$

T_1, T_2, T_3 = Base period of the various unit hydrograph in hours

T = Base period of the final average unit hydrograph in hours

The average unit hydrograph may be plotted taking into consideration the parameters of Q, T and t_p.

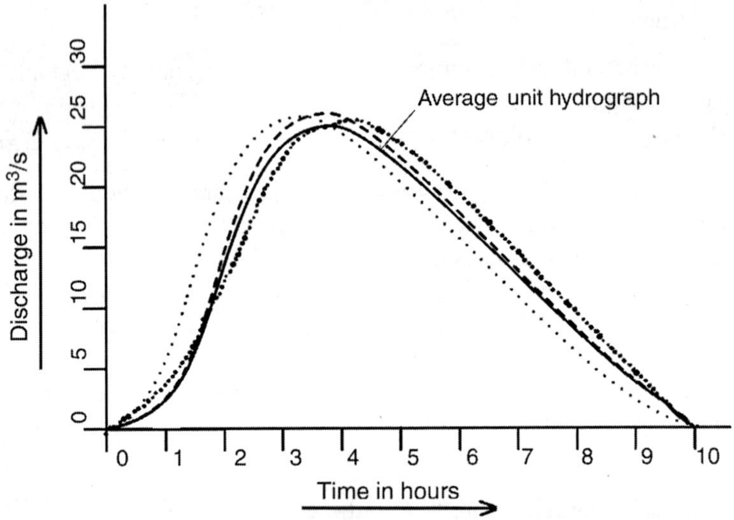

Fig. 6.32 Definition sketch of averaging unit hydrograph

An average unit hydrograph developed in a specific case is shown in Fig. 6.32.
It should be checked that the total area under the average unit hydrograph is unity.

Example 6.9

The unit hydrographs derived from different storms for a catchment are as follows:
Derive the average unit hydrograph.

Time (h)	0	1	2	3	4	5	6	7	8	9	10	Total
First Set	0	7	18	23	17	9	7	1	0	–		82
Second Set	0	11	22	22	13	7	4	2	1	1	0	83
Third set	0	6	14	21	21	11	4	3	2	0	–	82
Total	0	24	54	66	51	27	15	6	3	1	–	247
Average	0	8	18	22	17	9	5	2	1	0	–	82

$$\text{Average} \quad t_p = \frac{3+2+4}{3} = 3 \text{ h}$$

$$\text{Average} \quad T = \frac{8+10+9}{3} = 9 \text{ h}$$

Average maximum Discharge

$$Q_m = \frac{23+22+21}{3} = 22 \text{ m}^3/\text{s}$$

The resulting average unit hydrograph is as shown in Fig. 6.33.

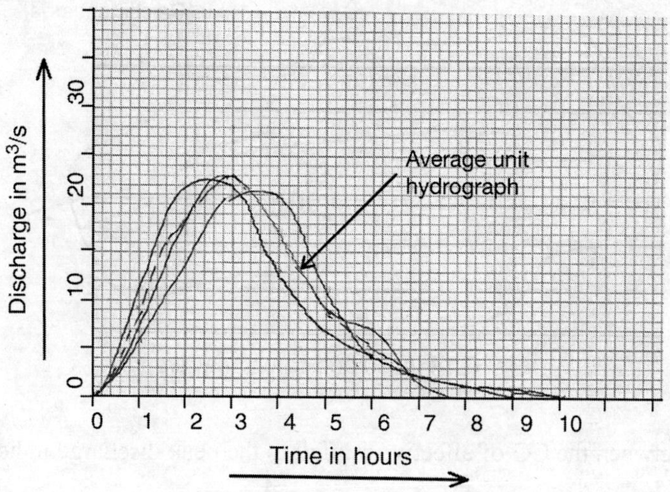

Fig. 6.33 Average unit hydrograph

Example 6.10

The peak of the flood hydrograph resulting from a 3-h 2.5-cm/hour storm is 320 m^3/s. If the ϕ value of infiltration is 5 mm/s, find the peak value of the 3 cm–3 h unit hydrograph. Assume the base flow to be 20 m^3/s.

Solution:

After deducting the base flow, the peak discharge of the flood hydrograph resulting due to the storm only will be $320 - 20 = 300$ m^3/s.

The excess uniform precipitation after deducting the infiltration rate will be $2.5 - 0.5 = 2$ cm/h for 3 h. The excess precipitation for a 3 cm–3 h unit hydrograph will have to be 1 cm/h for 3 h.

Therefore, the peak discharge for a 3 cm–3 h unit hydrograph will be $= \frac{1}{2} \times 300 = 150$ m^3/s.

6.8 SYNTHETIC UNIT HYDROGRAPH

When sufficient observed data are not available for the derivation of a unit hydrograph, then the unit hydrograph is framed based on the catchment characteristics.

6.8.1 Snyder Method

F. F. Snyder analysed a number of unit hydrographs in the Appalachian mountain region in USA and presented a set of equations for the synthetic unit hydrograph based on the following three catchment area characteristics viz. A, L & L_c (Fig. 6.34).

A = Catchment area in km^2

L = Length of main stream in km

L_c = Distance in km along the main stream from the outlet up to a point nearest to the centre of gravity (CG) of the catchment area

t = Duration of unit hydrograph in hours

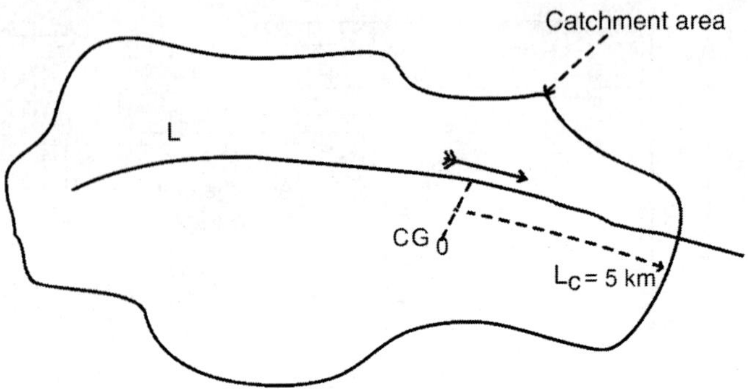

Fig. 6.34 Assumed catchment area

t_p = Time between the CG of effective rainfall to the peak discharge in hours (basin lag in hours)
Q_p = Peak discharge in m^3/s
T = Base period in hours
W_{75} = Width of unit hydrograph for 75% of Q_p
W_{50} = Width of unit hydrograph for 50% of Q_p
C_t = A constant depending upon the slope and S_D of the catchment area. (Value is normally between 1.35 to 1.65.)
C_p = A coefficient ranging from 0.56 to 0.69

The equations suggested are:

$$t_p = C_t (L \cdot L_c)^{0.3}$$
$$Q_p = 2.778 (C_p/t_p) A$$
$$T = 5.455\, t_p$$
$$W_{50} = 2.14 (Q_p/A)^{-1.08}$$
$$W_{75} = 1.22 (Q_p/A)^{-1.08}$$
$$t = 2\, t_p/11$$

The widths of unit hydrographs W_{50} and W_{75}, respectively, may be divided into two parts such that 1/3 lies in the rising curve and 2/3 in the recession curve.

The unit hydrograph can be framed, based on the above set of equations. As a check, the area of the unit hydrograph may be checked and corrected for a value of unit precipitation (Fig. 6.35).

Example 6.11

Derive a unit hydrograph for the catchment area shown in Fig 6.36

Solution:

Here A = 250 km^2, L = 20 km, L_c = 5.0 km
To suit the characteristics of the catchment, the values of the of constants assumed are C_p = 0.6, C_t = 1.50.

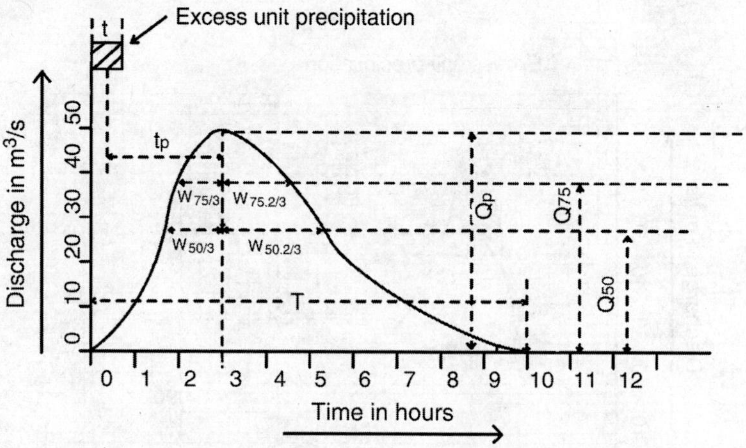

Fig. 6.35 Definition sketch of synthetic unit hydrograph

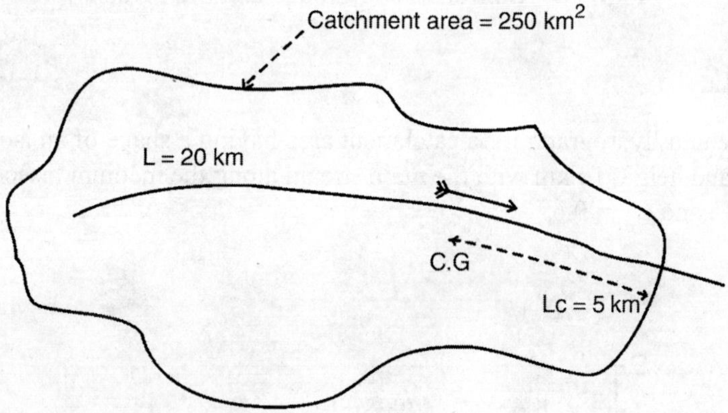

Fig. 6.36 Given catchment area

As per Synder's equations, the various parameters will be as follows:

$$t_p = C_t\,(L{\cdot}L_e)^{0.3} = 1.5\,(20 \times 5)^{0.3} = 5.97\text{ h} \approx 6.0\text{ h}$$

$$Q_p = (2.778 \times C_p \times A)/t_p = (2.778 \times 0.6 \times 250)/5.97 = 69.8\text{ m}^3/\text{s}$$

$$T = 5.455 \times t_p = 5.455 \times 6.0 = 32.73\text{ h} \approx 33\text{ h}$$

$$W_{50} = 2.14\,(Q_p/A)^{-1.08} = 2.14\,(69.8/250)^{-1.08} = 8.49\text{ h}$$

$$W_{75} = 1.22\,(Q_p/A)^{-1.08} = 1.22\,(69.8/250)^{-1.08} = 4.84\text{ h}$$

$$Q_{50} = 69.8 \times 0.5 = 34.9\text{ m}^3/\text{s} \quad \text{and} \quad Q_{75} = 69.8 \times 0.75 = 52.35\text{ m}^3/\text{s}$$

$$t = \text{ Unit time of excess precipitation} = 2\,t_p/11 = 2\,6/11 = 1.090\text{ h} \approx 1\text{ h}$$

The unit hydrograph thus derived is shown in Fig. 6.37.

After plotting the unit hydrograph, the area of the unit hydrograph was measured. The shape of the unit hydrograph was modified slightly so that it works to be unit.

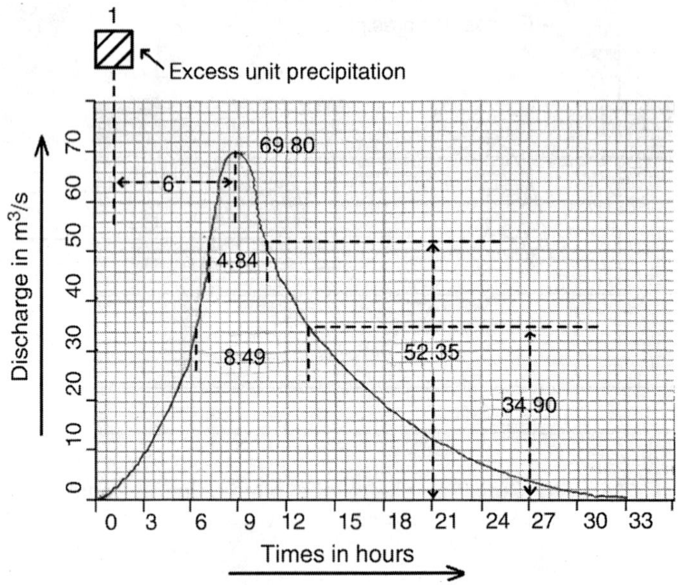

Fig. 6.37 Derivation of synthetic unit hydrograph

Example 6.12

Derive a synthetic unit hydrograph for a catchment area having a shape of an isosceles triangle of base width 6 km and height 10 km with the main stream along the medium bisecting the base. Assume $C_t = 1.5$ and $C_p = 0.6$

Solution:

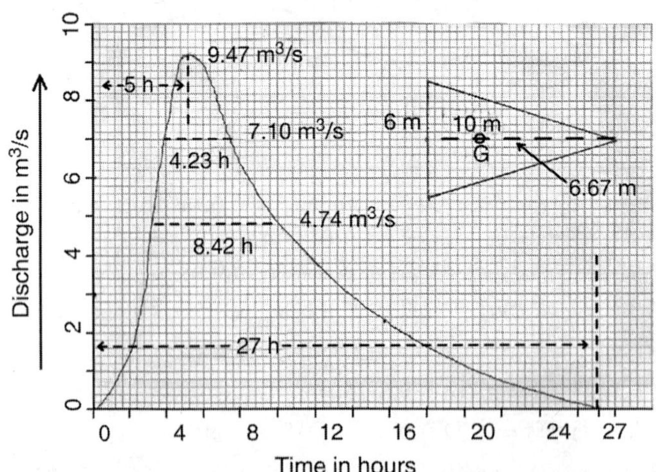

Fig. 6.38 Derivation of synthetic unit hydrograph for triangular catchment

Here, i. Area of catchment $= \frac{1}{2} \times 10 \times 6 = 30 \text{ km}^2$

ii. Length of the main stream = 10 km

iii. Length of the main stream from a point nearest to the CG up to outlet = 6.67 km

From these characteristics, the following parameters can be derived:

$$t_p = C_t (L L_e)^{0.3} = 1.5 (10 \times 6.67)^{0.3} = 5.28\text{h} \approx 5.3 \text{ h}$$
$$Q_p = 2.778 (C_p/t_p) A = 2.778 (0.6/5.0) \times 30 = 10.31 \text{ m}^3/\text{s}$$
$$T = 5.455 \, t_p = 5.455 \times 5.0 = 27.275 \approx 27 \text{ h}$$
$$t = 2 \, t_p/11 = 2 \times 5/11 = 0.909 \approx 1.0 \text{ h}$$
$$W_{50} = 2.14 (Q_p/A)^{-1.08} = 2.14 (10.31/30)^{-1.08} = 6.78 \text{ h}$$
$$W_{75} = 1.22 (Q_p/A)^{-1.08} = 1.22 (10.31/30)^{-1.08} = 3.86 \text{ h}$$

The shape of the catchment area and the unit hydrograph thus derived will be as shown is Fig. 6.38.

The area under the unit hydrograph measured = 42 cm^2
Volume of unit hydrograph = $42 \times 2 \times 3600 \times 1 = 30.24 \times 10^4$ m^3

This should be the volume due to unit uniform precipitation over the catchment.

Therefore, $$\frac{x \times A \times 10^6}{100} = 30.24 \times 10^4$$

$$x = 1.008 \approx 1.0 \text{ cm}$$

Therefore, the unit precipitation for the unit hydrograph is 1 cm.

6.8.2 Central Water Commission Method for Indian Catchments

The Central Water Commission (CWC), India, has recommended the following procedure for the construction of a synthetic unit hydrograph. This is based on the number of catchments in India of various shapes and sizes.

The recommended formulae are

$$Q_{pd} = 4.44 \, A^{3/4} \text{ for S} > 1 \text{ in } 360$$
$$Q_{pd} = 2.22 \, A^{3/4} \text{ for S} < 1 \text{ in } 360$$
$$t_p = \frac{3.95}{(Q_{pd}/A)^{0.9}}$$
$$D = 1.1 \, t_p$$

where, Q_{pd} = Peak discharge in m^3/s
A = Catchment area in km^2
S = Weighted mean slope of the catchment
t_p = Time in hours from the centroid of the effective rainfall to the peak discharge
D = Duration of effective rainfall

6.8.3 Dimensionless Unit Hydrograph

The United States, Soil Conservation Service (SCS) proposed a dimensionless unit hydrograph on the basis of the unit hydrographs derived for different basins having different catchment areas. This unit hydrograph is in the dimensionless form.

On the y-axis, Q_t/Q_p is expressed and on the x-axis, the ratio of t/t_p is expressed. The base period t/t_p is taken to be equal to 5.

Where, Q_t = Discharge at time t

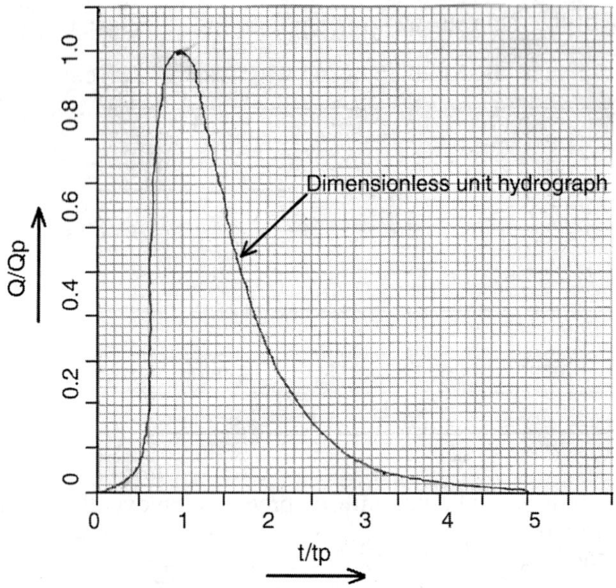

Fig. 6.39 Dimensionless unit hydrograph

Q_p = Maximum discharge
t = Time when Q_t is observed from the beginning of the unit hydrograph
t_p = Time when the maximum discharge is observed

The shape of the unit hydrograph thus derived is in agreement of the hydrograph expected for the catchment. For averaging the unit hydrographs derived for various catchments, this method is more effective. The unit hydrograph derived for the neighbouring catchments can be converted into a dimensionless form and transposed to an ungauged catchment.

The dimensionless unit hydrograph derived for a basin is shown in Fig. 6.39.

t/t_p	0	0.5	1.0	1.5	2.0	2.5	3.0	3.5	4.0	4.5	5.0
Q/Q_p	0	0.43	1.0	0.66	0.32	0.16	0.075	0.036	0.018	0.009	0

6.9 DISTRIBUTION GRAPH

Barnard proposed a unit hydrograph in this distribution form. It is a dimensionless representation of a unit hydrograph in the form of a bar chart. The base period of a unit hydrograph is divided into a number of equal time intervals. The runoff in each of this interval is first calculated and then its percentage with respect to the total runoff is worked out. The area during that interval is divided by the total area of the unit hydrograph. It is then represented on y-axis for that interval.

Figure 6.40 below shows a distribution graph in a specific case.

This unit hydrograph is in the form of percentage for each time interval. It is known as a *distribution graph*. The total of all percentages will be 100%. All unit storms irrespective of their intensity for a catchment have practically the same distribution graphs.

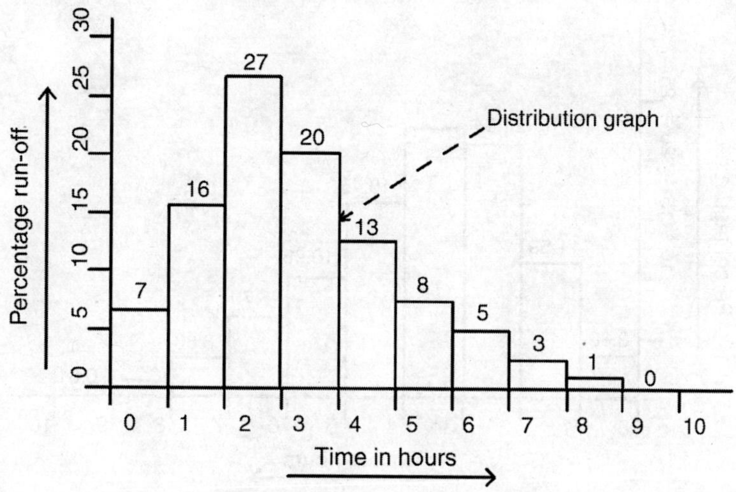

Fig. 6.40 Distribution graph

Example 6.13

The unit hydrograph coordinates of a basin are as under. Prepare the distribution graph.

Time (h)	0	1	2	3	4	5	6	7	8	9	10
Discharge (m^3/s)	0	6	13	22	16	11	7	4	2	1	0

Solution:

The percentage runoff during each hour is calculated as under. The table is self-explanatory.

1	2	3	4	5
Serial no.	Time (h)	Unit hydrograph (m^3/s)	Average runoff (m^3/s)	Runoff (%)
1	0	0	–	–
2	1	6	3	3.66
3	2	13	9.5	11.59
4	3	22	17.5	21.34
5	4	16	19.0	23.17
6	5	11	13.5	16.46
7	6	7	9.0	10.98
8	7	4	5.5	6.70
9	8	2	3.0	3.66
10	9	1	1.5	1.83
11	10	0	0.5	0.61
	Total	82	82	100

The graphical representation of the distribution will be as shown in Fig. 6.41.

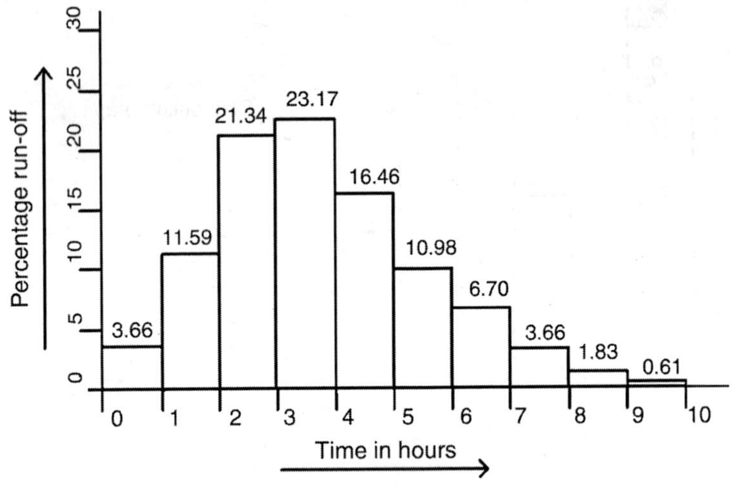

Fig. 6.41 Derived distribution graph

Example 6.14

The distribution graph coordinates as stated below are to be applied to a neighbouring catchment covering an area of 360 km^2. Evaluate the flood hydrograph when the excess rainfall is 2.0 cm/h for 2 h.

Time (h)	0	1	2	3	4	5	6	7	8	9	10
Distribution graph (%)	0	4	11	21	23	16	11	7	4	2	1

Solution:

The base period of the unit hydrograph = 10 h.

If precipitation of p cm/h occurs uniformly distributed over a catchment of A km^2, then the uniform discharge may be Q.

Time (h)	Distribution graph coordinates (%)	Discharge due to firstspell of 2 cm/h (m^3/s)	Discharge due to second of spell 2 cm/h (m^3/s)	Total discharge (m^3/s)
0	0	0	0	0
1	4	80	0	80
2	11	220	80	300
3	21	420	220	640
4	23	460	420	880
5	16	320	460	780
6	11	220	320	540
7	7	140	220	360
8	4	80	140	220
9	2	40	80	120
10	1	20	40	60
11	0	0	20	20
Total	100%	2000	2000	4000

Therefore, $Q = \frac{p}{100} \times \frac{A \times 10^6}{3600} = 2.77 \times p \times A$

Therefore, discharge with x% distribution $= \frac{x}{100} \times 2.77 \times p \times A$

$= 0.0277 \times x \times p \times A$

The flood hydrograph due to the precipitation of 2 cm/h for 2 h is tabulated above.

6.10 TRIANGULAR UNIT HYDROGRAPH

For simplicity, sometimes, a unit hydrograph is expressed in the form of a triangle with its base parallel to the x-axis. In this case, the base of the triangle represents the base period and the height of the triangle represents the peak discharge of the unit hydrograph as shown in Fig. 6.42.

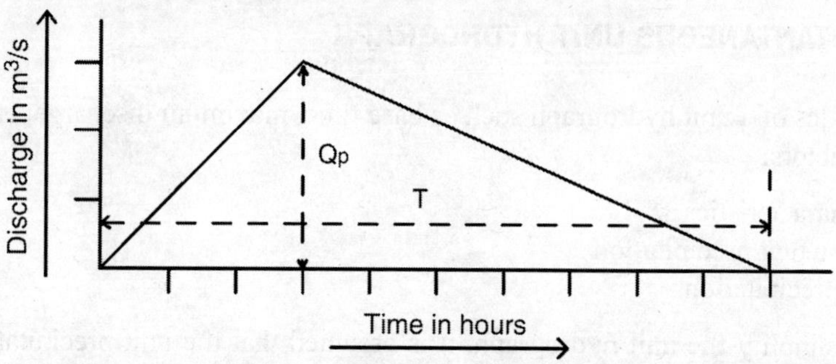

Fig. 6.42 Definition sketch of triangular unit hydrograph

Example 6.15

The 1 h–1 cm unit hydrograph of a catchment is in the form of a triangle of base width of 15 h and peak discharge of 80 m^3/s occurring after 5 h from the start. Find the catchment area.

Solution:

The unit hydrograph diagram will be as shown in Fig. 6.43.

Let the catchment area $= A$ km^2

i. The total runoff will be the area under the graph.

Area under the graph $= \frac{1}{2} \times 15 \times 80 = 600\ m^3/s \times 1\ h = 600\ m^3/s \times 3600\ s$

$= 216 \times 10^4\ m^3$.

ii. The runoff from the catchment area A km^2 from a precipitation of 1 cm/h for1h = $(A \times 10^6)\ 1/100 = A \times 10^4\ m^3$

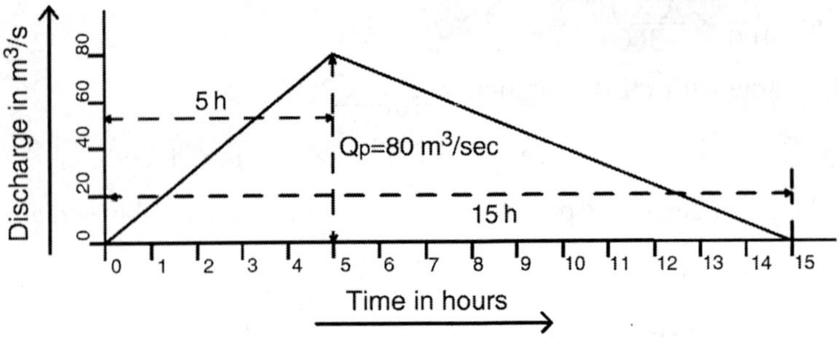

Fig. 6.43 Triangular unit hydrograph

The runoff calculated in (i) and (ii) should be equal.

i.e, $A \times 10^4 = 216 \times 10^4$

Therefore, $A = 216 \text{ km}^2$.

6.11 INSTANTANEOUS UNIT HYDROGRAPH

The characteristics of a unit hydrograph such as base time, maximum discharge, etc. depend upon the following factors:

- Catchment characteristics
- Intensity of the unit precipitation
- Time of unit precipitation

In order to simplify the unit hydrograph, it is assumed that the unit precipitation occurs uniformly over the entire catchment **instantaneously**, i.e. $t = 0$ or $t \to 0$.

It is a conceptual fictitious unit hydrograph.

This is a hypothetical case. It is incorrect to assume that the unit precipitation occurs instantly. However, further analysis of unit hydrograph theory is simplified. With this assumption, one factor (time of unit precipitation) influencing unit hydrograph is eliminated. Thus, a unit hydrograph of a catchment area, resulting from a unit precipitation occurring **instantly**, uniformly distributed over the entire catchment, is known as *instantaneous unit hydrograph* as shown in Fig. 6.44.

It is normally denoted as IUH and its shape is similar to a single peak hydrograph.

It may be noted that the basic principles of linearity and time invariance, assumed in the unit hydrograph theory, hold good for IUH also.

Indirectly, a unit hydrograph of time of precipitation of $t \to 0$ is an IUH.

6.11.1 Derivation of Instantaneous Unit Hydrograph

A lot of work and research is being done for the derivation of an instantaneous unit hydrograph. It can be derived from the unit hydrograph or from an S hydrograph. A number of conceptual models have been proposed. The most important are the following:

- Clark model
- Nash cascade model
- Chow–Kulandaiswami model
- Dooge model

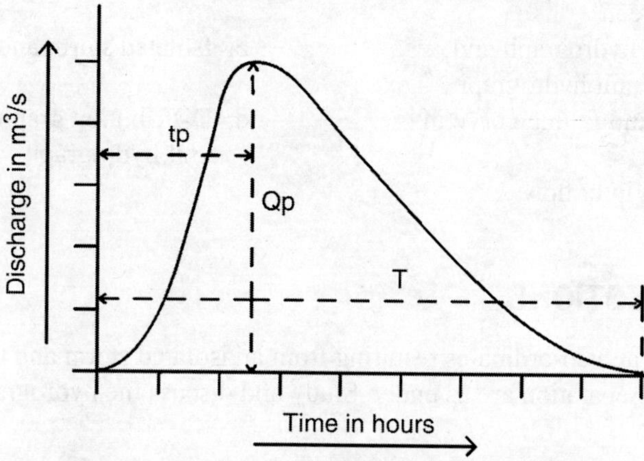

Fig. 6.44 Definition sketch of instantaneous unit hydrograph

REVIEW QUESTIONS

1. Define hydrograph. Discuss the normal units used to draw a hydrograph.
2. What is a complex storm? Sketch a complex storm and the resulting hydrograph.
3. Sketch a hydrograph resulting from an isolated storm and explain its features.
4. What is base flow? Why is it necessary to separate it from the hydrograph? Explain the different methods to separate it.
5. Explain with a neat sketch as to how the base flow is separated from a hydrograph resulting from a complex hydrograph.
6. Define unit hydrograph. Discuss the different definitions involved in this theory.
7. Explain the assumptions made in the unit hydrograph theory.
8. Explain how the unit hydrograph is used to find the flood hydrograph.
9. Discuss the limitations of unit hydrograph theory.
10. What is an S hydrograph? How is it derived from a unit hydrograph?
11. How is an S hydrograph used to change the period of the unit hydrograph?
12. Discuss the different methods followed to increase the time of the unit hydrograph.
13. Discuss the different methods to derive a unit hydrograph from observed data.
14. What is averaging of unit hydrograph? Why is it necessary?
15. What is a synthetic unit hydrograph? Why is it necessary?
16. Explain Snyder's method of framing a synthetic unit hydrograph.
17. Explain distribution graph.
18. Explain dimensionless unit hydrograph.
19. Write a detailed note on instantaneous unit hydrograph.
20. Write short notes on:
    a. Excess rainfall
    b. Master depletion curve
    c. Specifications of unit hydrograph
    d. Collin's method of derivation of unit hydrograph
    e. Ven Te Chow's forward substitution method
    f. Ven Te Chow's backward substitution method
    g. Dimensionless unit hydrograph
    h. Distribution graph
    i. Principle of linearity
    j. Principle of time invariance
    k. Time of concentration

21. Differentiate between
    a. Synthetic unit hydrograph and instantaneous unit hydrograph
    b. Isolated storm and complex storm
    c. Rising curve and falling curve in a hydrograph
    d. Distribution graph and dimensionless unit hydrograph
    e. Base flow and inter flow

NUMERICAL QUESTIONS

6.1. The flood hydrograph co-ordinates resulting from an isolated storm and the base flow co-ordinates after separation are as under. Study and discuss the hydrograph.

Time in Hours	0	1	2	3	4	5	6	7	8	9	10
Discharge in m^3/s	10	14	18	24	16	13	1	10	9	10	10
Base flow in m^3/s	10	8	5	2	0	2	4	6	7	9	10

6.2. The unit hydrograph co-ordinates of a 1 hour 1 cm unit hydrograph are as follows

Time in Hours	0	1	2	3	4	5	6	7	8	9	10	11	12
Discharge in m^3/s	0	5	13	25	40	29	20	13	8	3	2	1	0

Calculate the flood hydrograph for the following excess precipitation.
4 cm/hour for 1 hour followed by 7 cm/hour for 2 hours, no precipitation for next 1 hour followed by 2 cm/hour for 1 hour.
Assume a uniform base flow of 5 m^3/s

Ans: The flood hydrograph co-ordinates will be as follows

Time in Hours	0	1	2	3	4	5	6	7	8	9	10	11	12	13	14	15
Discharge in m^3/s	5	25	92	331	586	594	450	348	225	130	70	42	18	9	7	5

6.3. A 2 hours 1 cm, unit hydrograph co-ordinates are as under.

Time in Hours	0	1	2	3	4	5	6	7	8	9	10	11	12	13
Discharge in m^3/s	0	2.5	9.0	19.0	32.5	34.5	24.5	16.5	10.5	5.5	2.5	1.5	0.5	0

Derive a 3 hours, 1cm, unit hydrograph. Calculate also the catchment area.

Ans: The 3 hours 1 cm, unit hydrograph co-ordinates will be as follows.

Time in Hours	0	1	2	3	4	5	6	7	8	9	10	11	12	13
Discharge in m^3/s	0	1.67	6.00	14.34	26.00	31.34	29.67	20.67	13.67	8.00	4.34	2.00	1.00	0.34

The catchment area = 57.24 km^2

6.4. A 2 hours, 1 cm, unit hydrograph co-ordinates are as under.

Time in Hours	0	1	2	3	4	5	6	7	8	9	10	11	12	13
Discharge in m^3/s	0	2.5	9.0	19.0	32.5	34.5	24.5	16.5	10.5	5.5	2.5	1.5	0.5	0

Derive a 1 hour, 1cm, unit hydrograph. Draw the 'S' curve.

Ans: The 1 hour 1 cm, unit hydrograph co-ordinates will be as follows.

Time in Hours	0	1	2	3	4	5	6	7	8	9	10	11	12
Discharge in m^3/s	0	5	13	25	40	29	20	13	8	3	2	1	0

6.5. A catchment area covering 50 km^2 experienced the following storm.

(1) 5 cm/hr for 1 hour followed by
(2) 9 cm/hr for 1 hour followed by
(3) 6 cm/hr for 1 hours

The resulting hydrograph was as under

Time in Hours	0	1	2	3	4	5	6	7	8	9	10	11	12	13	14	15
Discharge in m^3/s	5	19	79	191	352	472	430	300	198	122	64	34	19	9	5	5

Assuming a base flow of 5 m^3/s and φ index equal to 2 cm/hr, derive the unit hydrograph.

Ans: The 1 hour, 1 cm unit hydrograph co-ordinates will be as follows

Time in Hours	0	1	2	3	4	5	6	7	8	9	10	11	12
Discharge in m^3/s	0	5	13	25	40	29	20	13	8	3	2	1	0

6.6. Derive a 1 hour, 1cm unit hydrograph for a catchment with the following details.

i) Catchment area ______________ 100 km^2
ii) Length of main stream ______________ 18 km
iii) Distance along the main stream from the outlet upto a point nearest to the C. G. of the catchment area. ______________ 5 km.

Assume Cp and Ct in Snyder's method to be '0.5 and 1.4 resp.

Ans: The required 1 hour, 1 cm unit hydrograph details will be as follows

(1) Maximum discharge :— Qp 2 42 m^3/s,
(2) Time of maximum discharge :— tp 5 4 hours.
(3) Base period :— T 5 22 hours.
(4) Time of unit precipitation :— 1 hour.
(5) Q50 5 21 m^3/s, W50 5 5.46 hours.
(6) $Q_{75} = 32\ m^3/s$, $W_{75} = 3.11$ hours.

6.7. A 1 hour, 1cm unit hydrograph co-ordinates are as under

Time in Hours	0	1	2	3	4	5	6	7	8	9	10	11	12
Discharge in m^3/s	0	5	13	25	40	29	20	13	8	3	2	1	0

Derive the distribution graph.

Ans: The distribution graph co-ordinates will be as under.

Time in Hours	0	1	2	3	4	5	6	7	8	9	10	11	12
Run-off in Percentage	0	1.57	5.66	11.94	20.44	21.69	15.40	10.30	6.60	3.46	1.57	0.94	0.31

6.8. The 1 hour, 1 cm unit hydrograph of a catchment area is in a triangular shape. The details as follows

(i) Catchment area ____________________ 25 km^2
(ii) Base period ____________________ 10 hours
(iii) Maximum discharge ____________________ 21 m^3/s
(iv) Time of max discharge ____________________ 3 hours

Calculate the flood hydrograph for the following storm.
4 cm / hour for 1 hour followed by 7 cm / hour for 2 hours followed by 2 cm/hour for 1 hour.
Assume a uniform base flow of 5 m^3/s

Ans: The flood hydrograph will be as under

Time in Hours	0	1	2	3	4	5	6	7	8	9	10	11	12	13
Discharge in m^3/s	5	33	110	236	336	366	326	266	206	146	86	38	11	5

MULTIPLE CHOICE QUESTIONS

1. Point of contraflexure on the recession curve of the hydrograph indicates the end of
   (a) The base flow (b) The precipitation
   (c) The overland flow (d) The interflow
2. The theory of unit hydrograph was first introduced by
   (a) Snyder (b) Ven Te Chow
   (c) Bernard (d) L. K. Sherman
3. The term 'unit' in unit hydrograph theory refers to
   (a) Unit area of basin (b) Unit precipitation
   (c) Unit discharge (d) Unit duration of precipitation
4. If the base period of a 4-h unit hydrograph is 48 h, then the hydrograph of a storm of 8 h, derived from this unit hydrograph, will be having a base period of
   (a) 44 h (b) 48 h
   (c) 52 h (d) 56 h
5. The upper limit of area of a basin for the application of unit hydrograph is
   (a) 1000 km^2 (b) 1500 km^2
   (c) 5000 km^2 (d) 10,000 km^2
6. The basic principles of unit hydrograph theory are
   (a) Linearity and time variance (b) Non-linearity and time variance
   (c) Linearity and time invariance (d) Non-linearity and time invariance

7. The S hydrograph is used to derive
   (a) Synthetic unit hydrograph
   (b) Unit hydrograph from a complex storm
   (c) To change the period of unit hydrograph
   (d) Flood hydrograph from a complex storm

8. The unit of time on x-axis of a hydrograph may be
   (a) Hours
   (b) Days
   (c) Months
   (d) Any of these

9. The discharge in the river before the storm is known as
   (a) Base flow
   (b) Basic flow
   (c) Sustained flow
   (d) Fair weather flow
   (e) Any of these

10. The recession curve of a hydrograph depends on
    (a) The period of precipitation
    (b) The intensity of precipitation
    (c) The basin characteristics of the basin
    (d) All the above

11. The master recession curve is also known as
    (a) Composite recession curve
    (b) Master depletion curve
    (c) Type curve
    (d) Any of the above

12. The base period of a unit hydrograph depends upon
    (a) Basin characteristics of the basin and the duration of the unit precipitation
    (b) Duration and the intensity of unit precipitation
    (c) Basin characteristics of the basin and the intensity of the unit precipitation
    (d) None of the above

13. The effect of the following physical characteristic of the basin is reflected in the shape of the unit hydrograph
    (a) Slope
    (b) Shape
    (c) Manning's coefficient
    (d) All the above

14. The slope to the S hydrograph curve is
    (a) Never negative
    (b) Horizontal at the beginning
    (c) Horizontal after the base period time
    (d) All the above

15. The peak discharge in case of the Snyder synthetic unit hydrograph is given by
    (a) $Q_p = \dfrac{2.778A}{C_p t_p}$
    (b) $Q_p = \dfrac{2.778AC_p}{t_p}$
    (c) $Q_p = \dfrac{2.778C_p t_p}{A}$
    (d) $Q_p = \dfrac{2.778t_p}{AC_p}$

16. The basic principle of instantaneous unit hydrograph is
    (a) Linearity and time variance
    (b) Non-linearity and time variance
    (c) Linearity and time invariance
    (d) Non-linearity and time invariance

17. In a dimensionless unit hydrograph, the base period ratio is generally taken to be
    (a) 3
    (b) 4
    (c) 5
    (d) 6

18. Collin's method is used to derive
    (a) Unit hydrograph
    (b) Synthetic unit hydrograph
    (c) Instantaneous unit hydrograph
    (d) Dimensionless unit hydrograph

19. Bernard proposed
    (a) Unit hydrograph
    (b) Synthetic unit hydrograph
    (c) Instantaneous unit hydrograph
    (d) Distribution graph

20. Conceptual models are proposed to derive
    (a) Unit hydrograph
    (b) Synthetic unit hydrograph
    (c) Instantaneous unit hydrograph
    (d) Distribution graph

21. The distribution graph for a catchment is practically the same irrespective of
    (a) Intensity of precipitation
    (b) Duration of precipitation
    (c) Catchment area
    (d) Characteristics of catchment

22. In the instantaneous unit hydrograph, the effect of the following parameter is eliminated.
    (a) Intensity of precipitation
    (b) Duration of precipitation
    (c) Catchment area
    (d) Characteristics of catchment

23. Principle of linearity means that the coordinates of the flood hydrograph are directly proportional to
    (a) Intensity of precipitation
    (b) Duration of precipitation
    (c) Catchment area
    (d) All of the above

24. The triangle representing a triangular unit hydrograph is a
    (a) Right angled triangle
    (b) Isosceles triangle
    (c) Equilateral triangle
    (d) None of the above

25. In Snyder's method, the widths of the unit hydrograph for 75% and 50% of maximum discharge are divided in the rising and falling curves in the ratio of
    (a) 1:1
    (b) 1:2
    (c) 1:3
    (d) 1:4

ANSWERS TO MULTIPLE CHOICE QUESTIONS

1. (c) 2. (d) 3. (b) 4. (c) 5. (c) 6. (c) 7. (c) 8. (d)
9. (e) 10. (c) 11. (d) 12. (a) 13. (d) 14. (d) 15. (b) 16. (c)
17. (c) 18. (a) 19. (d) 20. (c) 21. (a) 22. (b) 23. (a) 24. (d)
25. (b)

*** ——— ***

7 Runoff

Chapter Outline

7.1 DEFINITION

'Runoff' is that part of precipitation that appears in a drainage channel as surface flow in a perennial or an intermittent form. It is that part of water, which can be used for engineering purposes and hence is also known as *yield of catchment*.

The yield from a catchment is generally expressed in terms of volume, in a season or a year.

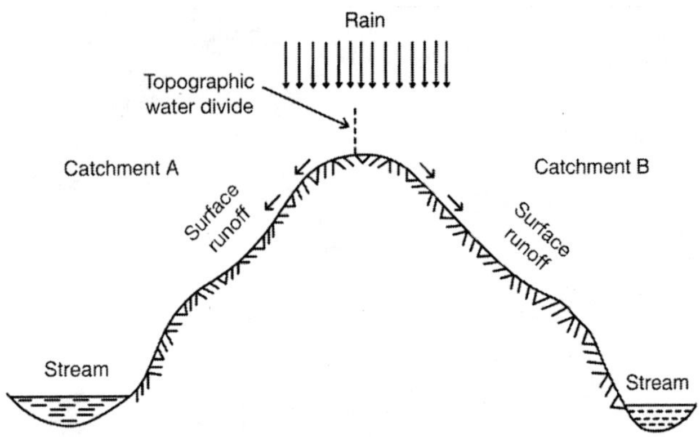

Fig. 7.1 Topographic divide

7.1.1 Units of Runoff

Runoff is expressed as the rate of flow during a specific period of flow, i.e. Q in time T and is indirectly a volume over the period. Thus, the unit for runoff is million cubic metres (10^6 m^3or million m^3), or ha m (denoting 1-m depth of water over an area of 1 ha).

Thus 1 million $m^3 = 10^2$ ha m

The runoff is also sometimes expressed in metres or millimetres as the depth of water spread uniformly over the entire catchment. For instance, 90.0 mm yield from a catchment of 50 km^2

$$= 50 \times 10^6 \times \frac{90}{1000} = 4.5 \text{ million m}^3 \text{ or } 450 \text{ ha m}.$$

7.1.2 Ridge Line

The line that demarcates the drainage area and hence the surface runoff of two adjacent rivers or drainage basins, is called the *ridge line*. It is also called the *topographic divide, topographic water divide, watershed divide* or simply *divide*. The precipitation on either side of the ridge line will flow in the opposite directions. Figure 7.1 shows a specific case of the topographic divide.

7.1.3 Groundwater Divide Line

The line that marks the direction of the groundwater movement of the two adjacent river basins is called the *groundwater divide*. It is also known as the *phreatic divide*.

It is very difficult to locate groundwater divide in hilly regions as well as in plains. Hence, for practical purpose, the groundwater divide is considered the same as topographic divide.

Figure 7.2 shows a specific case of a groundwater divide.

7.2 CATCHMENT AREA

Catchment area may be defined as *the area from which the surface runoff is derived.* It is also known as the *watershed area, drainage area, drainage basin* or simply *basin* or *catchment.*

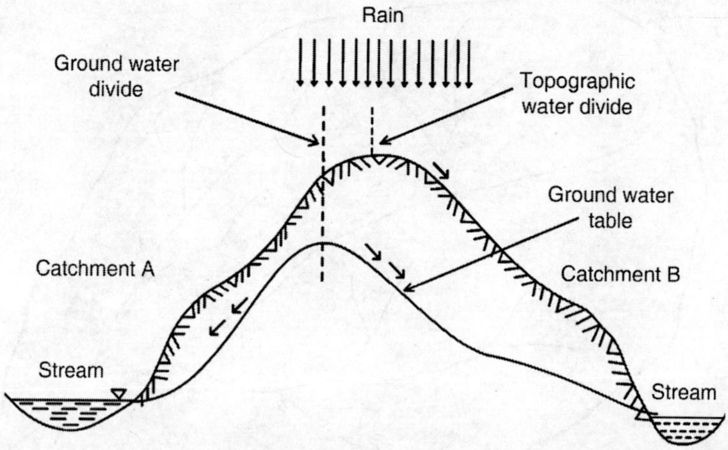

Fig. 7.2 Groundwater divide

The unit for the catchment area is km^2. If the catchment area is less than 25 km^2, it is mentioned in terms of hectares. The catchment areas formed by the divide lines at A and B are shown in Fig. 7.3.

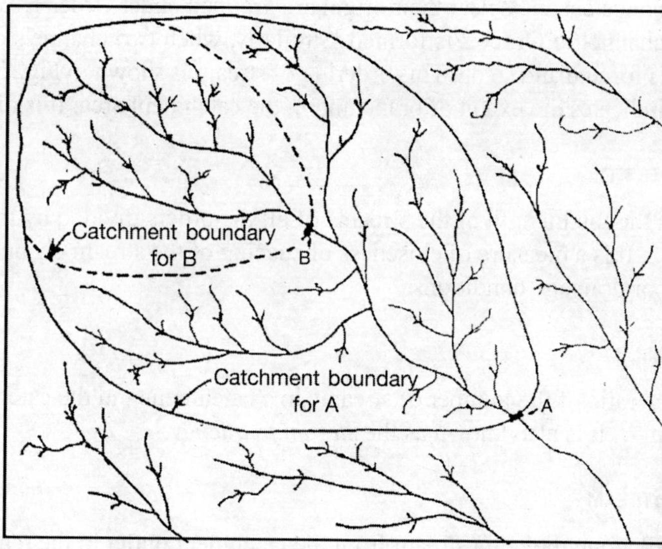

Fig. 7.3 Catchment areas formed by the divide lines at A and B

7.2.1 Stream Order

The classification reflecting the degree of branching or bifurcation of stream channels within a basin is known as *stream order*. Consider a map of the catchment area showing the channel network as shown in Fig. 7.4.

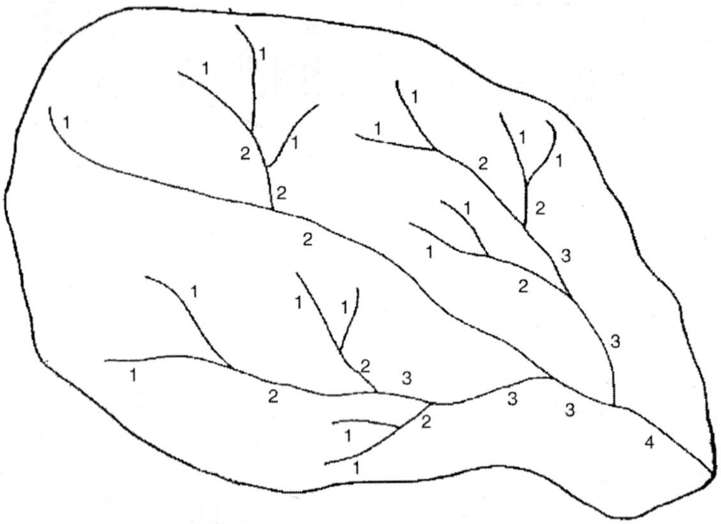

Fig. 7.4 Ordering of streams

The smallest drainage channels near their origin are grouped under Order 1. When two channels of Order 1 meet, a channel of Order 2 is formed. Similarly, when two channels of Order 2 meet, a channel of Order 3 is formed and so on. This ordering of stream is shown in Fig. 7.4. Thus, the order of the main stream indicates the extent of branching in the catchment area. It is dimensionless.

7.2.2 Drainage Density

Drainage density is the total length of the streams of all the orders divided by the catchment area. The unit will be m^{-1}. It is a measure of closeness of spacing of the stream channels. Low-drainage density indicates poor drainage conditions.

7.2.3 Stream Density

Stream density is the ratio of the number of streams in a catchment and the catchment area in km^2. The unit would be m^{-2}. It is also known as the *stream frequency*.

7.2.4 Length of Stream

The length measured along the main stream from the catchment outlet to the remotest point on the catchment boundary is called the *length of the stream* or *watershed length*. Naturally, the unit is expressed in metres.

7.2.5 Form Factor of Catchment Area

Form factor is defined as the ratio of the catchment area to the square of its length of the main stream. It is dimensionless.

7.2.6 Circularity Ratio

The circularity ratio is defined as the catchment area divided by the area of a circle whose perimeter is equal to the length of the ridge line of the basin. It is dimensionless.

7.2.7 Elongation Ratio of the Catchment Area

The elongation ratio of a catchment area is defined as the diameter of a circle whose area is equal to the catchment area divided by the length of the basin. It is dimensionless.

7.2.8 Shape Factor

The shape factor is the ratio of the square of the watershed length and the watershed area. It is dimensionless.

7.2.9 Relief of Catchment Area

The relief of a catchment area may be defined as the difference in elevations between the highest point on the ridge line and the basin outlet. The units would be expressed in metres.

7.2.10 Compactness Coefficient

The compactness coefficient is defined as the ratio of the perimeter of the basin to the perimeter of the circle whose area is equal to the catchment area. It is dimensionless.

Example 7.1

The details of a catchment collected from a toposheet are as follows.
Here the scale of the map is 1:50,000.
The required data were collected from the map and are given as under.

1. Area of the catchment: 28 cm^2 (on the map) 7.0 km^2 (actual)
2. Length of the main stream: 12 cm (on the map) 6 km (actual)
3. Length of the ridge line: 27 cm (on the map) 13.5 km (actual)
4. Level of the highest point on the ridge: 325 m
5. Level of the lowest point, i.e. at the outlet: 230 m
6. The order of all the streams was marked on the map and their lengths were measured as follows:

Order of the stream	No. of streams
1	22
2	11
3	6
4	1
Total	40

7. Total number of streams: 40
8. Length of all streams of all orders: 55.7 cm (on the map) 27.85 km (actual)

Calculate the characteristics of the catchment by studying the data given.

Solution:

Taking into consideration the scale of the map, 1:50,000, the corresponding prototype values are also mentioned.

The diameter of a circle whose perimeter is equal to the length of the ridge line

$$\pi D = 13.5$$

Therefore, $D = \frac{13.5}{\pi} = 4.29$ km

The diameter of a circle whose area is equal to the catchment area $\frac{\pi D^2}{4} = 7.0$

Therefore, $D = \left(\frac{28}{\pi}\right)^{1/2} = 2.98$ km

The characteristics of the catchments will be as follows:

a. Stream order of the catchment = 4

b. Drainage density $= \frac{27.85}{7} = 3.97\ \text{m}^{-1}$

c. Stream density $= \frac{40}{7} = 5.71\ \text{m}^{-2}$

d. Form factor $= \frac{7}{6^2} = 0.194$

e. Circularity ratio $= \frac{7}{\pi \times 4.29^2/4} = 0.482$

f. Elongation ratio $= \frac{2.98}{6} = 0.496$

g. Shape factor $= \frac{6^2}{7} = 5.14$

h. Relief of the catchment = 325 − 230 = 95 m

i. Compactness coefficient $= \frac{13.5}{2.98\pi} = 1.44$

7.2.11 Slope of a Catchment Area

It is noticed that the slope of a channel is normally steep at the early stage and it goes on reducing as it flows downstream. The average slope of the catchment area is calculated as in Fig. 7.5.

Draw the 'L' section of the main stream. Divide the length of the main channel into equal horizontal segments taking into consideration its slope.

Thus, $$S = \left[\frac{S_1^{1/2} + S_2^{1/2} + S_3^{1/2} + \ldots + S_n^{1/2}}{n}\right]^2$$

where, S = Slope of the catchment area

$S_1, S_2, S_3, \ldots$ = Slope of the main stream in various segments

n = Number of segments

Example 7.2

The L section along the main stream of a channel is shown in Fig. 7.6. Find the average slope of the channel.

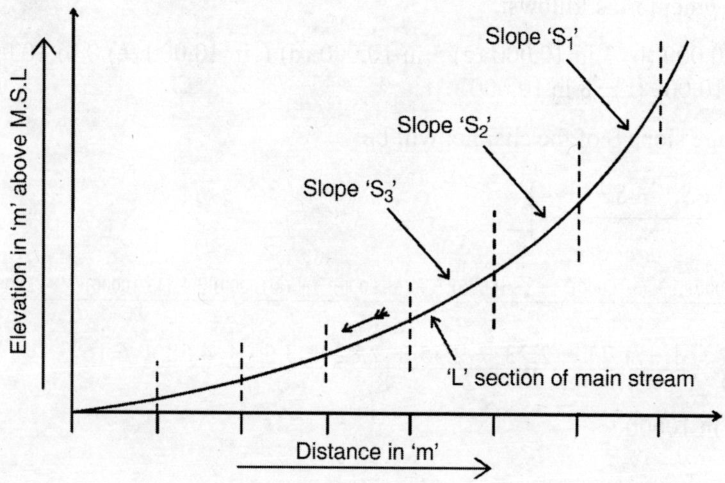

Fig. 7.5 Slope of a catchment area

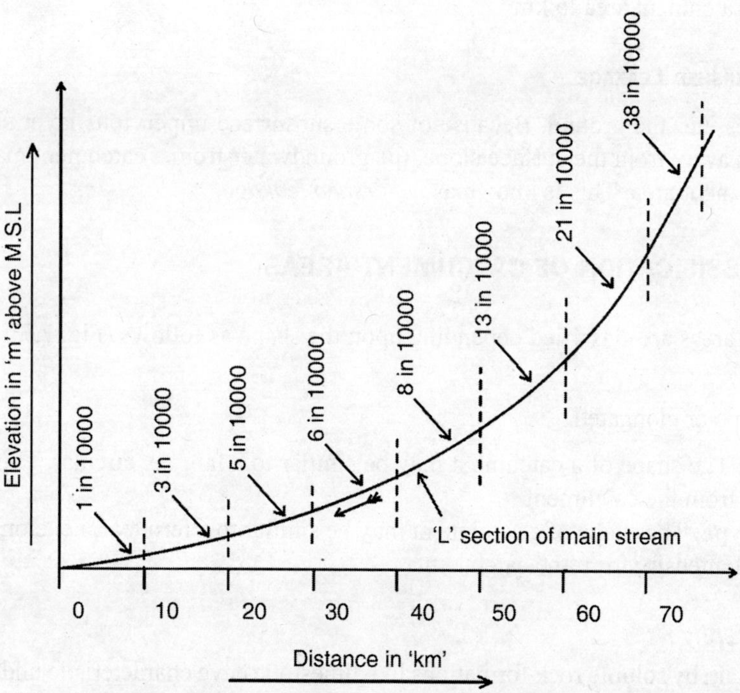

Fig. 7.6 Estimation of the slope of a catchment

Solution:

Divide the L section into eight equal horizontal segments. Each will be of 10 km. The bed slope of each segment is as follows:

(a) 1 in 10,000 (b) 3 in 10,000 (c) 5 in 10,000 (d) 6 in 10,000 (e) 8 in 10,000 (f) 13 in 10,000 (g) 21 in 10,000 (h) 38 in 10,000

The average slope S of the channel will be

$$S = \left[\frac{S_1^{1/2}+S_2^{1/2}+S_3^{1/2}+\cdots}{N}\right]^2$$

$$= \left[\frac{(1/10000)^{1/2} + (3/10000)^{1/2} + (5/10000)^{1/2} + (6/10000)^{1/2} + (8/10000)^{1/2} + (13/10000)^{1/2} + (21/10000)^{1/2} + (38/10000)^{1/2}}{8}\right]^2$$

$$= \left[\frac{1}{100} \times \frac{1 + 1.73 + 2.23 + 2.45 + 2.82 + 3.60 + 4.58 + 6.16}{8}\right]^2$$

$$= 9.43 \text{ in } 10000$$

7.2.12 Relation of Catchment Area with Its Length

After analysing a number of catchment areas, the following relation is noticed.

$$L = 1.2736\, A^{0.6}$$

where, L = Length in kilometre of the main channel
A = Catchment area in km^2

7.2.13 Watershed Leakage

Water infiltrates into the ground. Because of some subsurface impervious layer sloping in a different direction away from the surface slope, the groundwater from a catchment may flow into the adjoining catchment area. This is known as *watershed leakage*.

7.3 CLASSIFICATION OF CATCHMENT AREAS

The catchment areas are classified depending upon the shape as follows (Fig. 7.7):

1. Fan shape
2. Fern leaf-type or elongated

Fan shape: The shape of a catchment may be similar to a fan, i.e. circular. This will affect the flood intensity from the catchment.

Fern leaf-type: The shape of a catchment may be similar to a fern leaf, i.e. elongated. This will affect the flood intensity from the catchment.

Karst topography

Regions underlain by soluble rock formations like limestone have characteristic undulating surfaces with conical round hillocks and circular sinks. Such regions are said to have *Karst topography* and are known as *Karstic regions*. The runoff from such areas is excluded from the surface runoff

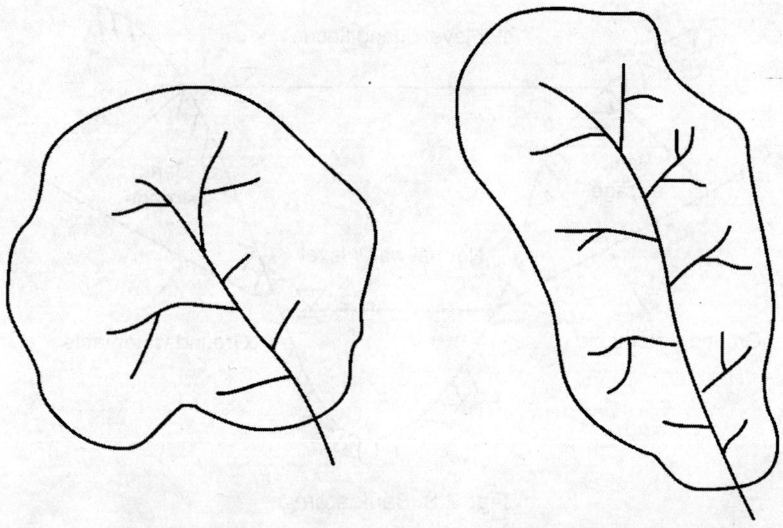

Fig. 7.7 Classification of catchments

estimation, since the precipitation over such areas may be lost into the underground passages to far distant streams, lakes or even sea.

7.4 CATCHMENT FLOW CHARACTERISTICS

7.4.1 Overland Flow

Precipitation falling on a ground surface, after satisfying the infiltration demand, is temporarily detained on the ground surface. When sufficient depth is built, it travels over the ground surface towards the stream. It is somewhat like a sheet flow. It is a laminar flow and may not continue for more than 50 m. Then it forms small streams. This is called *overland flow*.

The overland flow stops shortly after the precipitation stops.

7.4.2 Bank Storage

When the river is in flood, the water surface level in the river rises abruptly and is much higher than the groundwater table in the vicinity. Under such condition, water flows from the river into the banks of the river and is stored there temporarily.

The quantity of water thus stored by the banks temporarily is known as *bank storage*.

When the flood recedes, the water surface level in the river drops. The banks may retain substantial quantity of water, which may flow back into the river or enter the groundwater storage (Fig. 7.8).

7.4.3 Valley Storage

When heavy precipitation occurs, there is likelihood of flood in the river. However, this flood flow may reduce as it travels downstream because of the storage in channels and depressions. Such storage in channels and depressions is known as *valley storage*.

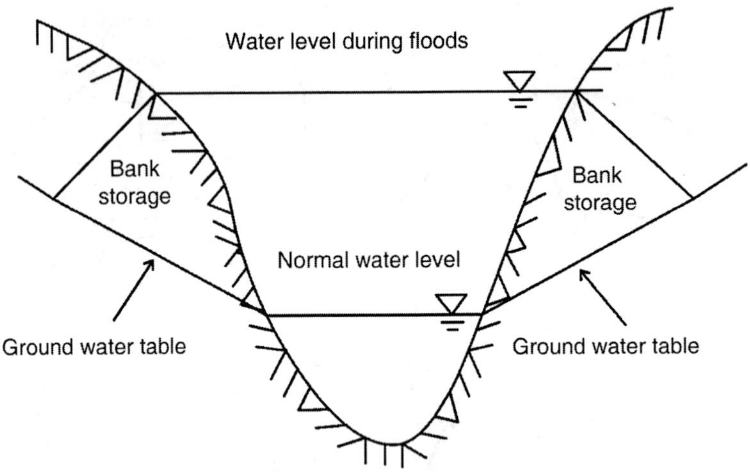

Fig. 7.8 Bank storage

7.4.4 Antecedent Precipitation Index

Precipitation occurring in late summer, when soil moisture is very low, may produce very low runoff and as against flood, due to equal precipitation occurring in rainy season, may contribute comparatively more runoff or high flood. This is because of the moisture present in the soil. This soil moisture depends on the precipitation, which has occurred previously, i.e. antecedent precipitation.

The Antecedent Precipitation Index (API) is a measure of the soil moisture condition existing on the day of precipitation. It is generally denoted by—

$$I_t = k\, I_{t-1}$$

where, I_t = API on the day of precipitation
I_{t-1} = API on the previous day of precipitation
k = A constant known as recession factor

The value of k ranges from 0.85 to 0.98. The API is taken into consideration while assessing the runoff or flood from a catchment area.

7.5 PROCESS OF RUNOFF

Whenever there is precipitation, the requirement of interception and depression storage is first met with. The infiltration also starts immediately. If the rate of precipitation is more than these abstractions, overland flow results. Otherwise, all precipitation water is lost in these abstractions. This overland flow is collected in small streams leading to a channel or a river.

Before and after this storm, there may be some discharge in the river. This may be due to interflow, delayed interflow or flow from the groundwater appearing into the channel. The surface flow observed in the river immediately after the precipitation is the combined flow due to the precipitation and the flow in the river before the precipitation.

7.6 CLASSIFICATION OF STREAMS

The streams may be classified as follows:

7.6.1 Influent Stream

When there is flow in a channel and the groundwater table is lower than the water level in the river, water from the channel may flow into the groundwater storage. A stream that feeds some flow to the groundwater storage is known as *influent stream.*

7.6.2 Effluent Stream

If the groundwater table level is higher than the water level in the channel then water may flow from the groundwater to the channel. A stream that receives some flow from the groundwater storage is known as *effluent stream*. Figure 7.9 shows these two types of flow.

A stream cannot be classified as influent stream or effluent stream for its entire length. It may change its nature depending upon the groundwater table.

7.6.3 Intermittent Stream

A stream that acts as an influent stream for some period in a year and for the rest of the period as effluent stream is called as *intermittent stream.*

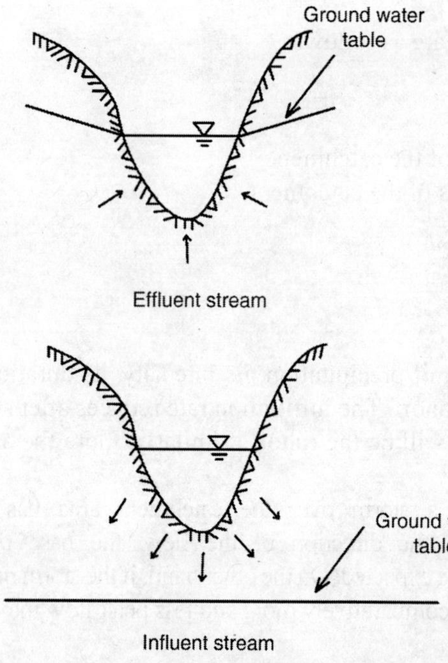

Fig. 7.9 Classification of streams

7.6.4 Ephemeral Stream

A stream that carries discharge in rainy days only and dries during hot weather is known as *ephemeral stream* or *seasonal stream*.

7.6.5 Perennial Stream

A stream that carries discharge throughout the year is known as *perennial stream*.

7.7 STREAM PATTERNS

The combined effect of climate, soil structure and geology of the catchment area is noticed in the *network of channel* or *stream pattern*. The usual patterns observed are as follows (Fig. 7.10):

- Dendrite type or tree like
- Radial pattern
- Trellis pattern
- Pinnate type
- Anastomising pattern
- Parallel
- Rectangular

These stream patterns have a significant effect on the draining period and the flow intensity of the surface runoff and indirectly on infiltration and total runoff.

7.8 FACTORS AFFECTING RUNOFF

The factors affecting the runoff are as follows:

- Precipitation
- Size and shape of catchment
- Geographical characteristics of the catchment
- Meteorological characteristics of the catchment
- Effect of drainage net
- Other factors

7.8.1 Precipitation

The runoff is clearly a function of precipitation, its intensity, its duration and its coverage. More the intensity, more will be the runoff. The infiltration rate reduces after some time; hence more the duration, proportionately more will be the runoff. Similarly, more the area covered by the storm, more will be the runoff.

Direction of movement of a storm over the catchment area has a definite effect on the runoff. If the storm moves in the direction of the flow, the base period of hydrograph will be less and more peak flow may be expected. On the other hand, if the storm moves against the flow direction, then the base period will be comparatively more and less peak flow may be expected (Fig. 7.11).

7.8.2 Size and Shape of the Catchment Area

The size of catchment has a definite effect on the runoff. More the area, more will be the runoff. So also, the shape will have a definite effect on the runoff.

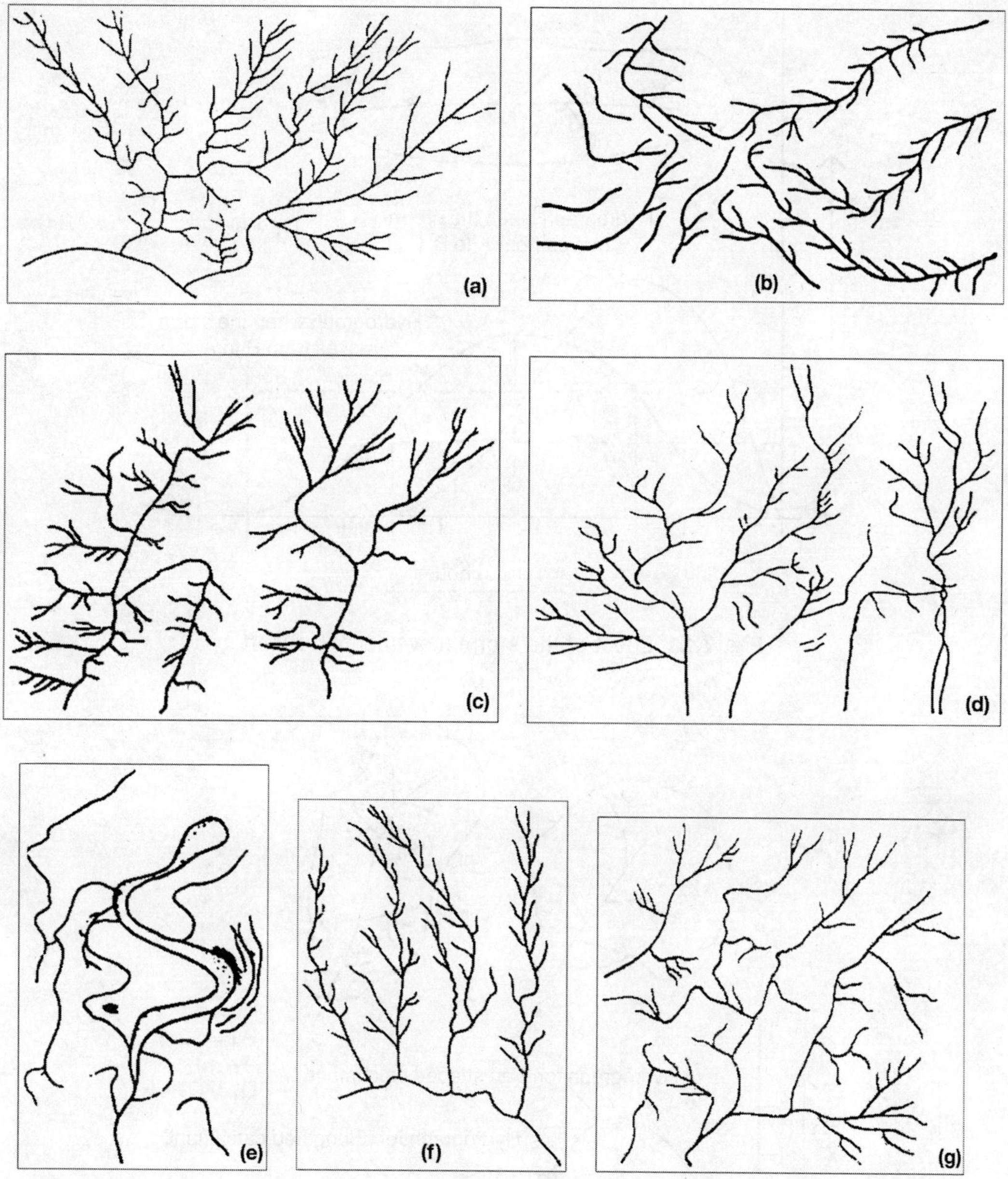

Fig. 7.10 Stream patterns

In case of a fan-shaped catchment area, the base period of the resulting hydrograph will be less and thus more peak flow may be expected. In case of an elongated catchment, the base period of the resulting hydrograph will be comparatively more and thus more will be the infiltration losses and less will be the runoff (Fig. 7.12).

The average value of (1) Manning's coefficient, (2) the shape factor, (3) stream order, (4) drainage density, (5) stream density, (6) circularity ratio, (7) elongation ratio, (8) compactness coefficient, (9) the slope of the channel and so on of the catchment thus definitely influence the runoff.

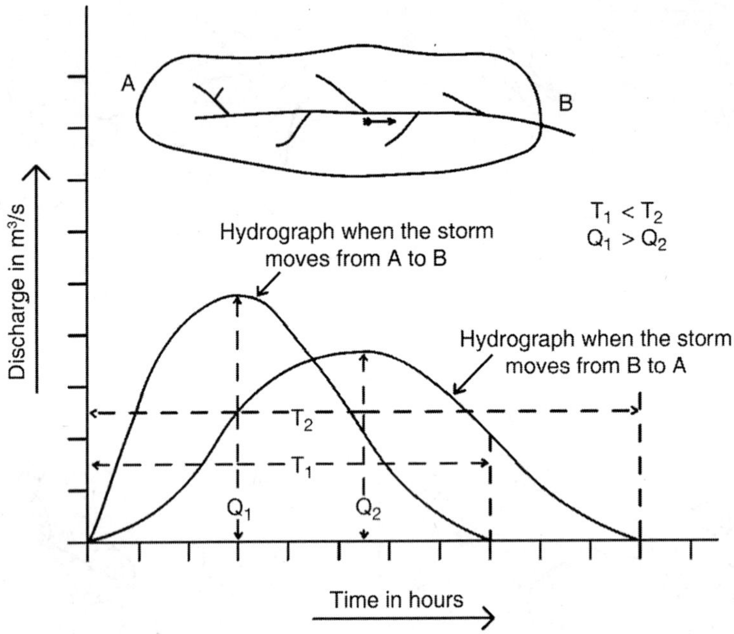

Fig. 7.11 Effect of the storm movement on runoff

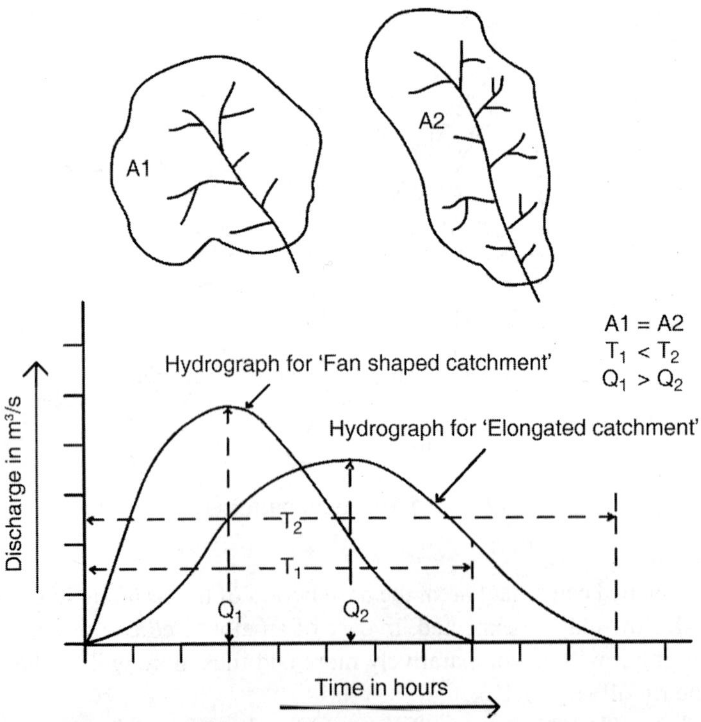

Fig. 7.12 Effect of the shape of catchment on runoff

7.8.3 Geographical Conditions

The nature of the soil, its permeability, has an effect on the infiltration rate and has indirect effect on the runoff. Impervious rock outcrops will increase the runoff. Also, impervious sub-surface layers at higher levels than groundwater table level increase the runoff.

Faults, fissures and cracks in the surface formations will allow more infiltration of the overland flow and the water thus infiltrated may find its opening in the adjoining catchment area or the same catchment somewhere else and may indirectly affect the runoff.

7.8.4 Meteorological Conditions

The temperature has an effect on the evaporation and infiltration and may indirectly affect the runoff. The barometric pressure, altitude and wind will not only affect the storm and its movement, but may also affect the runoff.

7.8.5 Drainage Net

The rainwater after meeting all abstractions, first flows through small rivulets and then flows to bigger ones. The pattern of the various tributaries normally known as drainage net or drainage pattern will affect drainage of the surface flow. In each case, the draining time will differ and will affect the infiltration and thus indirectly the runoff.

7.8.6 Other Factors

(1) Cultivation in an area, (2) contour bunding, (3) ploughing, (4) deforestation, (5) urbanization and so on have a direct effect on the runoff.

7.9 SOURCES OR COMPONENTS OF RUNOFF

The sources and components of runoff from a catchment area are as follows:

7.9.1 Direct Precipitation on the Stream Channel

When the river is in floods, there will be a spread of flood water. The precipitation occurring directly over this water surface contributes to the runoff without any abstractions therefrom.

However, this quantity is very small and normally neglected.

7.9.2 Surface Runoff

This is the major part of the runoff. The water quantity that reaches the stream from the overland flow and is carried by the stream on its surface is the *surface runoff*. It is also known as *quick flow*.

7.9.3 Interflow

All the water, which infiltrates into the ground, may not reach up to the groundwater table if it meets a local impervious layer. This water may move laterally as per the slope of this impervious local layer and may find an outlet into the stream. This component of runoff is known as *interflow*.

If this inflow is observed in a short period, it is called *prompt interflow*; otherwise it is called *delayed interflow*. This inflow moves slowly as compared to the overland flow and its contribution depends upon the orientation of the impervious layer as well as the soil characteristics. This is also known as *subsurface runoff, subsurface storm flow, storm seepage, secondary base flow, through flow* or *quick return flow*.

7.9.4 Groundwater Flow

The water that infiltrates may reach the groundwater storage if not obstructed by any impervious layers in between.

If the groundwater table level is higher than the water level in the stream or the bed level of the stream when it is dry, then the water will flow from the groundwater storage into the stream. The flow may lag behind the occurrence of rainfall by days or by weeks, or even by months. But the flow evens out and is regular over a longer period. This represents the long-term component of the total runoff and is important during the long dry spells when there is no precipitation and hence no surface runoff.

7.9.5 Regeneration

When water is supplied to the fields for irrigation, a small portion of it is utilized for the growth of the crops. Remaining portion infiltrates. A part of the infiltrated water finds its way into the adjoining stream and is called *regeneration*. The remaining flows to the groundwater.

For the regenerated water, the lag time is small. Normally, regeneration is considered to be 20% of the water applied for irrigation.

7.10 ESTIMATION OF RUNOFF

The runoff from a catchment is estimated by the following methods:

- Standard tables
- Empirical formulae
- Rainfall runoff correlation

All these methods give a rough estimate, since none of them is accurate.

7.10.1 Standard Tables

Observations of precipitation and the resulting runoff were taken for a number of catchments having different characteristics. Tables are prepared showing the relation between precipitation and the resulting runoff, taking into consideration the catchments. The following tables are in use.

7.10.1.1 Binnie's table

Sir Binnie suggested runoff as a percentage of total annual precipitation by observing rivers in Madhya Pradesh, India, as follows:

Table 7.1 Binnie's table

Serial no.	Average annual rainfall in the catchment (mm)	Runoff % of annual rainfall
1	500	15
2	600	21
3	700	25
4	800	29
5	900	34
6	1000	38
7	1100	40

The observations covered areas that were not having heavy rainfall but were having a precipitation up to 1100 mm. Hence, these percentages are not applicable for high-rainfall areas.

Example 7.3

The average annual precipitation over a catchment of 120 km^2 and its percentage is as follows:

Serial no.	% of area	Average annual precipitation (cm)
A1	20	50
A2	15	75
A3	40	90
A4	25	110

Find the runoff from the catchment using Binnie's table.

Solution:

The Binnie's coefficients corresponding to the average annual precipitation for each area and the corresponding runoff will be as follows:

Serial no.	Area (km^2)	Binnie's coefficient	Runoff (m^3)
1	$120 \times 0.20 = 24$	0.15	$24 \times 10^6 \times 0.50 \times 0.15 = 1.800 \times 10^6$
2	$120 \times 0.15 = 18$	0.27	$18 \times 10^6 \times 0.75 \times 0.27 = 3.645 \times 10^6$
3	$120 \times 0.40 = 48$	0.34	$48 \times 10^6 \times 0.90 \times 0.34 = 14.680 \times 10^6$
4	$120 \times 0.25 = 30$	0.40	$30 \times 10^6 \times 1.10 \times 0.40 = 13.200 \times 10^6$
			Total 33.325×10^6 m^3

7.10.1.2 Rainfall–Runoff correlation

Runoff is correlated to rainfall as follows:

$$\text{Runoff} = K \times (\text{annual rainfall})$$

where, K = Constant and its value for various types of catchment areas are shown in Table 7.2.

This method is applicable mainly to small urban catchment areas.

Table 7.2 Values of constant K for various types of catchment areas

Serial no.	Type of catchments	Value of K
1	Urban	0.05–0.5
2	Forest	0.2–0.5
3	Commercial and industrial	0.9
4	Parks and farms	0.3–0.5
5	Concrete pavements	0.85

Example 7.4

A catchment area of 150 km^2 has the following type of distribution:

Area	Type	%
A1	Urban	20
A2	Forest	35
A3	Commercial and industrial	30
A4	Concrete pavement	15

Find the annual runoff from the catchment when the average annual precipitation is 90 cm.

Solution:

The runoff coefficient K for the respective type of catchment and the annual runoff from each type of catchment will be as follows:

Serial no.	Area (km^2)	K	Runoff (m^2)
A1	$150 \times 0.20 = 30.0$	0.25	$30.0 \times 10^6 \times 0.9 \times 0.25 = 6.75 \times 10^6$
A2	$150 \times 0.35 = 52.5$	0.45	$52.5 \times 10^6 \times 0.9 \times 0.45 = 21.26 \times 10^6$
A3	$150 \times 0.30 = 45.0$	0.90	$45.0 \times 10^6 \times 0.9 \times 0.90 = 36.45 \times 10^6$
A4	$150 \times 0.15 = 22.5$	0.85	$22.5 \times 10^6 \times 0.9 \times 0.85 = 17.21 \times 10^6$
			Total 81.67×10^6

7.10.1.3 Strange's Coefficient

W. L. Strange suggested that rainfall–runoff relationship depends upon the nature of catchment as well as soil condition. The relation was developed on the basis of observations in the former Bombay State in India. The relation has been presented in the form of tables as well as curves.

The catchment is classified as (i) dry, (ii) damp and (iii) wet. It is also classified as (a) good, (b) average and (c) bad. The percentage coefficients for these classifications are given in Table 7.3.

It was also recommended to add 25% to the runoff figures for good catchments and deduct 25% for the bad catchments. The precipitation range as well as the type of catchment is covered in this approach. Still the results are not accurate. The daily runoff is calculated whenever there is rain and is added to calculate the seasonal or annual runoff.

Table 7.3 The percentage coefficient of different classes of the catchment

Serial no.	Daily rainfall (mm)	Runoff (%)		
		Dry	Damp	Wet
1	6.25	–	–	8
2	12.50	–	6	12
3	25.00	3	11	18
4	37.50	6	16	25
5	50.00	10	22	34
6	75.00	20	37	55
7	100.00	30	50	70

Example 7.5

The daily rainfall over a catchment of 140 km^2 is 5, 30 and 80 mm, respectively. The area is classified as follows:

1. Dry—30%
2. Damp—45%
3. Wet—25%

Find the runoff from this storm using Strange's coefficient. The area is classified as good catchment.

Solution:

The rainfall and the corresponding Strange's coefficient for the classification of the catchment will be as follows:

Serial no.	Day	Precipitation (mm)	Strange's coefficient (%)		
			Dry	Damp	Wet
1	1	5	–	–	0.08
2	2	30	0.035	0.13	0.20
3	3	80	0.230	0.40	0.62

The runoff, in mm, for each day for the three classifications will be as follows:

Serial no.	Day	Rainfall (mm)	Runoff (mm)		
			Dry	Damp	Wet
1	1	5	–	–	0.08 × 5 = 0.40
2	2	30	0.035 × 30 = 1.05	0.13 × 30 = 3.9	0.20 × 30 = 6.00
3	3	80	0.230 × 80 = 18.40	0.40 × 80 = 32.0	0.62 × 80 = 49.60
		Total	19.45	35.9	56.00

The runoff in m^3 from the three classifications will be as follows:

Serial no.	Day	Area (km^2)	Runoff (m)
1	1	$140 \times 0.30 = 42$	$42 \times 10^6 \times 19.45 \times 10^{-3} = 0.817 \times 10^6$
2	2	$140 \times 0.45 = 63$	$63 \times 10^6 \times 35.90 \times 10^{-3} = 2.262 \times 10^6$
3	3	$140 \times 0.25 = 35$	$35 \times 10^6 \times 56.00 \times 10^{-3} = 1.960 \times 10^6$
			Total $= 5.039 \times 10^6$ m^3

Since the area is classified as 'good', the runoff from the catchment will be increased by 25% and the runoff from the catchment will be $= 1.25 \times 5.039 \times 10^6$

$$= 6.30 \times 10^6 \text{ m}^3$$

7.10.1.4 Barlow's method

Barlow suggested percentage coefficients for small catchment areas depending on the nature of the catchment for different categories as given in Table 7.4.

These percentages were modified by multiplying them by a coefficient depending on the rainfall as given in Table 7.5.

Table 7.4 Percentage coefficient of small catchment areas depending on the nature of the catchment

Serial no.	Class	Nature of catchment	Runoff (%)
1	A	Flat, cultivated black cotton soil area	10
2	B	Flat partly cultivated consisting of different soils	15
3	C	Average land	20
4	D	Hills and plains with little cultivation	35
5	E	Very hilly and steep slopes with hardly any cultivation	45

Table 7.5 Coefficient depending on the rainfall

Serial no.	Nature of rainfall	Coefficient				
		A	B	C	D	E
1	Light rain	0.7	0.8	0.8	0.8	0.8
2	Average rain	1.0	1.0	1.0	1.0	1.0
3	Continuous downpour	1.5	1.5	1.6	1.7	1.8

Example 7.6

A catchment covering an area of 200 km^2 has the following classification:

A1. Flat cultivated land = 20%
A2. Hills and plain with little cultivation = 40%
A3. Very hilly and steep slopes with no cultivation = 40%

The average precipitation over the catchment during a storm for 3 days was as under

1. Light rain = 10 mm
2. Average rain = 50 mm
3. Heavy rain = 150 mm

Find the runoff from the catchment.

Solution:

The runoff is in metre for the classification of the catchment, as per Barlow's table, the coefficient being as follows:

Serial no.	Day	Nature of catchment	Runoff coefficient	Nature of Rainfall	Rainfall coefficient	Rainfall (m)	Runoff (m)
1	1	A1	0.10	Light	0.7	0.01	0.1 × 0.7 × 0.01 = 0.0007
2	2	A1	0.10	Average	1.0	0.05	0.1 × 1.0 × 0.05 = 0.005
3	3	A1	0.10	Heavy	1.5	0.15	0.10 × 1.5 × 0.15 = 0.0225
							Total = 0.0282 m
4	1	A2	0.35	Light	0.8	0.01	0.35 × 0.8 × 0.01 = 0.0028
5	2	A2	0.35	Average	1.0	0.05	0.35 × 1.0 × 0.05 = 0.0175
6	3	A2	0.35	Heavy	1.7	0.15	0.35 × 1.7 × 0.15 = 0.0893
							Total = 0.1096 m
7	1	A3	0.45	Light	0.8	0.01	0.45 × 0.8 × 0.01 = 0.0036
8	2	A3	0.45	Average	1.0	0.05	0.45 × 1.0 × 0.05 = 0.0225
9	3	A3	0.45	Heavy	1.8	0.15	0.45 × 1.8 × 0.15 = 0.122
							Total = 0.1481 m

The runoff in m^3 from the three classifications of the catchment will be as under.

Serial no.	Nature of catchment	Area (m^2)	Runoff (m)	Runoff (m^3)
1	A1	$200 \times 10^6 \times 0.2$	0.282	$20 \times 10^6 \times 0.0282 = 0.564 \times 10^6$
2	A2	$200 \times 10^6 \times 0.4$	0.1096	$40 \times 10^6 \times 0.1096 = 4.384 \times 10^6$
3	A3	$200 \times 16^6 \times 0.4$	0.1481	$40 \times 10^6 \times 0.1481 = 5.924 \times 10^6$
				Total = $10.872\ 10^6\ m^3$

7.10.2 Empirical Formulae

There are several empirical formulae in use. These involve the following parameters:

- Catchment area
- Annual rainfall

- Nature of catchment
- Average temperature

The formulae normally followed are as follows:

7.10.2.1 Sir Inglis formula

Sir Inglis suggested two formulae, based on the observations in the former Bombay State, India.

(A) Ghat-fed catchments:

$$R = 0.85\,P - 30.5$$

(B) Plain areas in water shadow regions:

$$R = \frac{P \times (P - 17.8)}{254}$$

where, P and R are in cm.

A catchment area may be partly in the ghat and partly in the plains. Runoff from both the types may be calculated separately and then added together.

Example 7.7

A catchment area of 200 km^2 has the following classification:

1. Ghat-fed—40%
2. Plain—60%

Find the runoff, if the annual average precipitation is 750 mm.

Solution:

The runoff is calculated by using the Inglis formula.

1. Ghat-fed area:

$$R = 0.85\,P - 30.5 = 0.85 \times \frac{750}{10} - 30.5 = 33.25 \text{ cm} = 0.34 \text{ m}$$

2. Plain area:

$$R = \frac{P \times (P - 17.8)}{254} = \frac{750/10(750/10 - 17.8)}{254} = \frac{75 \times 57.2}{254}$$

$$= 16.88 \text{ cm}$$

$$= 0.168 \text{ m}$$

$$\text{Total runoff} = 200 \times 0.4 \times 10^6 \times 0.34 + 200 \times 0.6 \times 10^6 \times 0.168$$

$$= 27.2 \times 10^6 + 20.16 \times 10^6$$

$$= 47.36 \times 10^6 \text{ m}^3$$

7.10.2.2 Khosla's formula

Dr Khosla suggested the following formula:

$$R = P - 5T$$

where, R = Runoff in mm
P = Average precipitation in mm
T = Average temperature in °C

The formula is modified further when the average temperature is less than 4 °C. The formula is used to calculate the runoff for each month and then the values are added to arrive at the annual runoff.

Example 7.8

The mean monthly temperature and monthly precipitation figures observed for the year 2004 for a catchment area, covering an area of 160 km^2, are as under.

Month	Jan.	Feb.	Mar.	Apr.	May	Jun.	Jul.	Aug.	Sept.	Oct.	Nov.	Dec.
Precipitation (cm)	5	5	2	0	3	13	33	30	16	2	1	1
Temperature (°C)	13	17	21	28	32	35	32	30	29	27	20	14

Find the annual runoff.

Solution:

Using Khosla's formula, the monthly precipitation, monthly temperature, monthly loss and monthly runoff, for the year 2004, are tabulated as follows:

Serial no.	Month	Monthly precipitation (cm)	Monthly temperature (°C)	Monthly loss (cm)	Monthly runoff (cm)
1	January	5	13	6.5	0
2	February	5	17	8.5	0
3	March	2	21	10.5	0
4	April	0	28	14.0	0
5	May	3	32	16.0	0
6	June	13	35	17.5	0
7	July	33	32	16.0	17
8	August	30	30	15.0	15
9	September	16	29	14.5	1.5
10	October	2	27	13.5	0
11	November	1	20	10.0	0
12	December	1	14	7.0	0
				Total	33.5

Therefore, total annual runoff $= 160 \times 10^6 \times \dfrac{33.5}{100} = 53.6 \times 10^6 \text{ m}^3$

7.10.2.3 Rational formula

The runoff from a catchment can be estimated by using a rational formula for small catchments as follows:

$$R = C\,A\,P$$

where, R = Runoff in million m^3
C = Constant
A = Area in km^2
P = Average annual precipitation in metres.

Table 7.6 gives the value of C as suggested by Richard.

The catchment area may be divided into sub-areas as per the type mentioned above. The runoff from each catchment area is evaluated separately and then added together to arrive at the total runoff.

Table 7.6 Value of C as suggested by Richard

Serial no.	Types of catchment	Value of C
1	Rocky and impermeable	0.8–1.0
2	Slightly permeable	0.6–0.8
3	Cultivated or covered with vegetation	0.4–0.6
4	Cultivated absorbent soil	0.3–0.4
5	Sandy soil	0.2–0.3
6	Heavy forest	0.1–0.2

Example 7.9

The classification of a catchment covering an area of 110 km^2 is as follows:

1. Rocky, impermeable—10%
2. Cultivated—60%
3. Forest—30%

Using the rational formula and the coefficients suggested by Richard, find the annual runoff from the catchment when the average annual precipitation is 800 mm.

Solution:

The catchment area, its classification, Richard's coefficients and the annual runoff is tabulated below.

Serial no.	Catchment area (m^2)	Richard's coefficient	Annual runoff (m^2)
1	$0.1 \times 110 \times 10^6 = 11 \times 10^6$	0.9	$0.9 \times 11 \times 10^6 \times 0.8 = 7.92 \times 10^6$
2	$0.6 \times 110 \times 10^6 = 66 \times 10^6$	0.5	$0.5 \times 66 \times 10^6 \times 0.8 = 26.40 \times 10^6$
3	$0.3 \times 110 \times 10^6 = 33 \times 10^6$	0.15	$0.15 \times 33 \times 10^6 \times 0.8 = 3.96 \times 10^6$
			Total 38.28×10^6

7.10.3 Deducting Abstractions from Precipitation

The runoff can also be estimated by deducting all abstractions from the precipitation. These abstractions are evaporation, surface retention and infiltration; however, it calls for a correct estimation of these abstractions.

7.10.4 Rainfall–Runoff Correlation

From the data available from past records, a graph can be plotted to correlate runoff and rainfall.

The graph will normally be a straight line having the following equation (Fig. 7.13):

$$R = m(p - x)$$

where, R = Runoff in million m^3
m = A constant
p = Rainfall in mm
x = Interception of the straight-line relationship on the x-axis.

The relation thus derived will be applicable for a specific catchment and is normally not used for other catchment areas. There are several changes carried out by humans every year and the relation has to be modified every year to account for them. However, it gives a rough estimate.

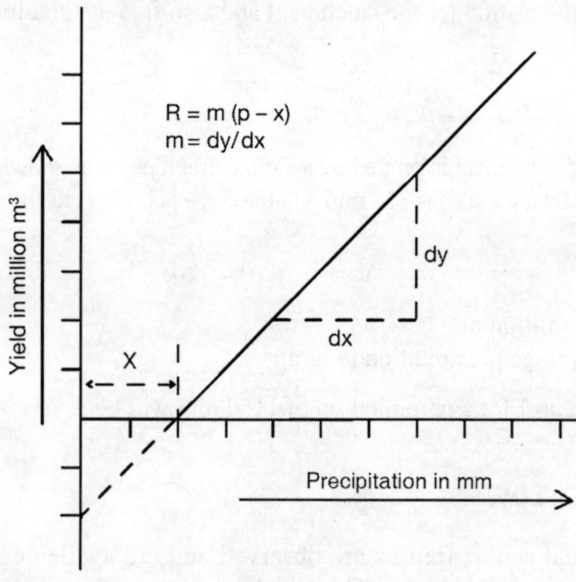

Fig. 7.13 Rainfall runoff co-relation

Example 7.10

The annual runoff and the annual precipitation observed over a catchment are as follows:

Serial no.	1	2	3	4	5	6	7	8	9	10	11	12
Precipitation (mm)	1000	1700	1200	1500	1360	1620	1100	1370	1630	1420	1400	1260
Runoff (million m^3)	553	1105	651	903	750	992	605	781	1005	875	854	751

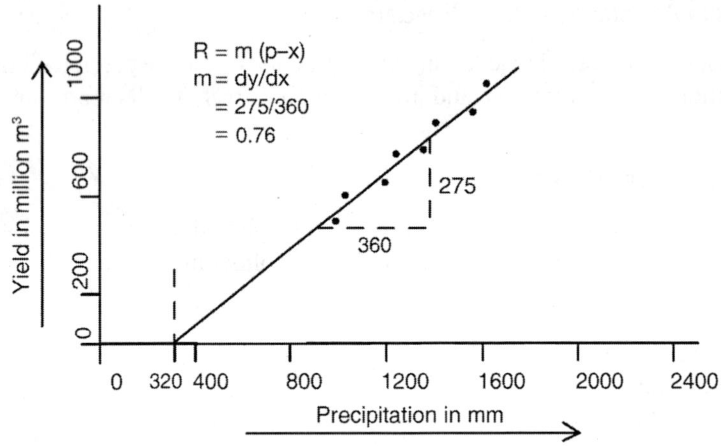

Fig. 7.14 Rainfall–runoff correlation

Derive the rainfall–runoff relation for the catchment and also find the possible runoff when the precipitation will be 1160 mm.

Solution:

The rainfall–runoff for the catchment is plotted on a simple graph paper as shown in Fig. 7.14. From the graph the interception on the x-axis is 320 mm and the slope is 0.76. Thus the relation works to be as follow:

$$R = 0.76\,(P - 320)$$

where, R = Runoff in million m^3

P = Annual average precipitation in mm

From this relation, the runoff for a precipitation of 1160 mm will be = 638×10^6 m^3.

7.11 DEPENDABLE FLOW

In the case when the annual runoff figures are observed and are available for n years, these figures are arranged in a descending order. The serial number m of a specific value of runoff is known as *ranking*. It can then be said that this specific value of runoff or more than that is available for m years out of n years and the percentage dependability of this value will be as follows:

$$\%\ \text{dependability} = \frac{m}{n} \times 100$$

For instance, if the data of annual runoff are available for 40 years and if these are arranged in a descending order, then the percentage dependability of the runoff value at Serial no. 30 will be $30/40 \times 100 = 75\%$.

Similarly, the percentage dependability of 50% will be the runoff figure at Serial no. 20.

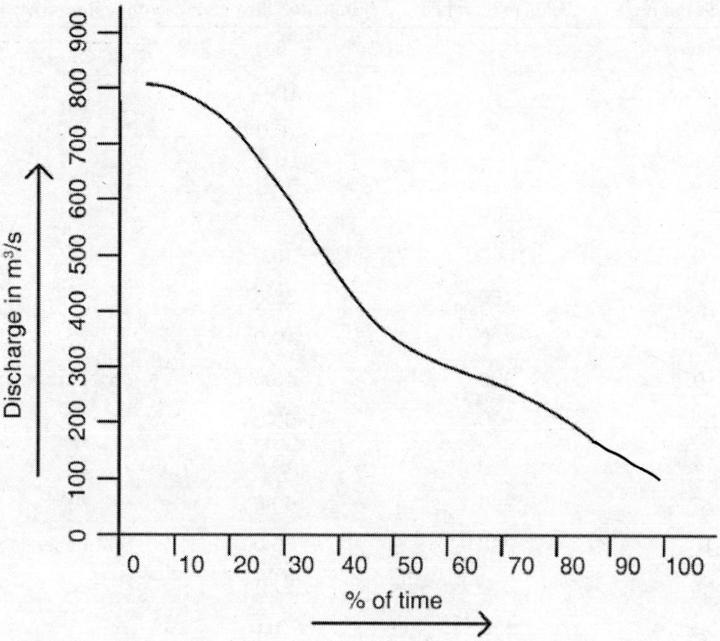

Fig. 7.15 Flow duration curve

7.12 FLOW DURATION CURVE

A graph of percentage dependability on x-axis and the corresponding runoff value on the y-axis are plotted and is known as *flow–duration curve.* It is also known as *discharge frequency curve.*

A flow–duration curve is shown in Fig. 7.15.

Example 7.11

The annual yield in million m^3 from a catchment for the last 20 years is as follows: 240, 350, 680, 400, 290, 160, 110, 105, 300, 600, 800, 540, 280, 130, 320, 810, 740, 220, 190 and 480.
Find the following:

(1) 80% dependable yield
(2) 45% dependable yield
(3) The % dependability of the yield of 540 million m^3
(4) Plot the flow–duration curve

Solution:

The annual yield in million m^3 for the 20 years is arranged in descending order. The yield for (1) 80% dependability will be at $20 \times 0.80 = 16$ (serial order) and (2) 45% dependability will be at $20 \times 0.45 = 9$ (serial order).

Serial no.	Yield (million m^3)	Dependability (%)	Remarks
1	810	5.0	
2	800	10.0	
3	740	15.0	
4	680	20.0	
5	600	25.0	
6	540	30.0	
7	480	35.0	
8	400	40.0	
9	350	45.0	45% dependable
10	320	50.0	
11	300	55.0	
12	290	60.0	
13	280	65.0	65% dependable
14	240	70.0	
15	220	75.0	
16	190	80.0	80% dependable
17	160	85.0	
18	130	90.0	
19	110	95.0	
20	105	100.0	

1. 80% dependable yield = 190×10^6 m^3
2. 45% dependable yield = 350×10^6 m^3
3. Since 540×10^6 m^3 is at Serial no. 6, its dependability will be $= \frac{6}{20} = 30.0\%$
4. The flow–duration curve is plotted in Fig. 7.15.

7.13 RUNOFF COEFFICIENT

All the water after precipitation will not flow to the river. There are some abstractions. After meeting the requirements of these abstractions, the balance water will flow to the river as runoff. Thus runoff will be less than the precipitation.

The runoff coefficient is the ratio of runoff and precipitation. It will be always be less than 1.

7.14 RUNOFF CYCLE

It is a descriptive term used for a part of the hydrological cycle. When precipitation occurs over the land area, some part of the precipitation flows into a river channel as overland flow, while part of the precipitation enters into the ground as infiltration. This quantity of infiltrated water finds its way into the river channel as interflow. Part of water precipitated returns to the atmosphere through evapo-transpiration.

This process, which is part of the hydrological cycle, is known as *runoff cycle*.

REVIEW QUESTIONS

1. Define runoff. Discuss the various components of runoff.
2. How is the average slope of a catchment area calculated?
3. Explain watershed leakage, with the help of a neat sketch. How does it affect the runoff from the catchment?
4. How are the catchment areas classified based on the shape of the catchment? Discuss their effect on the runoff.
5. Explain bank storage, with the help of a neat sketch.
6. What do you understand by Karst topography?
7. How are the streams classified based on the flow consideration?
8. Discuss the factors affecting runoff.
9. What are the sources of runoff?
10. How is runoff from a catchment estimated?
11. State and explain the different formulae used to estimate runoff from a catchment. Discuss their limitations also.
12. Explain Binnie's table to estimate the runoff from a catchment. What are its limitations?
13. What are Strange's coefficients? Discuss their limitations.
14. What is a dependable yield? How is it calculated for a given dependability?
15. What is a flow-duration curve? How is it constructed?
16. Write short notes on the following:
    a. Ridge line
    b. Valley storage
    c. Groundwater divide line
    d. Antecedent precipitation index
    e. Drainage area
    f. Process of runoff
    g. Stream order
    h. Stream pattern
    i. Drainage density
    j. Rainfall runoff correlation
    k. Steam density
    l. Runoff cycle
    m. Length of stream
    n. Ephemeral stream
    o. Form factor
    p. Compactness coefficient
    q. Circulatory ratio
    r. Relief of catchment area
    s. Elongation ratio
    t. Shape factor
17. Is there any relation between area of the catchment and its length? Discuss.
18. Distinguish between the following:
    a. Topographic divide and groundwater divide
    b. Shape of the catchment and pattern of the catchment
    c. Influent stream and effluent stream
    d. Intermittent stream and perennial stream
    e. Drainage density and stream density
    f. Form factor and shape factor
    g. Circulatory ratio and elongation ratio
    h. Fan-shaped catchment and elongated catchment
    i. Overland flow and interflow
    j. Binnie's table and Barlow's table for the estimation of runoff

NUMERICAL QUESTIONS

7.1. The levels taken along the cross-section of a channel at a regular interval of 1000 m are as follows.

Section	0	1	2	3	4	5	6	7
Level in m	245	246	248	251	255	260	266	273

Calculate the average slope of the channel.

Ans: The average slope of the channel = 3.69 in 1000

7.2. The annual average precipitation in mm of a catchment covering 1000 km^2 is 1000 mm . Out of this catchment 60% is ghat-fed and the rest is plain. Find the run-off from the catchment.

Ans: $45.69 \times 10^6 m^3$

7.3. The mean average annual precipitation and the run-off from a catchment observed over the last 10 years is as follows

Sr. No.	1	2	3	4	5	6	7	8	9	10
Precipitation in mm	541	581	739	942	838	1022	703	903	781	603
Run-off in mm	263	341	438	705	581	743	442	603	526	408

Derive a relation between precipitation and run-off, for the catchment. Also calculate the possible run-off when the precipitation is 720 mm

Ans:

(1) The relation between run-off and precipitation is R = 0.909 (P − 203)

(2) The run-off when the precipitation is 720 mm = 470 mm

7.4. The annual yield in million m^3 from a catchment for the last 20 years is as follows

1410	1310	1100	935	1370	1020	1203	930	1440	1290
980	1400	1176	1425	1340	1261	1130	1300	1480	1280

Calculate

(1) 75 % dependable yield

(2) 50 % dependable yield

(3) the % dependability for 1300×10^6 m^3

Ans:

(1) 75 % dependable yield ______________ 1130×10^6 m^3

(2) 50 % dependable yield ______________ 1290×10^6 m^3

(3) the % dependability for 1300×10^6 m^3 ______________ 45 %

MULTIPLE CHOICE QUESTIONS

1. Normally the regeneration of irrigated water is considered to be

(a) 30% (b) 25%

(c) 20% (d) 15%

2. Interflow is also known as
   (a) Subsurface runoff
   (b) Subsurface storm flow
   (c) Storm seepage
   (d) Secondary base flow
   (e) All the above
3. Catchment area is also known as
   (a) Watershed area
   (b) Drainage area
   (c) Drainage basin
   (d) Catchment
   (e) Basin
   (f) All the above
4. A stream that carries discharge in rainy season and dries during hot weather is known as
   (a) Intermittent stream
   (b) Influent stream
   (c) Ephemeral stream
   (d) Effluent stream
5. A stream that carries discharge throughout the year is known as
   (a) Seasonal stream
   (b) Perennial stream
   (c) Influent stream
   (d) Effluent stream
6. The graph showing the relation of rainfall runoff has the following relation with usual notations
   (a) $R = mp - x$
   (b) $R = mx - p$
   (c) $R = m(p - x)$
   (d) $R = mpx$
7. For calculating the dependable yield, the annual runoff figures are arranged in
   (a) Descending order
   (b) Ascending order
   (c) Chronologically
8. Flow–duration curve is normally plotted on
   (a) Normal graph paper
   (b) Semi-log paper
   (c) Log-log paper
9. Ridge line is also known as:
   (a) Topographic water divide
   (b) Watershed divide
   (c) Topographic divide.
   (d) All the above
10 Groundwater divide line is also known as
   (a) Topographic water divide
   (b) Watershed divide
   (c) Phreatic divide
   (d) All the above
11. The runoff is expressed in
   (a) m of depth of water uniformly spread over the entire area
   (b) Million m^3
   (c) ha m
   (d) All the above
12. The peculiarity of the Karst topography is that it
   (a) Does not produce any runoff
   (b) Produces only direct runoff and no base flow
   (c) Produces only base flow and no direct runoff
   (d) None of the above
13. The runoff from a catchment area of 50 km^2 receiving an excess rainfall of 100 mm will be equal to
   (a) 5 million m^3
   (b) 5×10^2 ha m
   (c) 0.1 m of water
   (d) All are correct

14. The slope of a catchment area with usual notations is given by

(a) $\dfrac{S_1 + S_2 + S_3 + \ldots + S_n}{n}$

(b) $\dfrac{S_1^{0.5} + S_2^{0.5} + S_3^{0.5} + \ldots + S_n^{0.5}}{n}$

(c) $\left(\dfrac{S_1^{1/2} + S_2^{1/2} + S_3^{1/2} + \ldots + S_n^{1/2}}{n}\right)^2$

(d) $\dfrac{S_1L_1 + S_2L_2 + S_3L_3 + \ldots + S_nL_n}{L_1 + L_2 + L_3 + \ldots + L_n}$

ANSWERS TO MULTIPLE CHOICE QUESTIONS

1. c	2. e	3. e	4. c	5. b	6. c	7. a	8. a
9. d	10. c	11. d	12. a	13. d	14. c		

*** ———— ***

8 Floods

Chapter Outline

8.1 DEFINITION

Any flow of water that is relatively high and that may overtop the natural or the artificial banks in a reach of a channel may be called as *flood.*

Water may overtop the bank at some location where it may be treated that the channel is in floods. However, at some other location of the same channel, water may be contained within the banks of the channel. Thus, the same channel may not be considered in floods at that location.

In large rivers, an arbitrary level is decided keeping in mind the habitation or the farmland and other properties so that when water level rises above this level, the river is designated to be in flood.

8.1.1 Features of Flood

The features of flood are mentioned as follows:

- When the river is in flood, normally the maximum discharge in m^3/s it carries, is the consideration of the flood.
- In some cases, the maximum level reached by the flood water is the consideration.
- The spread of water, i.e. the area inundated during the flood is also a consideration.
- The duration of the high water spread may be a consideration.

However, all these considerations are interdependent.

8.1.2 Causes of Flood

Flood may be due to the following reasons:

- Whenever there is heavy precipitation over the catchment in terms of intensity, duration and spread, the river will carry high discharge and thus this is the main reason of a river to be in flood.
- Whenever there is a heavy landslide in the river, it may cause flood on the u/s side due to arrest of flow and consequent rise in water level.
- Sometimes, because of earthquake, it may so happen that the river bed is raised. This may cause flood on the u/s due to the arrest of flow as well as on the d/s, for some distance for a short period.
- In case there is breach of a dam, the reservoir water may rush towards the d/s side causing heavy flood for a short period.
- Floods may be caused due to heavy melting of snow and ice.

8.1.3 Water Year

The 12 months of the year selected for maintaining or presenting records of flow is known as a *water year*. It is also known as *hydrological year*.

8.2 FACTORS AFFECTING FLOOD

The factors that may affect the flood intensity are discussed below:

8.2.1 Precipitation

It is obvious that precipitation will affect the flood. More the intensity, more the duration and more the coverage of the storm, more will be the flood.

8.2.2 Size and Shape of Basin

It is obvious that larger the size of the catchment more will be the flood.

The shape of the basin also definitely affects floods. For a fan-shaped catchment, the time of concentration will be less and hence the storm hydrograph base period will be less and the peak discharge will be more. As against this, in the case of an elongated-shaped catchment of equal area

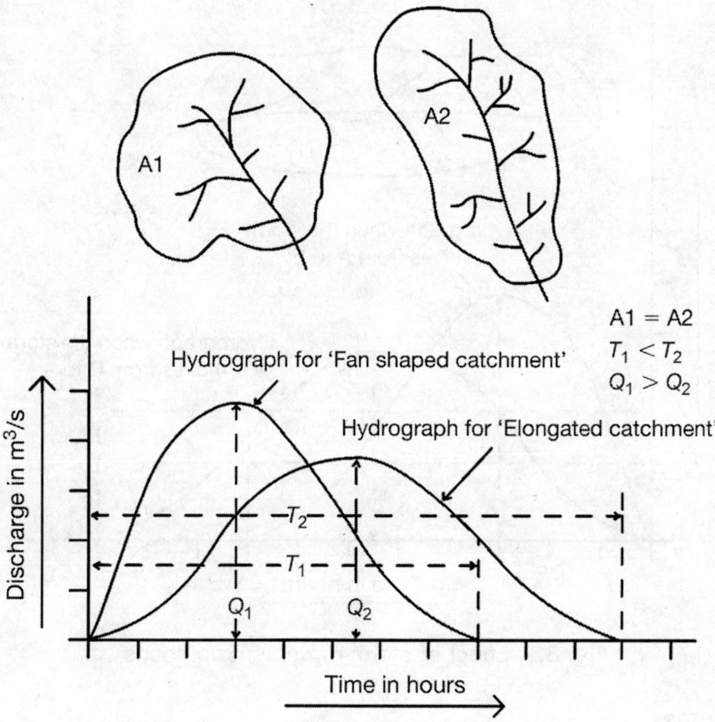

Fig. 8.1 Effect of shape of catchment on floods

and the same storm, the time of concentration will be more. Thus, the storm hydrograph base period will be more and the peak discharge will be comparatively less.

Figure 8.1 shows the two flood hydrographs in the case of a fan-shaped catchment and an elongated catchment.

8.2.3 Movement of Storm

When the catchment is large, the storm may not cover the entire catchment area, but may cover a part of the area and then move towards other parts. If the storm moves along the flow direction, the time of concentration will be less, the base period will be less and peak discharge will be more. If the storm moves against the flow direction, for the same storm and the same basin, the time of concentration will be more, the base period will be more and the peak discharge will be less.

Figure 8.2 shows the two flood hydrographs when the storm moves along the flow and against the flow.

8.2.4 Antecedent Precipitation Index

When precipitation occurs, the soil may be wet because of the precipitation on the previous day. In this case, the antecedent precipitation index (API) may be high and hence the infiltration rate of the soil will be low resulting into high value of surface runoff. This may result in higher value of flood discharge also. Thus, higher value of API results in higher flood.

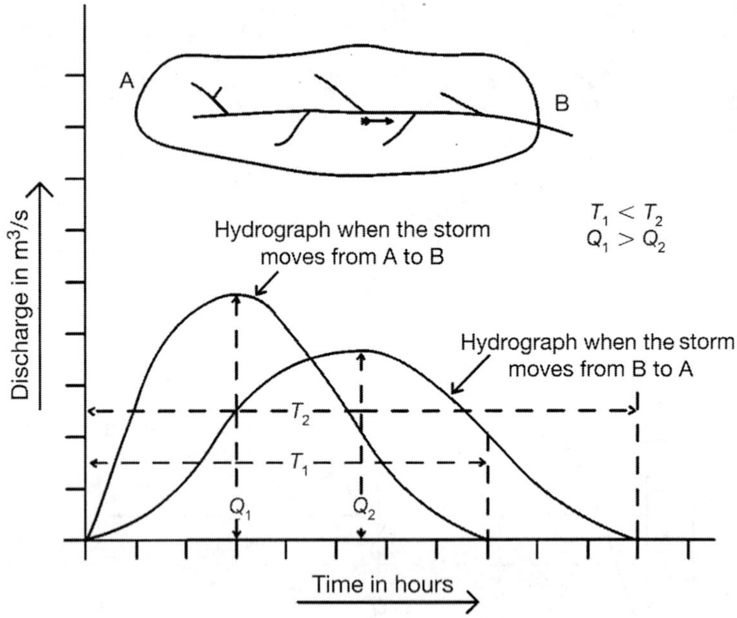

Fig. 8.2 Effect of storm movement on floods

8.2.5 Other Factors

The following factors, which affect the infiltration rate, also indirectly affect the flood.

- **Physical factors:** Manning's coefficient, slope of catchment, form factor, drainage density, drainage pattern, etc.
- **Geographical factors:** Nature of soil, permeability of soil, rock outcrops, impervious layers in the ground at higher levels, faults, fissures, cracks, vegetation, conservation measures such as contour bunding.
- **Climatic factors:** Temperature, barometric pressure, altitude, wind speed and so on.

8.3 CLASSIFICATION OF FLOODS

Floods may be classified as under:

- **Annual flood:** The maximum discharge observed in a water year is known as '*Annual flood.*'
- **Maximum observed flood:** The maximum flood observed during the period for which data is available is known as '*Maximum observed flood.*'
- **Mean annual maximum flood:** The average of all the annual floods recorded is known as 'Mean annual maximum flood.'
- **Maximum-probable flood:** The maximum probable flood can be defined as the flood that may be expected from the most severe combination of meteorological conditions. It is normally a very large flood. It is also known as *maximum possible flood.* The normal short form used is MPF.
- **Design flood:** The discharge adopted for the design of a structure is known as 'Design flood.' Design flood is finalized considering the nature of the structure, etc. For less important structures, the design flood may be less than the MPF.

- **Flash flood:** A flood of short duration and abrupt rise with a relatively high-peak rate of flow, usually resulting from a high intensity of rainfall, is known as 'Flash flood'.
- **N-year flood:** A flood that has a probability of being equalled or exceeded once in *N* years is known as an 'N-year flood'.

8.4 ESTIMATION OF FLOODS

Floods may be estimated by the following methods:

- Estimating the observed flood and increasing it by certain percentage
- Envelope curves
- Empirical formulae
- Unit hydrograph application
- Statistical methods

8.4.1 Estimation and Increase of Observed Floods

The maximum observed flood may be increased by some percentage depending upon the importance of the structure.

If sufficient observed data are not available, then, by local inquiry, the maximum water level reached or recorded on some building or bridge or any other structure may be noted or may be obtained from the old persons residing in the locality. The flood discharge for this maximum water level can be calculated from the following data about the river:

- River cross section at the flood level observed
- Manning's coefficient (an appropriate judgement)
- The river bed slope

The flood discharge thus calculated may be increased or decreased suitably for design purposes, depending upon the importance of the structure.

Example 8.1

In a village, the maximum water level reached in the last 32 years was recorded on the compound wall of the Town hall as 105.00 m corresponding to the bed of the river at 100.00 m.

A survey was conducted at this cross section and levels were taken. These are as follows:

Chainage (m)	0.0	5.00	10.0	17.5*	27.5	32.5	35.0	40.0
Level (m)	106.00	104.00	101.00	100.00	101.00	103.00	105.00	106.0

*River bed

The river bed slope at this location was observed to be 1 in 2500. The Manning's coefficient corresponding to the topography of the river was 0.030.

Find the flood discharge corresponding to the maximum water level reached.

Solution:

The cross section of the river is plotted to scale as shown in Fig. 8.3.

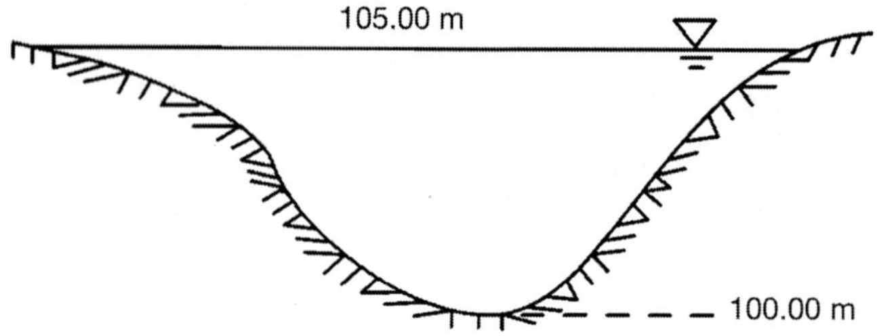

Fig. 8.3 Cross Section of a river

The area of flow corresponding to this HFL calculated by a plannimeter was 112.5 m^2.
The wetted perimeter corresponding to this HFL was 35.6 m.

Therefore, hydraulic mean depth $= R = \dfrac{112.5}{35.6} = 3.16$ m
Using Manning's formula

$$\text{Now,}\quad Q = A \times V = A \times \frac{R^{2/3} \times S^{1/2}}{n} = 112.5 \times \frac{3.16^{2/3} \times (1/2500)^{1/2}}{0.03}$$

$$= 162.0 \text{ m}^3/\text{s}$$

8.4.2 Envelope Curve

The observed or the calculated flood discharges of catchment areas, hydrographically and meteorologically homogenous regions are considered and plotted on a simple graph paper. A curve covering all these points is drawn. This is known as *envelope curve*.

Figure 8.4 shows such a curve drawn for a catchment area.

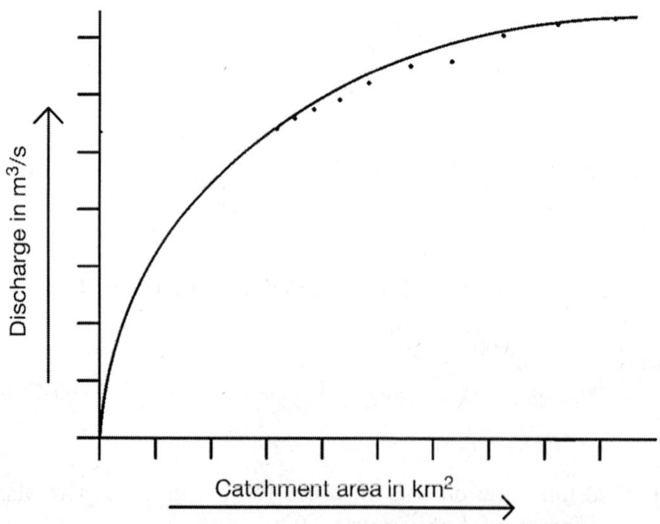

Fig. 8.4 Envelope curve for flood discharge

The flood discharge from this curve for the required basin may be calculated. Following equation for such an envelope curve is derived.

$$Q = \frac{3010\,A}{(277 + A)^{0.78}} \tag{8.1}$$

where, Q = Flood discharge in m^3/s

A = Catchment area in km^2

The flood discharge used for plotting the curve are all observed ones and hence the flood discharge calculated from such an envelope curve has to be increased suitably depending upon the importance of the structure.

Example 8.2

The catchment area of a proposed bandhara is 100 km^2. The maximum average annual flood discharges in the adjoining area with hydrologically homogeneous catchments are as follows:

Catchment	A	B	C	D	E	F	G
Catchment area (km^2)	32.0	120.0	60.0	90	155	80.0	140.0
Maximum flood discharge (m^3/s)	1600	3300	2100	2900	3700	2800	3500

Find the maximum annual flood discharge at the site of the bandhara.

Solution:

On a simple graph paper, catchment area vs discharge, is plotted as shown in Fig. 8.5.

A curve by free hand covering all these points is drawn (envelope curve).

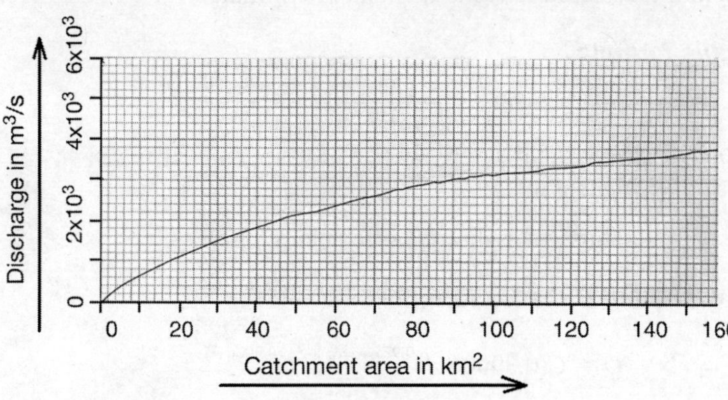

Fig. 8.5 Derived envelope curve for flood discharge

The maximum annual flood discharge at the bandhara site, with a catchment area equal to 100 km^2, was read from this envelope curve. It is 3400 m^3/s.

Example 8.3

For a catchment area of 100 km^2, find the annual average flood discharge by using the envelope curve equation.

Solution:

The envelope curve equation is as follows:

$$Q = \frac{3010\,A}{(277 + A)^{0.78}}$$

Substituting the value of A,

$$Q = \frac{3010 \times 100}{(277 + 100)^{0.78}} = \frac{3010 \times 100}{102.22}$$

$$Q = 2944 \text{ m}^3/\text{s}.$$

8.4.3 Empirical Formulae

There are innumerable empirical formulae derived to suit the topographical, climatological and geological conditions of the catchment area.

In most of these empirical formulae, the catchment area is the main factor and to suit the location of the catchment area as well as its characteristics, one or two constants are used.

Some of the formulae used for estimating the flood discharge are as follows:

8.4.3.1. Dickens formula

$$Q = CA^{3/4} \tag{8.2}$$

where, Q = Flood discharge in m^3/s
A = Area of catchment in km^2
C = A constant. Its value ranges from 11 to 20 for plain areas and 22 to 25 for ghat-fed catchments.

Normally, this formula is used for central and northern India.

8.4.3.2. Sir Inglis formula

$$Q = \frac{124\,A}{\sqrt{(A + 10.4)}} \tag{8.3}$$

where, Q = Flood discharge in m^3/s
A = Area of catchment in km^2

Normally, this formula is used for ghat-fed catchments.

8.4.3.3. Ali Nawaz Jung Bahadur formula

$$Q = C(0.39006\,A)^{0.925 - \frac{1}{14}\log_{10}(0.3906\,A)} \tag{8.4}$$

where, Q and A have the usual notations as in Eqs. (8.2) and (8.3) and C is a constant whose value ranges from 48 to 60.

Normally, this formula is used for old Hyderabad state and South India.

8.4.3.4. Rational formula

Originally, it was derived for British system without any conversion factor. Hence, it was called rational formula. However, in SI system, it is modified as under.

$$Q = 2.778\,C\,A\,I \tag{8.5}$$

where, Q = Flood discharge in m^3/s
C = Coefficient ranging from 0.1 to 0.5 depending upon the nature of the catchment
A = Area in km^2
I = Maximum intensity of rainfall in cm/h

Example 8.4

Find the maximum flood discharge by using the empirical formulae for a catchment area of 80 km^2 having a maximum precipitation intensity of 4 cm/h.

Assume
C in Dicken's formula = 22
C in Ali Nawaz Jung Bahadur formula = 50
C in rational formula = 0.35

Solution:

The flood discharge Q by using the various empirical formulae will be as follows:

1. Dicken's formula

$$Q = CA^{3/4} = 22 \times 80^{3/4} = 22 \times 26.74$$
$$= 588 \text{ m}^3/\text{s}$$

2. Inglis formula

$$Q = \frac{124 \times A}{\sqrt{(A + 10.4)}} = \frac{124 \times 80}{\sqrt{(80 + 10.4)}} = \frac{124 \times 80}{9.51}$$
$$= 1043 \text{ m}^3/\text{s}$$

3. Ali Nawaz Jung Bahadur formula

$$Q = C(0.39006\,A)^{0.925 - \frac{1}{14}\log_{10}(0.3906 \times A)}$$
$$= 50\,(0.39006 \times 80)^{0.925 - \frac{1}{14}\log_{10}(0.3906 \times 80)}$$
$$= 50 \times 31.20^{.819}$$
$$= 837 \text{ m}^3/\text{s}$$

4. Rational formula

$$Q = 2.778\,CAI = 2.778 \times 0.35 \times 80 \times 4$$
$$= 311 \text{ m}^3/\text{s}$$

8.4.4 Unit Hydrograph Procedure

The unit hydrograph theory and procedure can be used to estimate the flood. For this purpose, the following data are required:

1. **Possible maximum precipitation:** The possible maximum precipitation (PMP) that can be expected over the basin can be worked out from the available data. India Meteorological Department (IMD) has divided India in various zones for this purpose, PMP in each zone has been indicated. This data can also be used.

2. **Unit hydrograph:** The unit hydrograph derived for the catchment area up to the basin outlet is necessary. If such a unit hydrograph is not available, then synthetic unit hydrograph up to the basin outlet may be derived.
   By using the PMP and the unit hydrograph, the flood hydrograph up to the basin outlet can be estimated.

The peak discharge in the flood hydrograph thus derived may be considered as the design flood. With the use of this unit hydrograph procedure, not only the flood discharge is estimated but the flood hydrograph thus derived can also be used for the design of a spillway of a dam for flood routing purposes.

Example 8.5

For a catchment area of 50 km^2, the 1 h–1 cm unit hydrograph coordinates are as follows:

Time (h)	0	1	2	3	4	5	6	7	8	9	10
Discharge (m^3/s)	0	6	28	35	27	20	14	9	5	2	0

The worst possible storm covering the entire catchment area is 2 cm/h for 3 h. Find the maximum flood discharge.

Solution:

Since the unit hydrograph is for 1 h, the storm will be divided into three parts each one of 1-h duration. Discharge due to these three parts of the storm with a time lag of 1 h will be as follows:

Serial no.	Time (h)	Unit hydrograph coordinates		Discharge due		Total
			First part	Second part	Third part	
1	0	0	0	—	—	0
2	1	6	12	0	—	12
3	2	28	56	12	0	68
4	3	35	70	56	12	138
5	4	27	54	70	56	180
6	5	20	40	54	70	164
7	6	14	28	40	54	122
8	7	9	18	28	40	86
9	8	5	10	18	28	56
10	9	2	4	10	18	32
11	10	0	0	4	10	14
12	11	—	—	0	4	4
13	12	—	—	—	—	0

Therefore, the maximum flood discharge is 180 m^3/s.

8.4.5 Statistical Methods

The observed data can be used to predict future flood of a particular probability or return period. For this purpose, adequate reliable flood data should be available for a minimum of 20–25 years. The longer the period, more reliable will be the results. The data should also be correct. If required the data may be corrected or adjusted before use.

It is based on the assumption that the combination of numerous factors, which affect flood, is a pure chance and therefore is subject to analysis as per the theory of probability.

There are two methods of collecting the flood discharge data.

1. **Annual flood series**
2. **Partial duration flood series**

In the case of annual flood series, only the highest flood in each year is taken for analysis, whereas in the case of partial duration flood series, all the flood discharges above a selected peak are used. In this case, two or more peaks in a year may have to be used.

However, the results obtained by using both the approaches do not vary much.

The following statistical terms are generally referred to:

1. **Recurrence interval:** If the annual flood discharges of n (say 100) years are arranged in a descending order, then there will be some discharge figure (say 1000 m^3/s) at serial order at m (say 25). Then it can be said that this discharge of 1000 m^3/s at m or more than this will occur in m (25 years) out of n (100 years), i.e. its recurrence interval is 25/100 = 1/4, i.e. once in 4 years.

   It is also known as *return period* and generally represented by T_r.
   Here, $T_r = 4$. This does not mean that in every 4 years it will occur once, but on an average in 40 years it will occur 10 times or in 80 years it will occur 20 times.

2. **Probability:** In the example stated above, the probability of flood discharge of 1000 m^3/s or more, occurring in 100 years is 1/4. It is generally denoted as p. Probability is reciprocal of recurrence interval, i.e. $p = 1/T_r$. In this case, it is 0.25.
   Thus, probability is always a fraction, i.e less than 1.

3. **Frequency:** Probability expressed in terms of percentage is *frequency*. In the above example, the frequency is $= p \times 100 = 1/T_r \times 100 = 25\%$.

   From the available data, the recurrence interval and the probability of a specific discharge can be calculated as follows:

8.4.5.1 Probability plotting

Arrange the available discharge figures (say n in total) in descending order. Let the serial order of this specific discharge Q be m. This serial order m is called *rank* or *order of the observation*. The recurrence interval and the probability of the specific discharge Q in the series can be calculated by the following methods.

1. **California formula,**

$$T_r = \frac{n}{m}$$

2. **Hazen's formula,**

$$T_r = \frac{n}{m - 0.50}$$

3. **Weibull's formula,**

$$T_r = \frac{n + 1}{m}$$

4. **Gumbell's formula,**

$$T_r = \frac{n}{m + C - 1}$$

Here, C is called *gumbell's correction*. Its value depends on m and n as given in the following table.

m/n	1	0.9	0.8	0.7	0.6	0.5	0.4	0.3	0.2	0.1	0.04
C	1	0.95	0.88	0.845	0.78	0.73	0.68	0.59	0.52	0.4	0.28

Once recurrence interval T_r is calculated, the probability p can be calculated, since $T_r = 1/p$. Weibul's formula is generally used.

Example 8.6

The maximum annual flood discharges in m^3/s observed at a proposed dam site from 1973 to 2004 (32 years) are as follows:

395, 619, 766, 422, 282, 887, 705, 528, 520, 436, 697, 624, 496, 489, 598, 359, 696, 726, 527, 310, 408, 721, 814, 459, 440, 632, 343, 634, 464, 373, 289, 371

Find the return period for all the discharges by using the following formulae:

(1) California formula, (2) Hazen's formula, (3) Weibul's formula and (4) Gumbell's formula.

Solution:

The four formulae referred to are as follows.

1. California formula, $T_r = \frac{n}{m}$
2. Hazen's formula, $T_r = \frac{n}{m - 0.50}$
3. Weibul's formula, $T_r = \frac{n + 1}{m}$
4. Gumbell's formula, $T_r = \frac{n}{m + C - 1}$

The series of 32 discharges was arranged in the descending order. The return period calculated by the first three formulae are tabulated in Table A. Since the procedure for the fourth method is slightly different, the return period worked out by this method is tabulated in Table B.

Table A					
Serial no.	Discharge(m^3/s)	Discharge in descending order (m^3/s)	T_r		
			California formula	Hazen's formula	Weibull's formula
1	395	887	32	64	33
2	619	814	16	21.33	16.5
3	766	766	10.67	12.80	11
4	422	726	8	9.14	8.25
5	282	721	6.40	7.11	6.6
6	887	705	5.33	5.81	5.5
7	528	697	4.57	4.92	4.71
8	705	686	4	4.27	4.12
9	520	634	3.55	3.76	3.67
10	436	632	3.2	3.36	3.3
11	697	624	2.9	3.04	3
12	624	619	2.67	2.78	2.75
13	496	598	2.46	2.56	2.54
14	589	589	2.28	2.37	2.36
15	598	528	2.13	2.2	2.2
16	359	527	2	2.06	2.06
17	686	520	1.88	1.94	1.94
18	726	496	1.77	1.82	1.83
19	527	464	1.68	1.73	1.74
20	310	459	1.6	1.64	1.65
21	408	440	1.52	1.56	1.57
22	721	436	1.45	1.49	1.5
23	814	422	1.39	1.42	1.43
24	459	408	1.33	1.36	1.38
25	440	395	1.28	1.31	1.32
26	632	373	1.23	1.25	1.27
27	343	371	1.18	1.21	1.22
28	634	359	1.14	1.16	1.18
29	464	343	1.1	1.12	1.13
30	373	310	1.06	1.08	1.1
31	289	289	1.03	1.05	1.06
32	371	282	1	1.02	1.03

Table B

Serial no.	Discharge (m^3/s)	Discharge in descending order (m^3/s)	$\frac{m}{n}$	C	$m + C - 1$	T_r
1	395	887	0.030	0.27	0.27	118.5
2	619	814	0.0620	0.29	1.29	24.8
3	766	766	0.09	0.38	2.38	13.44
4	422	726	0.125	0.42	3.42	9.35
5	282	721	0.156	0.46	4.46	7.17
6	887	705	0.187	0.51	5.51	5.8
7	528	697	0.218	0.52	6.52	4.9
8	705	686	0.25	0.53	7.53	4.25
9	520	634	0.289	0.595	8.55	3.74
10	436	632	0.312	0.58	9.58	3.34
11	697	624	0.344	0.59	10.59	3.02
12	624	619	0.375	0.64	11.64	2.75
13	496	598	0.406	0.65	12.65	2.53
14	589	589	0.437	0.69	13.69	2.33
15	598	528	0.469	0.71	14.71	2.17
16	359	527	0.5	0.73	15.73	2.03
17	686	520	0.531	0.75	16.75	1.91
18	726	496	0.562	0.77	17.77	1.86
19	527	464	0.593	0.78	18.78	1.71
20	310	459	0.625	0.79	19.79	1.62
21	408	440	0.656	0.81	20.81	1.54
22	721	436	0.689	0.84	21.84	1.46
23	814	422	0.718	0.85	22.85	1.4
24	459	408	0.75	0.86	23.75	1.35
25	440	395	0.781	0.87	24.87	1.29
26	632	373	0.812	0.89	25.89	1.24
27	343	371	0.843	0.91	26.91	1.19
28	634	359	0.875	0.93	27.93	1.14
29	464	343	0.906	0.95	28.95	1.11
30	373	310	0.937	0.967	29.97	1.07
31	289	289	0.969	0.99	30.99	1.03
32	377	282	1.0	1.0	31.00	1.02

All these formulae will give the recurrence interval and probability for any value covered in the series. But if the recurrence interval and the probability of any value more than the values in the series are required, then this can be obtained by extrapolation by using the following graphs:

1. Natural scale
2. Semi-log
3. Log-log

The results obtained by using log-log plot are considered more accurate.

Example 8.7

For the data in Example 8.6, find the discharge with a return period of 50.

Solution:

The return periods for all the 32 discharges have been worked out by four different formulae in Example 8.6. These results will be extrapolated on graph paper. The following three types of graph papers will be used.

1. Natural scale graph paper (Fig. 8.6)
2. Semi-log scale graph paper (Fig. 8.7)
3. Log-log scale graph paper (Fig. 8.8)

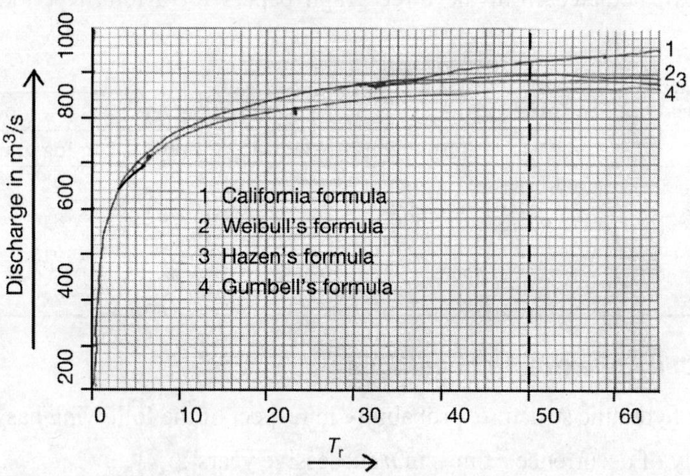

Fig. 8.6 A simple graph of the discharge for the return period of 50

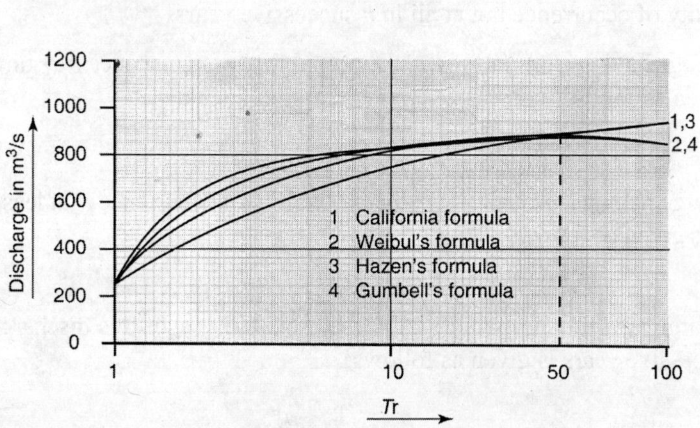

Fig. 8.7 A semi-log graph of discharge for the return period of 50

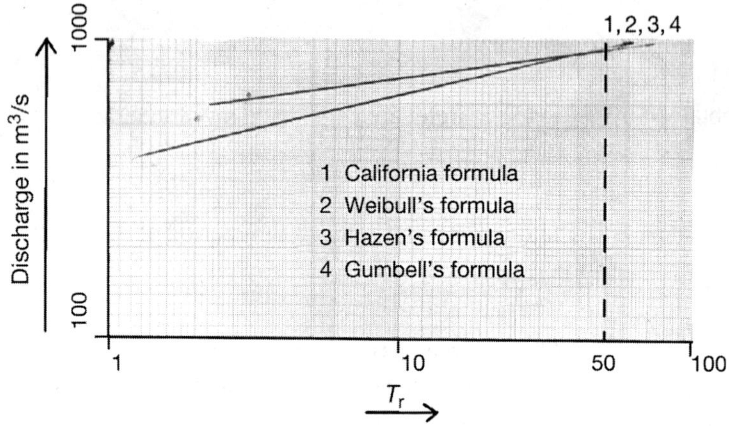

Fig. 8.8 A log-log graph of the discharge for the return period of 50

The discharges calculated from the three graph papers for a return period of 50 years are tabulated below.

Serial no.	Formula	Discharge (m^3/s)		
		Simple graph	Semi-log graph	Log-log graph
1	California	932	948	920
2	Hazen's	880	865	840
3	Weibul's	928	920	910
4	Gumbell's	854	850	820

8.4.5.2 Probability of extreme conditions

In the case of any hydraulic structure, probability in respect of the following has to be considered.

(A) Probability of occurrence r times in n successive years

(B) Probability of occurrence at least once in n successive years

(C) Probability of occurrence not at all in n successive years

(A) For a discharge having a probability p, the probability of occurrence p_r, r times in n successive years is given as follows:

$$p_r = \frac{n! \times p^r (1-p)^{n-r}}{(n-r)! \times r!}$$

(B) For a discharge having a probability p, the probability of occurrence p_r at least once in n successive years is given as follows:

$$p_r = 1 - (1-p)^n$$

(C) For a discharge having a probability of p, the probability p_r of this discharge not occurring at all in n successive years is given as follows:

$$p_r = (1-p)^n$$

Example 8.8

For a dam, the designed discharge of 1000 m^3/s has a return period of 50 years. Find the probability that

(a) This discharge occurs twice in 20 years.

(b) This discharge occurs once in 15 years.

(c) This discharge will not occur at all in 25 years.

Solution:

$$p = \frac{1}{50} = 0.02$$

$$\text{(a) } p_r = \frac{20! \times 0.02^2 \times (1 - 0.02)^{(20-2)}}{(20-2)! \times 2!} = 0.0528$$

$$\text{(b) } p_r = 1 - (1 - 0.02)^{15} = 0.262$$

$$\text{(c) } p_r = (1 - 0.02)^{25} = 0.603$$

Note: The procedure discussed in this sub-chapter is applicable to all hydrological parameters, e.g. precipitation, runoff, etc.

8.5 GUMBELL'S DISTRIBUTION

For the estimation of floods of higher recurrence intervals and probabilities, Gumbell's distribution is widely used. Gumbell considered the annual flood series.

Let $Q_1, Q_2, Q_3, \ldots Q_n$ be the discharge series of n figures arranged in descending order. As per this distribution, the probability of occurrence of a discharge figure is given by

$$p = 1 - e^{-e^{-y}} \tag{8.6}$$

where, p = Probability,
y = Reduced variate that is a dimensionless number

The value of $y = a(Q - Q_f)$ (8.7)

Here, a and Q_f are the parameters of the distribution.

$$a = \sigma n / sx$$

and

$$Q_f = \overline{Q} - (yn / \sigma n) \times sx$$

where, $\overline{Q}$ and sx are the mean and standard deviation, respectively, of the discharge series and σn and yn depend on n, i.e. the number of discharge series. These values are given in following table.

n	σn	yn	n	σn	yn	n	σn	yn
10	0.9497	0.4952	32	1.1193	0.5380	52	1.1638	0.5493
12	0.9833	0.5035	34	1.1255	0.5396	54	1.1667	0.5501
14	1.0095	0.5100	36	1.1313	0.5410	56	1.1696	0.5508
16	1.0316	0.5157	38	1.1363	0.5424	58	1.1721	0.5515
18	1.0493	0.5202	40	1.1413	0.5436	60	1.1746	0.5521
20	1.0629	0.5235	42	1.1458	0.5448	62	1.1770	0.5527
22	1.0754	0.5268	44	1.1499	0.5458	64	1.1793	0.5533
24	1.0864	0.5296	46	1.1538	0.5468	66	1.1814	0.5538
26	1.0961	0.5320	48	1.1574	0.5477	68	1.1834	0.5543
28	1.1047	0.5343	50	1.1607	0.5485	70	1.1854	0.5548
30	1.1124	0.5362				72	1.1873	0.5552

Yn = reduced mean

σn = reduced standard deviation

Consider Eq. 8.6,

$$p = 1 - e^{-e^{-y}}$$

$$1 - p = e^{-e^{-y}}$$

But,

$$p = \frac{1}{T_r}$$

Therefore, $1 - p = 1 - \dfrac{1}{T_r} = \dfrac{T_r - 1}{T_r}$

Therefore, $e^{-e^{-y}} = \dfrac{T_r - 1}{T_r}$

Taking logarithms of both sides,

$$-e^{-y} = \log_e \frac{(T_r - 1)}{T_r}$$

$$e^{-y} = -\log_e \frac{(T_r - 1)}{T_r} = -2.303 \log_{10} \frac{(T_r - 1)}{T_r} = 2.303 \log_{10} \frac{T_r}{(T_r - 1)}$$

Again, taking logarithms of both sides,

$$-y = \log_e 2.303 + \log_e \left(\log_{10} \frac{T_r}{T_r - 1}\right)$$

$$\therefore \quad y = -\left[\log_e 2.303 + \log_e \left(\log_{10} \frac{T_r}{T_r - 1}\right)\right]$$

Therefore, $$y = -\left[0.834 + 2.303 \log_{10}\left(\log_{10} \frac{T_r}{T_r - 1}\right)\right] \tag{8.8}$$

For $Q = Q_T$, y will be y_T, probability will be p_T and return period will be T_r.

Equation 8.7 can be written as

$$y_T = a\,(Q_T - Q_f)$$

$$= \left(\frac{\sigma n}{sx}\right)\left[Q_T - \left\{\overline{Q} - \left(\frac{yn}{\sigma n}\right) \times sx\right\}\right]$$

$$y_T \times \left(\frac{sx}{\sigma n}\right) = Q_T - \overline{Q} + \left(\frac{yn}{\sigma n}\right) \times sx$$

Therefore,

$$Q_T - \overline{Q} = y_T \times \left(\frac{sx}{\sigma n}\right) - \left(\frac{yn}{\sigma n}\right) \times sx$$

$$Q_T = \overline{Q} + sx\left[\left(\frac{y_T}{\sigma n}\right) - \left(\frac{yn}{\sigma n}\right)\right]$$

$$= \overline{Q} + \left(\frac{y_T - yn}{\sigma n}\right) sx \qquad (8.9)$$

$$= \overline{Q} + K_T sx.$$

K_T is known as *frequency factor* $= (y_T - yn)/\sigma n$.

The step-by-step procedure to follow Gumbell's distribution is as follows:

1. Arrange the discharge series for n years in the descending order.
2. Find out the mean discharge $\overline{Q}$ and standard deviation sx of the discharge series.
3. From Table 8.2, find the values of yn and σn corresponding to n.
4. From Eq. 8.8, find the value of y_T for the desired value of T_r.
5. From Eq. 8.9. find the value of Q_T.

Thus, Q_T will be the discharge having a return period of T_r.

Gumbell's distribution procedure can also be used by following a graphical solution. A special graph *Gumbell probability paper*, generally known as *Powell's probability paper*, can be used. This graph paper is shown in Fig. 8.9.

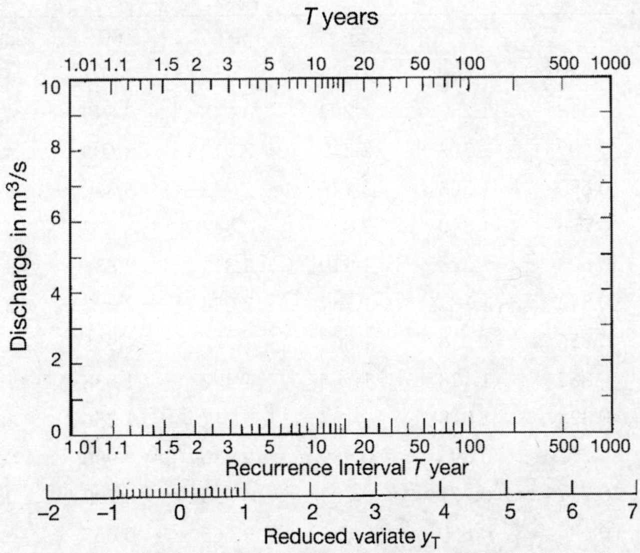

Fig. 8.9 Gumbell's probability paper

On the x-axis, the return period T_r is plotted and on the y-axis, the discharge is plotted. This graph is a straight line. From this graph, the discharge for any return period can be calculated.

8.6 LOG-PEARSON TYPE III DISTRIBUTION

This type of distribution is used for anticipating flood of higher return period. It is commonly used in the USA.

Let the observed flood discharge series be Q_1, Q_2, Q_3, ..., Q_n of n figures, arranged in the descending order. This discharge series is converted into a new log series, viz. log Q_1, log Q_2, log Q_3, ... Of course, the number of terms, i.e. n will remain unchanged. The base for the logarithms should be 10.

This new log series will now be considered further. The value of the variate y_T for a return period T_r is given as follows:

$$y_T = \bar{y} + K\,sx \tag{8.10}$$

where, y_T = Reduced variate

$\bar{y}$ = Mean of the log series

K = Frequency factor

sx = Standard deviation of the log series

The coefficient of skew g will be as follows:

$$g = \frac{n\sum (y - \bar{y})^3}{(n-1)(n-2)sx^3}$$

The frequency factor K is a function of T_r and g.

The following table gives the values of K for various values of T_r and g.

Different values of K for various values of T_r and g

g	T_r							
	2	5	10	25	50	100	200	1000
3.0	−0.396	0.420	1.180	2.278	3.152	4.501	4.970	7.250
2.5	−0.360	0.518	1.250	2.262	3.048	3.845	4.652	6.600
2.0	−0.307	0.609	1.302	2.219	2.912	3.605	4.298	5.910
1.5	−0.240	0.690	1.333	2.146	2.743	3.330	3.910	5.250
1.0	−0.164	0.758	1.340	2.043	2.542	3.022	3.489	4.540
0.5	−0.083	0.808	1.323	1.910	2.311	2.686	3.041	3.815
0.0	0.000	0.842	1.282	1.751	2.054	2.326	2.576	3.090
−0.5	0.083	0.856	1.216	1.567	1.777	1.955	2.108	2.400
−1.0	0.164	0.852	1.128	1.366	1.492	1.588	1.664	1.880
−1.5	0.240	0.825	1.018	1.157	1.217	1.256	1.282	1.373
−2.0	0.307	0.777	0.895	0.959	0.980	0.990	0.995	1.000
−2.5	0.360	0.711	0.771	0.793	0.798	0.799	0.800	0.802
−3.0	0.396	0.636	0.660	0.666	0.666	0.667	0.667	0.668

From the value of g calculated for the log series and T_r, the value of y_T can be calculated from Eq. 8.10.

From y_T, the value of Q_T can be calculated as follows:

$$Q_T = \text{Antilog } (y_T) \tag{8.11}$$

The step-by-step procedure to follow the Log-Pearson type III distribution is as follows:

1. Arrange the flood discharge series of n terms in descending order.
2. Convert this series into logarithmic series, taking the base as 10.
3. For this log series, find out the mean, $\bar{y}$; standard deviation, sx and the skew coefficient, g.
4. For the required return period T_r and g, find out the value of K from Table 8.3.
5. Find y_T from Eq. 8.10.
6. Obtain the value of Q_T from equation 8.11.

Q_T is the discharge with the required return period T_r.

Example 8.9

For the data given in Example 8.6, find the discharge with a return period of 50 years using 1. Gumbell's distribution 2. Log-Pearson Type III distribution.

Solution:

1. **Gumbell's distribution**

From the data, the following were computed:

(a) Number of terms in the series $n = 32$
(b) Mean discharge $\bar{Q} = 535 \text{ m}^3/\text{s}$
(c) Standard deviation $sx = 162.93$
(d) From Table 8.2, for $n = 32$ $\sigma n = 1.1193$
and $yn = 0.5380$

Now, $$y_T = -\left\{0.834 + 2.303 \log_{10}\left(\log_{10}\frac{T_r}{T_r - 1}\right)\right\}$$

Substituting the value of $T_r = 50$,

We get,

$$y_T = -\left\{0.834 + 2.303 \log_{10}\left(\log_{10}\frac{50}{50-1}\right)\right\}$$
$$= 3.9028$$

Now, $$Q_T = \bar{Q} + sx\left\{\frac{(y_T - yn)}{\sigma n}\right\}$$

Substituting the values of sx, y_T, yn and σn,

We get, $$Q_T = 535 + 162.93\left\{\frac{(3.9028 - 0.5380)}{1.1193}\right\} = 535 + 490$$
$$= 1025 \text{ m}^3/\text{s}.$$

2. **Log-Pearson Type III distribution**

The original series of 32 discharges was arranged in the descending order. This was then converted into logarithm series taking the base as 10 and tabulated in Table C.

Table C

Serial no.	Q (m^3/s)	Q (m^3/s) in descending order	$\log_{10}Q$ (m^3/s)
1	395	887	2.9479
2	619	814	2.9106
3	766	766	2.8842
4	422	726	2.8609
5	282	721	2.8579
6	887	705	2.8481
7	528	697	2.8432
8	785	686	2.8363
9	520	634	2.8020
10	436	632	2.8007
11	697	624	2.7951
12	624	619	2.7916
13	496	598	2.7767
14	589	589	2.7701
15	598	528	2.7226
16	359	527	2.7218
17	686	520	2.7160
18	726	496	2.6954
19	527	464	2.6665
20	310	459	2.6618
21	408	440	2.6434
22	721	438	2.6394
23	814	422	2.6253
24	459	408	2.6106
25	440	395	2.5965
26	632	373	2.5717
27	343	371	2.5693
28	634	359	2.5550
29	464	343	2.5352
30	373	310	2.4913
31	289	289	2.4608
32	377	282	2.4502

Mean $\log_{10} Q = 2.7080$, Standard deviation, $sx = 0.1377$

From Table A, we get,

(a) Number of terms, $n = 32$

(b) Mean log Q, $\overline{Q} = 2.7080$

(c) Standard deviation, $sx = 0.1377$

(d) Coefficient of skew,

$$g = \frac{n\Sigma\,(y - \bar{y})^3}{(n - 1)\,(n - 2)\,(sx)^3}$$

$$= \frac{32 \times (-0.0148)}{31 \times 30 \times 0.002610}$$

$$= -\,0.195$$

For $T_r = 50$ and $g = -0.195, K = 1.946$

Now,

$$y_T = \bar{y} + K \times sx = 2.7080 + (1.946 \times 0.1377) = 2.9759$$

$$Q_T = \text{Antilog } 2.9759 = 946 \text{ m}^3/\text{s}.$$

8.7 RISK

Any hydraulic structure is designed for a discharge having a probability of p. So also this hydraulic structure is expected to have a useful life of N years. There is a probability that this discharge or more than that may occur in the lifetime of the structure. Thus, there is some sort of risk involved in this.

This risk R is given as follows:

$$R = 1 - (1 - p)^N$$

or,
$$R = 1 - \left(1 - \frac{1}{T_r}\right)^N$$

On percentage basis, $R = [1 - (1 - p)^N] \times 100$

or,
$$R = \left[1 - \left(1 - \frac{1}{T_r}\right)^N\right] \times 100$$

Naturally, in practice the acceptable risk is decided by the economic and policy considerations.

And reliability will be $= 1 - R = (1 - p)^N = \left(1 - \frac{1}{T_r}\right)^N$

Example 8.10

A cofferdam is designed for a return period of 30 years. It will require 3 years to complete the dam. Find

1. The risk that the cofferdam may get washed away.
2. If the risk is expected to be only 5%, what will be the return period for which the cofferdam should be designed?

Solution:

1. $R = 1 - \left(1 - \frac{1}{T_r}\right)^N = 1 - \left(1 - \frac{1}{30}\right)^3 = 8.74\%.$

2. $0.05 = 1 - \left(1 - \frac{1}{T_r}\right)^3$

Therefore, $T_r = 59$ years

REVIEW QUESTIONS

1. What is flood? Discuss the various features of flood.
2. Discuss the various causes of floods.
3. What are the factors affecting flood?
4. How floods are classified?
5. When there are no observed data available, how the flood is estimated at a river outlet?
6. What is an envelop curve? How is it used to estimate for estimating flood at a site?
7. Which are the different empirical formulae used to estimate the flood discharge?
   Write a note on the rational formula used to estimate the flood at a site.
9. Explain how the unit hydrograph theory is used to estimate flood discharge from a catchment area.
10. Differentiate between annual flood series and partial duration flood series.
11. Explain:
    a. Recurrence interval
    b. Probability
    c. Frequency of floods
12. Explain the procedure to evaluate the probability of a specific discharge in an observed discharge series. Discuss the various formulae used to asses these.
13. The probability of a specific discharge, which is more than the observed discharge series, is to be calculated. Discuss the various methods.
14. Describe the method of estimating a T_r-year flood using Gumbell's distribution.
15. Describe the procedure of estimating a T_r-year flood using Log-Pearson type distribution.
16. How is the risk in the design that a structure may fail in its lifetime calculated?
17. Write short notes on:
    a. Frequency factor
    b. Rank of observation
    c. N-year flood
    d. Effect of the shape of catchment on the flood
    e. Gumbell probability paper

NUMERICAL QUESTIONS

1. In a village, the maximum water level reached on the wall of the school building is 105 m. The cross section of the river is in the form of a symmetrical triangle with vertical angle 90° and apex at a level of 100 m.
   Assuming the slope of the river to be 1 in 3600 and the Manning's coefficient to be 0.035, find the discharge corresponding to this maximum water reached.

Ans: 17.38 m^3/s

2. The catchment area of a proposed weir is 80 km^2. The maximum flood discharges in the adjoining area with hydrological homogeneous catchments are as follows:

Catchment	A	B	C	D	E	F
Catchment area (km^2)	103	176	48	141	72	220
Flood discharge (m^3/s)	1122	1478	380	1362	1007	1645

Evaluate the maximum flood discharge at the proposed weir.

Ans: 1020 m^3/s

3. Using the empirical formulae, find the maximum flood discharge for a catchment of 100 km^2, having the maximum intensity of precipitation of 4.0 cm/s.
Ans: Inglis formula = 1180 m^3/s; Dicken's formula = 790 m^3/s;
Rational formula = 388 m^3/s; Ali Nawaz Jung Bahadur formula = 979 m^3/s

4. For a weir, the design discharge of 500 m^3/s has a return period of 30 years.
Find: (i) This discharge occurs twice in 25 years
(ii) This discharge occurs once in 25 years
(iii) This discharge will not occur at all in 20 years

Ans: (i) 12.7%, (ii) 56%, (iii) 51%

5. The maximum flood discharges in m^3/s observed at a proposed dam site for 1985–2004 (20 years) are as follows:

825, 480, 713, 671, 451, 870, 462, 648, 601, 425, 890, 620, 540, 580, 705, 465, 790, 680, 733, 540

Find the discharge with a return period of 25 years using
(i) Gumbell's distribution
(ii) Log-Pearson type III distribution

Ans: (i) 1020 m^3/s, (ii) 931 m^3/s

6. A cofferdam is a design period of 25 years. It will require four rears to complete the dam.
Find: (i) The risk that the cofferdam may get washed away
(ii) If the risk is expected to be only 5%, the return period for which the cofferdam should be designed

Ans: (i) 15% (ii) T_r = 76 years

MULTIPLE CHOICE QUESTIONS

1. As per Weibull's formula, the return period is
(a) $\frac{n}{m}$
(b) $\frac{n}{(m+1)}$
(c) $\frac{n}{(m-1)}$
(d) $\frac{(n+1)}{m}$

2. As per Hazen's formula, the return period is
   (a) $\frac{n}{m}$ (b) $\frac{2n}{(2m-1)}$
   (c) $\frac{(2n+1)}{m}$ (d) $\frac{n}{(2m+1)}$
3. As per California formula, the return period is
   (a) $\frac{n}{m}$ (b) $\frac{2n}{(2m-1)}$
   (c) $\frac{(2n+1)}{m}$ (d) $\frac{n}{(2m+1)}$
4. As per Gumbell's formula, the return period is
   (a) $\frac{(n+C+1)}{m}$ (b) $\frac{(n+C-1)}{m}$
   (c) $\frac{n}{(m+C-1)}$ (d) $\frac{n}{(m+C+1)}$
5. The most commonly used formula for calculating the return period is
   (a) California formula (b) Hazen's formula
   (c) Weibul's formula (c) Gumbell's formula
6. A flood with a return period of 100 years is the flood that occurs
   (a) After every 100 years (b) Once in 100 years
   (c) Only after 100 years in the future (d) None of these
7. The probability that a flood of return period of T_r occurs in a year is
   (a) $\frac{1}{T_r} - 1$ (b) $\frac{1}{T_r}$
   (c) $\frac{1}{T_r} + 1$ (d) $1 - \frac{1}{T_r}$
8. A structure having a life of N years is designed for a return period of T_r years. The probability that it will not fail during its life period is
   (a) $1 - \frac{1}{T_r}$ (b) $1 + \frac{1}{T_r}$
   (c) $(1 - \frac{1}{T_r})^N$ (d) $(1 + \frac{1}{T_r})^N$
9. Dicken's formula for calculating the maximum flood discharge from a catchment is given by
   (a) $Q = CA^{4/3}$ (b) $Q = CA^{3/4}$
   (c) $Q = CA^{2/3}$ (d) $Q = CA^{3/2}$
10. Inglis formula for calculating the maximum flood discharge from a catchment is given by
   (a) $Q = \frac{124\,(A + 10.4)}{\sqrt{A}}$ (b) $Q = \frac{124\,(A - 10.4)}{\sqrt{A}}$
   (c) $Q = \frac{124\,A}{\sqrt{(A + 10.4)}}$ (d) $Q = \frac{124\,A}{\sqrt{(A - 10.4)}}$

11. As per Gumbell's distribution, the probability of occurrence of a discharge figure is given by

(a) $p = 1 + e^{-e^{+y}}$ (b) $p = 1 - e^{-e^{+y}}$

(c) $p = 1 + e^{-e^{-y}}$ (d) $p = 1 - e^{-e^{-y}}$

12. For a hydraulic structure designed for a return period of T_r years and having a useful life of N years, the risk that this flood may occur during the useful life is

(a) $R = \left[1 - \left(1 + \frac{1}{T_r}\right)^N\right] \times 100$ (b) $R = \left[1 - \left(1 - \frac{1}{T_r}\right)^N\right] \times 100$

(c) $R = \left[1 + \left(1 + \frac{1}{T_r}\right)^N\right] \times 100$ (d) $R = \left[1 + \left(1 - \frac{1}{T_r}\right)^N\right] \times 100$

ANSWERS TO MULTIPLE CHOICE QUESTIONS

1. (d) 2. (b) 3. (a) 4. (c) 5. (c) 6. (b) 7. (b) 8. (c)
9. (b) 10. (c) 11. (d) 12. (b)

*** ——— ***

9

Discharge Measurement

Chapter Outline

9.1 DEFINITION

Discharge is defined as *the rate of flow,* i.e. *volume flowing per unit time.*

9.1.1 Importance of Discharge Measurement

All over the world, there is always a demand for surface water resource and hence it is essential to assess the available water resource as accurately as possible. It can be assessed theoretically by following some empirical formulae, but this will not be accurate. Even for the derivation of an empirical formula, observed discharge data are required.

The surface water resource can be evaluated by measuring discharge in a stream continuously over a period. The surface water resource available at a site can be evaluated by scanning the observed data. This will give more realistic estimate and may involve less error.

9.1.2 Units of Discharge

The normal unit of discharge is m^3/s. The term *Cumecs* is also used instead of m^3/s. If the discharge is small as in a laboratory, then the unit lit/s or cc/s is used. The discharge from a specific catchment is sometimes expressed as *Discharge Per Unit Catchment Area* as $m^3/s/km^2 = m/s$, per unit area of the catchment.

Thus, a discharge of 1 mm/h from a catchment area of 25 km^2 will be

$$\frac{(1/1{,}000)}{3600} \times (25\times10^6) = 6.9 \text{ m}^3/\text{s}$$

9.1.3 Objectives of Discharge Measurement

The objectives of discharge measurement are as follows:

1. To assess the available surface water accurately from the catchment area of a project—may be for irrigation, hydroelectricity generation, flood control or water supply
2. To study the relation between precipitation and runoff
3. To study the variation in runoff
4. To evaluate the maximum water level that may be reached in the case of a bridge or a culvert for the largest possible flood by extrapolation from the observed stage-discharge record
5. To study the regeneration of water due to irrigation on the banks of the river

9.1.4 The Basic Principle

The basic principle followed in all the discharge measurement methods is the continuity equation

$$Q = A\times V$$

where, Q = Discharge in m^3/s

A = Area of cross section in m^2 at right angles to the velocity of flow

V = Average velocity of flow in m/s

The two parameters A and V are measured and the discharge is calculated.

In recent years, some modern techniques are used so that the discharge flowing in a stream is calculated without measuring the velocity and the area of flow.

9.1.5 Water Potential

Water potential is the volume of water at a site during a specific time, say, a year, a season or even a month.

9.1.6 Units of Water Potential

The normal units of water potential are m^3 or million m^3 (10^6 m^3), normally denoted by million m^3 or ha m (10^4 m^3).

$$1 \text{ m}^3/\text{s a day} = 86{,}400 \text{ m}^3 = 0.0864 \text{ million m}^3 = 8.64 \text{ ha m}$$

The water potential from a catchment area is sometimes mentioned as m/km^2, i.e. the potential per unit area of the catchment area. It means that so many metres of water depth are uniformly spread over the entire catchment. If the water potential from a catchment area of 100 km^2 is 2.00 m, the potential in usual volume terms will be:

$$2 \times 100 \times 10^6 = 200 \times 10^6 \text{ m}^3$$

$$= 200 \text{ million m}^3$$

$$= 2 \times 10^4 \text{ ha m}$$

9.1.7 The Stream

A stream is that through which water flows naturally. When the stream is small, it is also called *runnel, bróok, bourne, nalla* or *rivulet.* A river is termed for a stream when the discharge is comparatively more. Channel is a term also used for a river. Canal is a term normally used for a man-made channel.

9.2 THE STAGE OF A RIVER

The depth of water flowing in a river is known as *The Stage of the River* at that time. The stage of the channel is measured because from this observation, the area of flow can be calculated. The stage of the river is expressed in metres with datum as mean sea level. However, sometimes it is also mentioned with datum as lowest bed level.

9.2.1 Measurement of a Stage

Different methods are followed to measure the stage of the channel depending upon the channel, whether small or large. These are as follows:

- Actual wading through the stream with a staff gauge
- Fixing gauges along the cross section
- By suspending a gauge from a structure
- Recording type of gauge
- Automatic water-stage recorder

9.2.1.1 Wading the stream

A person with a gauge in hand may move across the stream and record the depth of flow at a location by the gauge in his hand. This method is crude and may be adopted if the stream is small. The person standing in the stream may cause an obstruction to the flow and may affect the gauge reading. The stage thus recorded will be at a specific location and at a specific time.

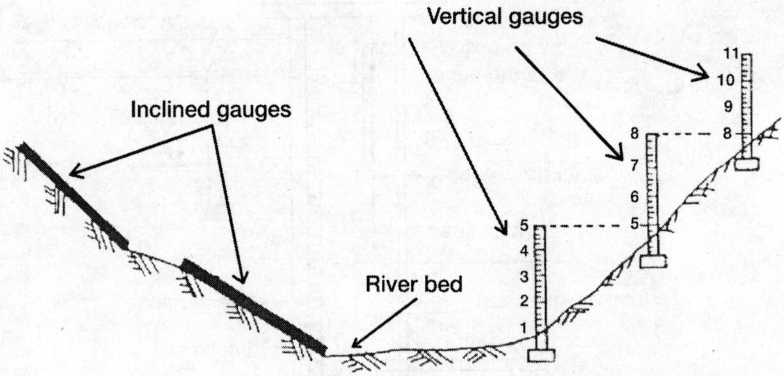

Fig. 9.1 Vertical staff gauges and inclined gauges.

9.2.1.2 Fixing gauges

The stage of a stream can be measured by installing gauges in the stream.

Vertical staff gauges may be fixed along the cross section of the river on one bank and may be correlated to each other with a minimum overlap of 0.5 m between two successive staves as shown in Fig. 9.1.

However, these gauges form an obstruction to the flow. Alternately the gauges may be fixed or painted on a bridge pier, a culvert or any other hydraulic structure, if available.

Inclined gauges

Gauges along the slope of the cross section of the stream may be fixed as shown in Fig. 9.1. While marking the depth of flow on the gauge, the side slope of the stream has to be taken into consideration. Such gauges will not cause any obstruction to the flow.

In both the cases, the water level will have to be observed from the banks of the stream. Knowing the water level and the bed level of the stream, the depth of flow can be evaluated.

By following this procedure, the stage of the river can be found out at that moment. However, the maximum and the minimum stages reached cannot be found out unless observed at those instances.

9.2.1.3 Suspension weight method

In this method, a weight attached to a rope may be lowered from a bridge, a culvert or any hydraulic structure, until it touches the water surface. The length of the rope touching the water surface from a specific reference point may be read.

This will give the vertical length of the water surface from the specific suspension point. Knowing the vertical length of the rope up to the bed of the stream when it is dry, the stage of the stream can be found out.

9.2.1.4 Water-stage recorder

A stilling well may be constructed on one of the banks of the stream as shown in Fig. 9.2. The well may be square or circular, uniform in cross section of suitable cross-sectional area connected to the stream by horizontal pipes at different levels at right angles to the flow to avoid the effect of the velocity of flow. The diameter of these pipes be such that the floating material will not choke the pipes.

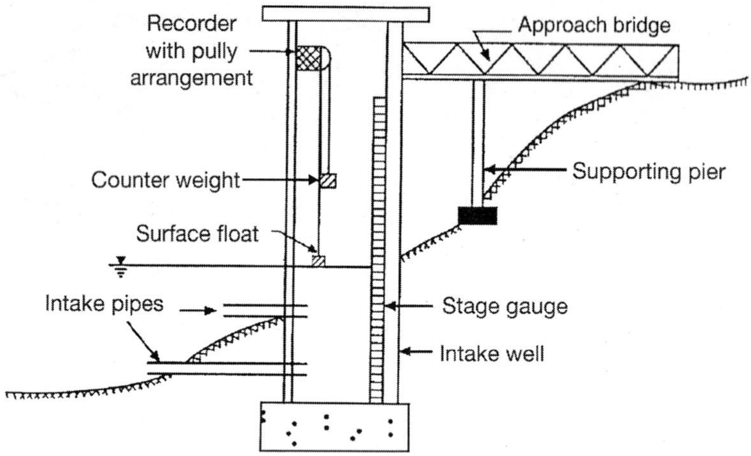

Fig. 9.2 Water stage recorder

- These pipes should be cleaned regularly.
- A vertical gauge may be painted or fixed on the inside of one of the walls.

Thus, the water level in the stream will be transferred to the water column in the well through the pipes without any effect of the velocity of flow. The water level in the well can be comfortably read since the oscillations in the river will be dampened and a steady water level in the well be noticed. Knowing the water level and the bed of the stream, the stage of the river can be worked out.

9.2.1.5 Automatic water level recorder

The water level in a stilling well is automatically recorded by the following methods:

1. Float method
2. Electrical resistance method

Float-type automatic water level recorder

In this method, a float is used to identify the water level in the well. The level indicated by the float and the time is transferred by a suitable mechanism by a pointer on a graph paper wound on a drum. The drum is kept rotating uniformly by a clock work so that it makes one rotation a day or a week. This type of mechanism is similar to the one used in automatic rain gauge. This graph paper has to be changed every day or week as the case may be. The minimum and maximum water levels reached and the time when reached are automatically recorded.

Electric resistance-type water level recorder

In this method, two electrodes are dipped vertically in the observation well keeping the distance between these two electrodes constant. An electric current flows between these two electrodes. The resistance between these two electrodes varies due to the change in water level.

The change in resistance will be a measure of the water level and is transferred by a suitable mechanism to a graph paper mounted on a drum by a pen pointer. This drum is kept rotating uniformly by a clock work and it makes a rotation a day or a week. This graph paper will have to be changed every day or a week as the case may be. The maximum and minimum water levels reached and the time when reached are automatically recorded.

In recent years, the water levels thus recorded are transferred by a wireless arrangement to a central recording station, where all meteorological data in that region are collected.

Crest gauge

When it is not possible to install an automatic water level recorder due to any reason, a crest gauge may be installed in the observation well. This crest gauge will record the maximum water level reached. This can be done by two ways:

1. It consists of a small float that rises as the water level rises in the well but will not move down as the water level in the well reduces. Thus, the maximum water level reached in the well is recorded.
2. One of the sides of the well or a bridge pier may be painted by a water-soluble paint, of course, well protected from rain. As the water level rises, the portion of the paint coming in contact with the water will get washed away and thus the maximum water level reached can be obtained.

9.3 MEASUREMENT OF DISCHARGE

The different methods followed to measure discharge in a stream are as follows:

- Area–slope method
- Area–velocity method
- Salt titration method
- Measurement by hydraulic structures
- Hydraulic model method
- Modern methods

However, in spite of the development in different sciences and in instrumentation, it is still not possible to measure the discharge in a stream accurately.

9.3.1 Selection of a Site

For the measurement of discharge by any method, certain field observations are required to be taken at a suitable site on the stream. Following are the requirements of a site for the measurement of discharge:

- The reach of the stream should be a straight one. No other stream should join the stream in this reach nor some discharge withdrawn from it.
- The reach should be preferably three to four times the width of the river during flood.
- The cross section of the stream should be uniform with no rock outcrops, vegetal growth or any other obstruction.
- The site should be accessible in all seasons.
- The site should be free from any disturbance.

9.3.2 Cross-sectional Area of a Stream

For the evaluation of discharge, the cross-sectional area of the stream is required to be calculated. The cross-sectional area of a stream is calculated by the following two methods:

(1) accurate survey and (2) echo-sounder.

9.3.3 Evaluation of a Cross-Sectional Area

After the site for the measurement of discharge is finalized, two cross sections of the stream are selected at a distance of 50–100 m. Precise survey is carried out for both the cross-sections when the stream is dry or the depth of flow is negligible.

If the depth of flow is substantial, then hydrographic survey is carried out. Based on this precise survey done, two graphs as shown in Figs. 9.3 and 9.4 are plotted. From these two graphs, the average values of area and wetted perimeter of the two cross sections for a known observed water level can be read.

These two graphs are further used to evaluate the area of flow and wetted perimeter for any known depth of flow. There is quite a possibility that the cross sections may change due to silting or scouring. Hence, precise survey is repeated regularly and the two graphs modified, if required.

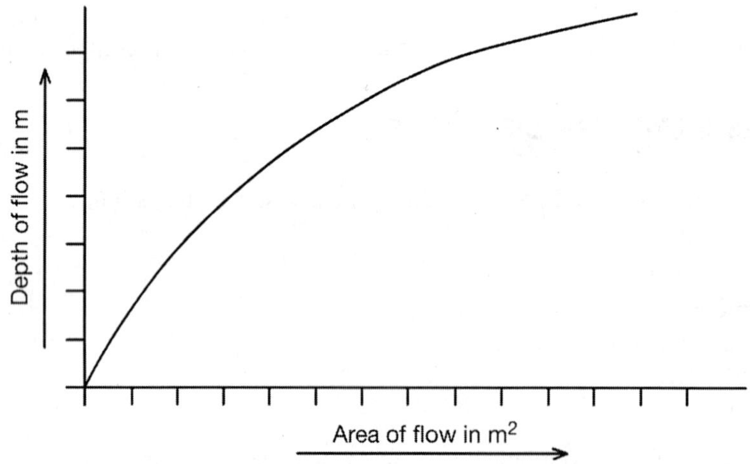

Fig. 9.3 Depth versus area flow

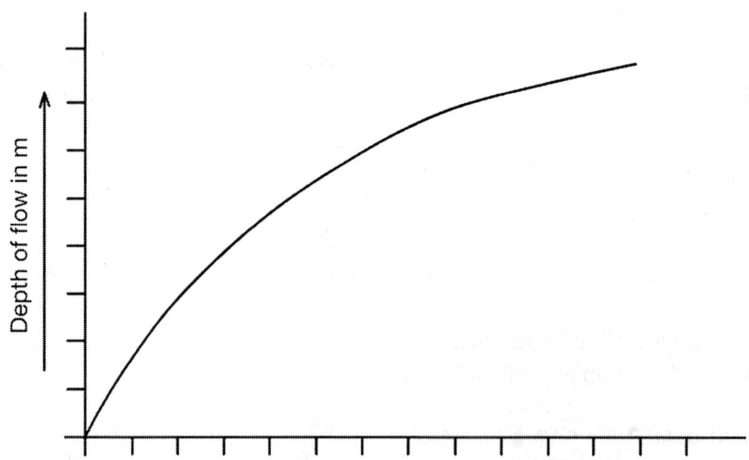

Fig. 9.4 Depth versus wetted perimeter

Example 9.1

After doing precise survey, the cross section of the stream at a gauging site is as shown in Fig. 9.5. Prepare the two graphs: (1) depth vs area and (2) depth vs hydraulic mean depth.

Solution:

The cross section was divided into five segments each of 1-m height and the area of each segment was measured accurately. So also the length of the cross section coming in contact with the water on either side was measured and tabulated.

Serial no.	Depth (m)	Segment	Area (m^2)	Σ Area (m^2)	P_L (m)	P_R(m)	$\Sigma (P_L + P_R)$ (m)	HMD(m)
1	1.0	A1	5.0	5.0	—	9.9	9.9	0.505
2	2.0	A2	9.0	14.0	2.2	2.0	14.1	0.992
3	3.0	A3	13.5	27.5	1.8	1.7	17.6	1.560
4	4.0	A4	20.5	48.0	1.9	1.7	21.2	2.264
5	5.0	A5	24.0	72.0	1.8	1.6	24.6	2.926

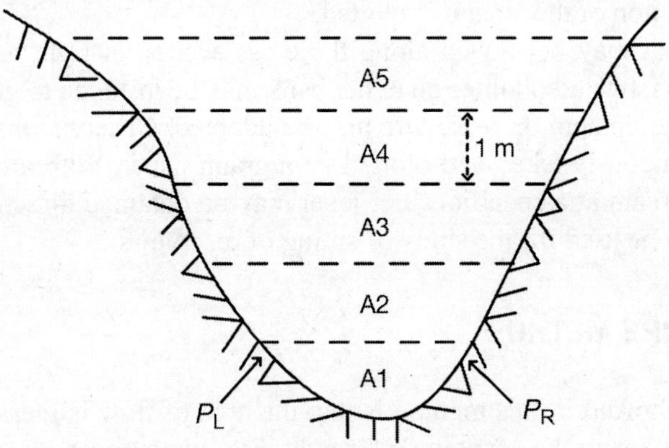

Fig. 9.5 Cross section of river

$$\text{Here, HMD} = \frac{\Sigma \text{Area}}{\Sigma(P_L + P_R)}$$

Two graphs are drawn as (1) Depth vs Hydraulic Mean Deapth (HMD) and (2) Depth vs area, on simple graph papers as shown in Fig. 9.6.

9.3.4 Echo Sounder

When the stream is large and the depth of flow is substantial, it may not be possible to carry out precise survey nor hydrographic survey, then an *Echo Sounder* is used to evaluate the depth of flow and in turn the entire cross section.

An echo sounder is an equipment normally fitted to a boat. Through this equipment, sonic pulses of short duration are transmitted from the water surface by electrodes dipped in the water.

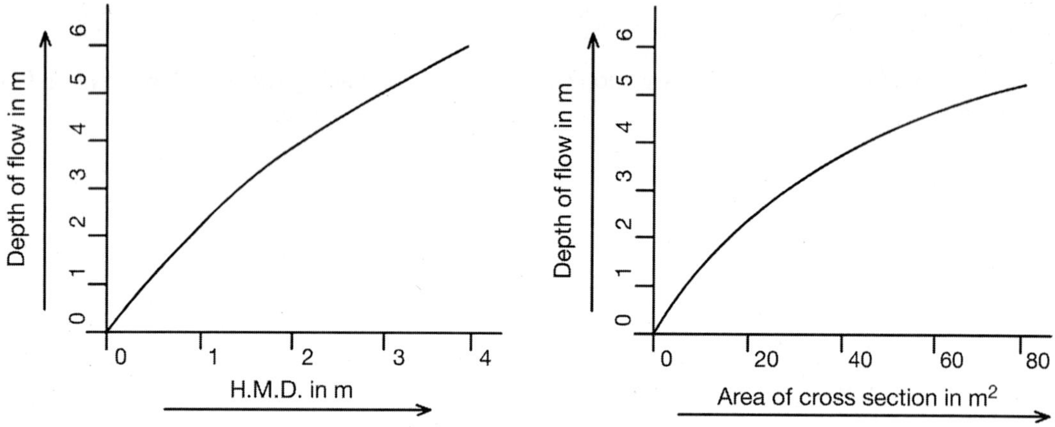

Fig. 9.6 Details of river cross section

These pulses are reflected back by the bed of the channel. The time of dispatch and receipt of pulses is recorded. The depth of flow then can be worked out from this time interval and the speed of the sonic pulse in water. Hence, the name of the equipment is 'Echo sounder'.

Such observations are taken at different locations on the cross section preferably at equal distance and the cross section of the stream is plotted.

Alternately, the boat may be moved along the cross section at right angles to the flow at a constant speed. For this, two theodolites on either bank may be installed to guide the boat properly, or any other ranging equipment or procedure may be adopted. Observations of depth by the echo sounder may be continuously taken and plotted on a graph paper. With a continuous plotter, the cross section of the stream at a specific water level may be obtained directly on the graph paper. Echo sounder can also be used for the study of silting of reservoirs.

9.4 AREA–SLOPE METHOD

The basic principle followed in this method is that the area of flow is measured and the velocity of flow is evaluated by using the Manning's formula. For this purpose, the water surface slope is observed and the Manning's 'n' is assumed. Then the continuity equation, $Q = A \times V$ is used.

where, Q = Discharge in m^3/s
A = Area of cross section in m^2 at right angles to the velocity of flow
V = Velocity of flow in m/s

The procedure to be followed is as follows:

1. Select two cross sections of the stream at a distance of about 150 m. For the selection of this reach, the criteria as explained in Section 9.3.1 are followed.
   It is assumed that steady uniform flow exists between these two cross sections.
2. Observe the depths of flow at these two cross sections as well as the water surface level accurately between these two cross sections.
3. Find the area of flow and also the wetted perimeter at these two cross sections from the predetermined stage vs area and stage vs wetted perimeter curves.

4. Find the mean area of flow Am and the mean wetted perimeter Wp.
5. From the water levels at the two cross sections, find accurately S, i.e. the slope of the water surface. (This will be the slope of energy also, since 'steady uniform flow' is assumed between these two cross sections.)
6. Calculate the mean hydraulic radius R_m as

$$R_m = \frac{A_m}{W_p}$$

7. Assume a suitable value of the Manning's n for the reach of the stream between these two cross sections.
8. Calculate the average velocity of flow using the Manning's formula as

$$V_m = \frac{(R_m^{2/3} \times S^{1/2})}{n}$$

where, V_m = Average velocity of flow in m/s
R_m = Mean hydraulic radius thus calculated in m
S = Energy slope
n = Assumed value of Manning's coefficient

9. Calculate the discharge as

$$Q = V_m \times A_m$$

The discharge is calculated by using the area and the slope of the water surface and hence it is known as *Area–slope Method.* To achieve accuracy, the procedure may be repeated and the average value of discharge of all the trials may be finalized.

9.4.1 Merits and Demerits of the Area–Slope Method

The merits of the area–slope method are as follows:

(a) This method is based on only two or three observations at site.
(b) Time required to complete the procedure as compared to other methods is less.
(c) A team of two or three observers is adequate to conduct the test.

The demerits of the area–slope method are as follows:

(a) The assumed value of n may not be very correct and may lead to erroneous result.
(b) The method is based on the assumption that 'steady uniform flow' exists between the two cross sections. This may not be correct, since during flood, the flow in a river is neither steady nor uniform.
(c) It is very difficult to observe the water surface levels accurately, since there are oscillations in the water surface.

Example 9.2

While taking discharge observations on a stream by area–slope method, the following observations were taken:

1. Distance between two observation sections of the stream = 250 m
2. Depth of flow at the u/s cross section A = 3.50 m

3. Depth of flow at the d/s cross section B = 3.50 m
4. Water level at the u/s cross section A = 236.62 m
5. Water level at the d/s cross section B = 236.37 m
6. Manning's coefficient (assumed) = 0.030

Find the discharge flowing in the stream.

Solution:

From the graphs of depth vs area of cross section and depth vs wetted perimeter, the area of cross section and the wetted perimeter at both A and B were evaluated as follows:

Cross section	Area (m²)	Wetted perimeter (m)
A	110	76.2
B	108	75.8

$$\text{Therefore, Mean area} = \frac{110 + 108}{2} = 109 \text{ m}^2$$

$$\text{Mean wetted perimeter} = \frac{76.2 + 75.8}{2} = 76.0 \text{ m}$$

$$\text{HMD} = R_m = \frac{109}{76} = 1.434 \text{ m}$$

$$\text{Therefore, } R_m^{2/3} = 1.434^{2/3} = 1.273$$

$$\text{Water surface slope} = \frac{236.62 - 236.37}{250} = \frac{0.25}{250} = 1 \text{ in } 1000$$

$$\text{Therefore, } Q = \frac{A_m \times R_m^{2/3} \times S^{1/2}}{n} = \frac{109 \times 1.273 \times 1}{0.030 \times \sqrt{1000}} = 146.26 \text{ m}^3/\text{s.}$$

9.5 AREA–VELOCITY METHOD

In this method also the basic continuity equation, $Q = A \times V$, is used to estimate the discharge flowing in a stream. The values of A and V are assessed separately and then the discharge is calculated.

9.5.1 Area of Flow

The area of flow is assessed in a similar way as that for area–slope method discussed earlier.

9.5.2 Velocity of Flow

The velocity of flow is not uniform over the depth. It is very low (theoretically zero) at the bottom. It increases with the depth. However, it slightly reduces at the surface. The normal velocity profile is shown in Fig. 9.7.

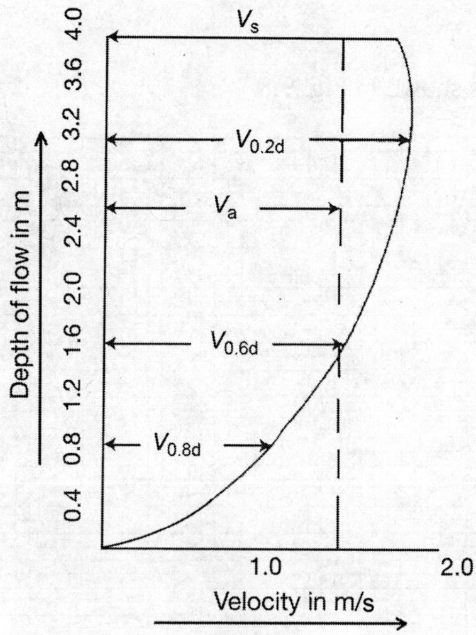

$$V_a = 0.85\ Vs$$
$$= V_{0.6d}$$
$$= \frac{V_{0.2d} + V_{0.8d}}{2}$$
$$= \frac{V_{0.2d} + 2V_{0.6d} + V_{0.8d}}{4}$$

Fig. 9.7 Normal vertical velocity profile in a stream

From this velocity profile, the average velocity is correlated as under:

$$V_a = \text{Average velocity} = 0.85\ V_s\ (V_s = \text{Velocity at the surface})$$
$$= V_{0.6D}{}^*$$
$$= \frac{(V_{0.2D} + V_{0.8D})}{2}$$
$$= \frac{(V_{0.2D} + 2V_{0.6D} + V_{0.8D})}{4}$$

*0.6D from the water surface, i.e. 0.4D from the bed of the channel.

Example 9.3

Observations for velocity of flow taken over a vertical of a stream at various depths are as follows.

Depth from water surface (m)	0	0.5	1.0	1.5	2.0	2.5	3.0	3.5	4.0	4.5	5.0
Velocity of flow (m/s)	1.47	1.49	1.52	1.48	1.41	1.34	1.24	1.14	1.00	0.82	0.00

Find the average velocity. Plot the velocity profile, and derive an equation for the velocity profile.

Solution:

The velocity profile will be as shown in Fig. 9.8.

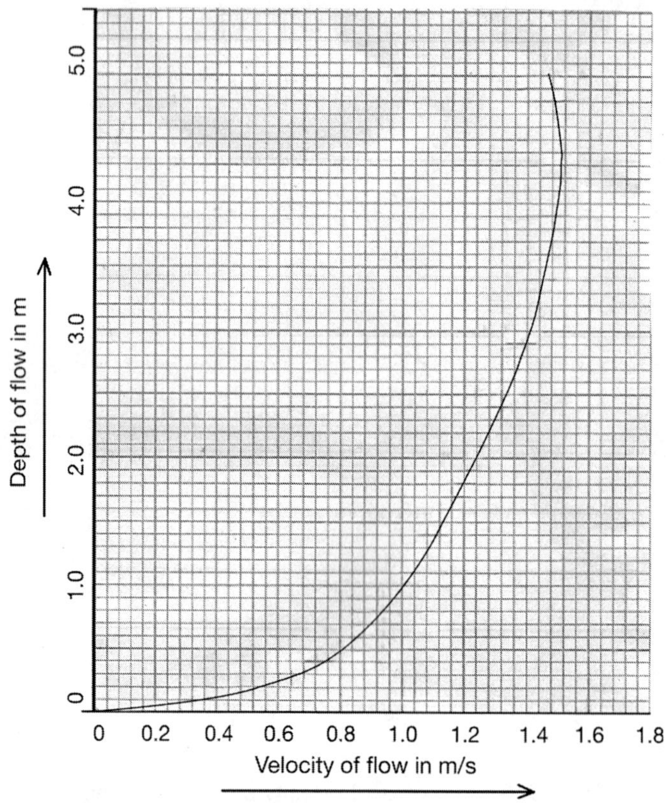

Fig. 9.8 Velocity profile

1. $V_m = \sum \frac{V}{n}$

$$= \frac{1.47 + 1.49 + 1.52 + 1.48 + 1.41 + 1.34 + 1.24 + 1.14 + 1.00 + 0.82 + 0.00}{11}$$

$$= \frac{12.91}{11} = 1.17 \text{ m/s}$$

$$V_m = \frac{(V_{0.2D} + V_{0.8D})}{2} = \frac{1.52+1.00}{2} = \frac{2.52}{2} = 1.26 \text{ m/s}$$

$$V_m = \frac{(V_{0.2D} + 2 \times V_{0.6D} + V_{0.8D})}{4} = \frac{1.52+(2\times1.24)+1.00}{4} = \frac{5.00}{4} = 1.25 \text{ m/s}$$

$$V_m = V_{0.6D} = 1.24 \text{ m/s}$$

2. Let the equation of the velocity profile be $V = d^n$.

Therefore, $\log V = n \log d$

or, $$n = \frac{\log V}{\log d}$$

The value of n for three observations will be as follows.

Serial no.	V (m/s)	d (m)	log V	log d	n
1	0.82	0.5	−0.1984	−0.6930	0.286
2	1.14	1.5	0.1310	0.4054	0.323
3	1.48	3.5	0.3920	1.2527	0.313
				Total	0.922

Therefore, the average value of $n = 0.922/3 = 0.307$.
The equation of the velocity profile will be $V = d^{0.307}$

9.5.3 Evaluation of Velocity of Flow

The velocity of flow is evaluated by anyone of the following methods:
(1) float method (2) current meter method and (3) pitot tube method

9.5.3.1 Float method

The velocity of flow is evaluated by a float floating over the water surface, known as *Surface Float*. When a float is dropped on a flowing water surface, it moves with the same velocity at the surface, i.e. V_s (Fig. 9.9).

The procedure followed to assess the surface velocity by a float is as follows:

- Two cross sections PP′ and QQ′ are selected at a distance L from each other (say about 150 m). The criteria for selection of site given in Section 9.3.1 also applies here.
- Two vertical posts PP′ and QQ′ are erected on either bank of the two cross sections at right angles to the flow.
- The water surface along the cross section P-P′ on the u/s side is divided into different equal segments depending upon:
  (a) Width of the river
  (b) Variation in depth
  (c) Accuracy of discharge measurement required
- The cross sections are plotted on a paper and the segments thus decided are marked on each of the cross section. The area of each of the segments from the two cross sections is measured. The average area of the corresponding two segments from the two cross sections is calculated such as $A_1, A_2, A_3, A_4, \ldots$
- Two observers having stopwatches are assigned to one of the banks of the two cross sections. The stopwatches with each of the observers are started simultaneously, and then the observers move to their respective appointed cross sections.
- Floats are dropped from the bank sufficiently u/s of cross section P-P′ such that the float reaches the centre of the desired segment as far as possible.

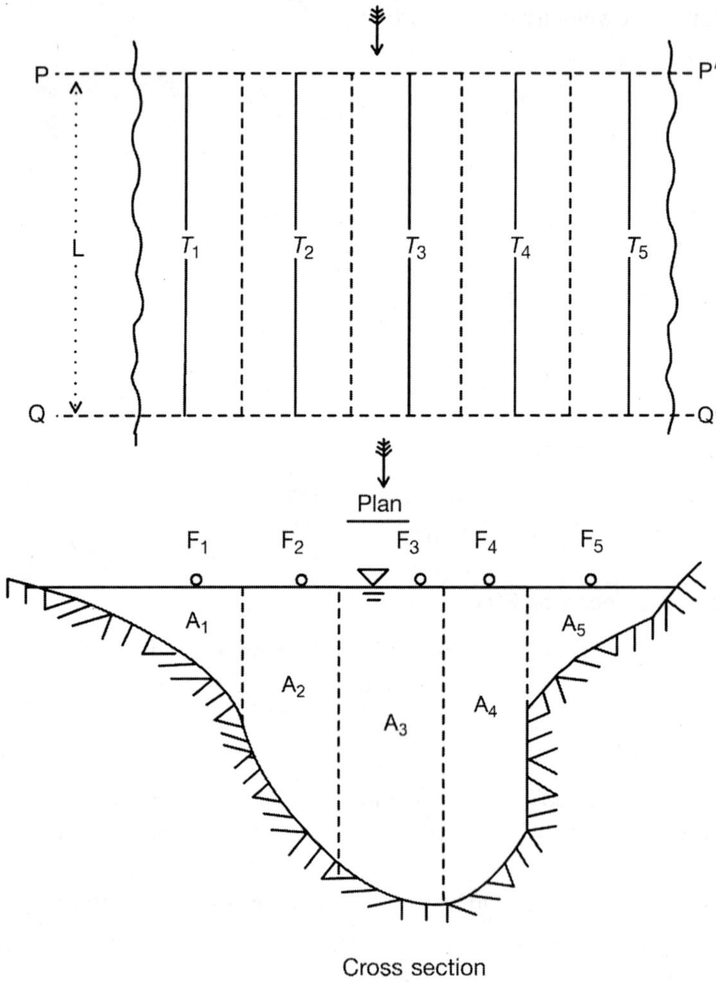

Fig. 9.9 Area–velocity method by float

- As the float crosses the cross sections P-P′ and Q-Q′, the observers at these cross sections should stop their respective stopwatches. Thus, the time of travel from the first cross section P-P′ to the next Q-Q′, i.e. to travel a distance L can be calculated.

Let the time taken by the float to travel each of the segment be T_1, T_2, T_3,.... Then the surface velocity in each of the segment will be

$$V_{s1} = \frac{L}{T_1}; \quad V_{s2} = \frac{L}{T_2}; \quad V_{s3} = \frac{L}{T_3}$$

It may be noted that T_1, T_2, T_3,... will not be equal, since the surface velocity over a cross section is maximum at the centre and goes on decreasing towards the banks.

Normally, the average velocity over a depth is less than the surface velocity, i.e.

$$V_a = 0.85\ V_s.$$

- The average velocity in each segment, V_{a1}, V_{a2}, V_{a3},... is calculated and then the discharge in the stream is calculated as:

$$Q = Q_1 + Q_2 + Q_3 + \ldots$$

$$= 0.85\,(A_1 V_{s1} + A_2 V_{s2} + A_3 V_{s3} + \ldots)$$

9.5.3.2 Types of floats

The floats normally used are:(a) surface float, (b) canister float and (c) rod float. These are shown in Fig. 9.10.

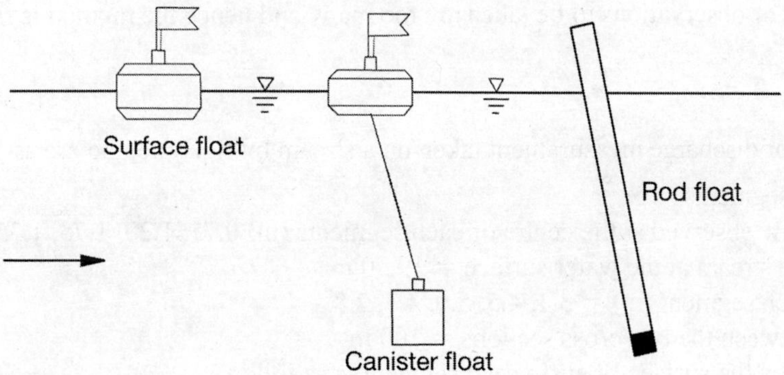

Fig. 9.10 Types of floats

9.5.3.3 Characteristics of the surface float

The characteristics for the surface floats are as follows:

- It should be heavy or weighed properly so that most of its part should remain under water so that its movement with flow is not affected by the wind velocity. The length under water should also be minimum so that the float records only the surface velocity.
- It should be suitably painted or a flag may be fixed on the float so that it can be easily located.
- It should not be of such shape that it drifts laterally.

9.5.3.4 Rod float

It is expected that a float should not drift laterally. To avoid this lateral drifting, a rod float is used. It is a wooden float or a lightweight metallic rod having a weight fixed at the lower end. A coloured flag may be fixed to the other end. It should float vertically. Since some part of this rod float is under water, it may not move with the surface velocity.

The average velocity of the segment in such a case may be calculated as follows:

$$V_a = V\left(1.02 - 0.0116\sqrt{\frac{D}{d}}\right)$$

where, V_a = Surface velocity of the segment in m/s
V = Observed velocity of the rod float in m/s
D = Clearance of the rod float from the bed of the stream in m
d = Depth of flow at the location of float in m

9.5.3.5 Merits and demerits of the float method

The merits of the float method are as follows:

- The method is very simple to follow.
- No sophisticated equipment is required.
- When the velocity of flow is high and other methods cannot be used, this method is followed.

The demerits of the float method are as follows:

- A team of observers is required.
- Wind velocity may affect the velocity of float.
- Dropping the float exactly at the centre of the segment is a difficult job.
- The float may not travel along the centerline of the segment.
- Number of observations to be taken are too many and hence the method is time-taking.

Example 9.4

Observations for discharge measurement taken on a stream by float method are as follows:

1. Number of segments = 5
2. Depth of flow observed at the centre of each segments (m) 0.75, 1.20, 1.75, 1.35, 0.60
3. Width of the stream at the water surface = 19.70 m
4. Width of each segment (m) = 3.1, 4.5, 5.0, 4.3, 2.8
5. Distance between the two cross sections = 100 m
6. Time taken by the surface floats in each segment is as follows:

Segment	Minutes	Seconds
1	2	11
2	1	45
3	1	40
4	1	55
5	2	09

Find the discharge flowing in the stream.

Solution:

Assuming $V_a = 0.85\ V_s$, the average velocity in each segment and the area of flow are calculated as under.

Segment	Surface velocity (m/s)	Average velocity (m/s)	Area of flow (m^2)	Discharge (m^3/s)
1	100/131 = 0.763	0.649	0.75×3.1 = 2.325	1.508
2	100/105 = 0.952	0.809	1.20×4.5 = 5.400	4.369
3	100/100 = 1.000	0.850	1.75×5.0 = 8.750	7.437
4	100/115 = 0.869	0.739	1.35×4.3 = 5.805	4.287
5	100/129 = 0.775	0.659	0.60×2.8 = 1.680	1.106
			Total	18.707

9.5.4 Current Meter Method

A current meter is an instrument used to measure velocity of flow. Current meters are of two types: (1) cup-type current meter and (2) propeller-type current meter.

9.5.4.1 Cup-type current meter

A cup-type current meter has a series of cone-shaped cups (normally six in number) fixed on a wheel freely rotating about a vertical axis mounted on the periphery of a circle as shown in Fig. 9.11.

The current meter is lowered below the water surface at the desired depth and the velocity of flow is assessed at that specific depth. The cup wheel rotates with a speed in proportion with the velocity of water at that depth. More the velocity, more will be the rotations of the cups. These rotations of the cup wheel are proportional to the velocity of water. The rotations of the cups are recorded by an electric *Make and Break* circuit. For each rotation, there is a click and these clicks are either recoded by a counter or are counted by using a stopwatch. The number of clicks, i.e., rotations, is measured over a period of 1 or 2 min and then the rotations per second are calculated.

The velocity of flow then can be calculated as follows:

$$V = a + bN$$

where, V = Velocity of flow in m/s
N = Number of rotations in one second
a, b = Calibration constants

Normally, the current meter is calibrated by the manufacturer and the formula with these numerical constants is supplied along with the instrument. A fish weight with a streamlined shape is fixed at the bottom to hold the current meter at a specific location. Fins are provided at the tail end so that the current meter will always align itself along the direction of flow.

This type of current meter is also known as *Price Current Meter* named after the designer. The range of velocity that can be measured by this type of current meter is 0.15–4.0 m/s. Because of the construction of this type of current meter, it cannot be used to measure the surface velocity nor the bed velocity.

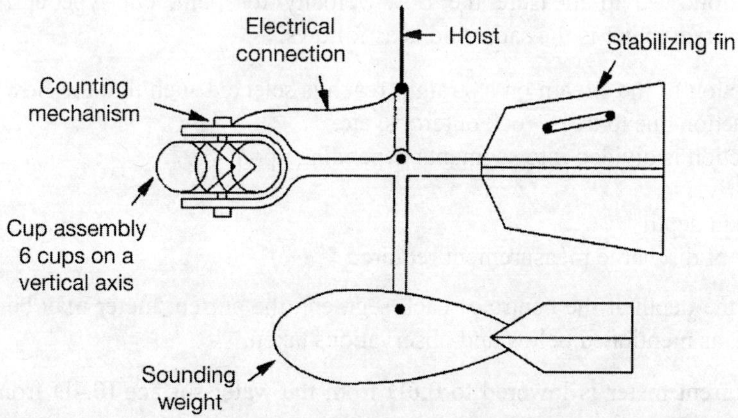

Fig. 9.11 Cup-type current meter

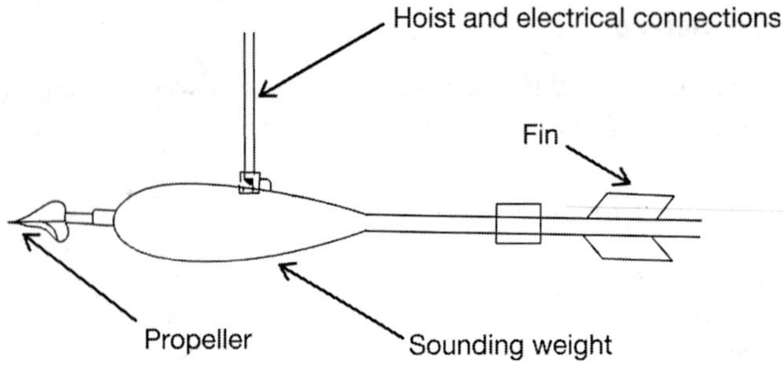

Fig. 9.12 Propeller-type current meter

9.5.4.2 Propeller-type current meter

In this type, a propeller facing the flow direction, rotating about a horizontal axis, is used as shown in the Fig. 9.12. The current meter is lowered below the water surface at the desired depth.

The rotations of the horizontal axis are proportional to the flow velocity at that depth. More the velocity, more will be the rotations of the axis. The rotations made by this axis are recorded by the electric make and break circuit. For each rotation, there is a click and these clicks are recorded by a counter or counted by using a stopwatch. The flow velocity then can be calculated by using the formula,

$$V = a + bN$$

Normally, the current meter is calibrated by the manufacturer and the formula along with the numerical constants is supplied along with the instrument. A fish weight of streamlined shape is mounted on the horizontal axis to keep the current meter in position. Fins are provided at the tail end so that the current meter will always align itself in the direction of flow.

The range of velocity that can be measured by this type of current meter is 0.15–4.00 m/s. The current meter cannot be used to measure surface velocity nor the bed velocity because of its construction.

9.5.4.3 Observations by a current meter

The procedure followed to measure the flow velocity for both, cup-type current meter and propeller-type current meter is the same and is as follows:

- One cross section of the stream on a straight reach is selected such that the flow is undisturbed by any obstruction due to trees, rock outcrops, etc.
- This cross section is divided into segments depending upon
  (a) Width of river
  (b) Variation in depth
  (c) Accuracy of discharge measurement required

Knowing the depth at the centre of each segment, the current meter may be lowered at the desired depth, as mentioned below and observations taken.

(i) The current meter is lowered to 0.6D from the water surface (0.4D from the bed) and velocity observations are taken.
This method is known as *Single-point Method* or *Six-tenth Method*.

(ii) The velocity observations may be taken at 0.2D and 0.8D also. The average velocity may be worked out as

$$V_a = \frac{(V_{0.2D} + V_{0.8D})}{2}$$

This is known as *Two-point Method.*

(iii) The velocity observation may be taken at 0.2D, 0.6D and 0.8D from the water surface and the average velocity can be worked out as

$$V_a = \frac{(V_{0.2D} + 2V_{0.6D} + V_{0.8D})}{4}$$

This is known as *Three-point Method* and normally followed for deep flow streams.

(iv) Observations may be taken by lowering the current meter in each of the segment with a uniform speed from the water surface till it reaches the bottom and lifted gradually upwards with the same uniform speed up to the water surface.

The total time for the movement of the current meter as well as its speed is recorded. The average velocity in each segment is calculated as follows:

$$V_a = a + b\left(\frac{N}{T}\right)$$

where, V_a = Average velocity in each segment in m/s
N = Number of rotations in time T required for downward and upward movement of the current meter
a, b = Calibration constants

This method is known as *Integrated Method of Velocity Observation.*

9.5.4.4 Lowering the current meter

The current meter may be lowered up to the desired depth at the centre of each segment by using the following methods.

- When the stream is shallow, the observer should wade through the stream with a current meter mounted on a rod and take the observations. In such case, the observer himself may cause obstruction to the flow. Normally, one observation is taken for each segment with minimum obstructions.
- The current meter may be lowered from an existing bridge and velocity observations are taken on the d/s side between two successive piers. The bridge piers cause obstruction to the flow.
- The current meter may be lowered from a trolley specifically erected for this purpose from a cableway as shown in the Fig. 9.13.
- The current meter may be lowered from a boat. The boat will have to be directed from the banks and kept steady at the centre of the segment. This can be done by two observers on either bank directing the boat. Alternatively, the boatman may keep the boat steady by ranging with the help of two fixed points on either bank.

9.5.4.5 Air-line and wet-line corrections

When a current meter is lowered by a chain from a bridge or a cableway, the current meter will be drifted d/s due to the flow velocity. The chain may take the shape as shown in Fig. 9.14.

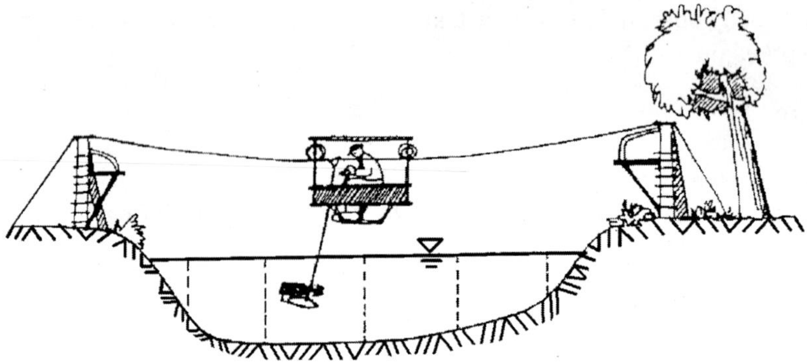

Fig. 9.13 Current meter lowered from a trolley

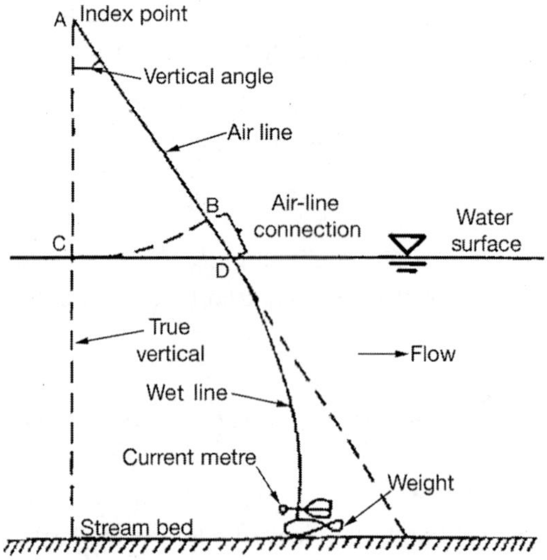

Fig. 9.14 Air-line and wet-line corrections

Because of the suspended inclined nature of the chain, the depth of water where the observations are taken may go wrong. The correct depth is calculated by applying two corrections as

(a) Air-line correction, K_a
(b) Wet-line correction, K_w

Air-line correction

- The air-line correction is the correction for the portion of the chain from the point of suspension up to the water surface.
- This correction depends on the vertical angle of drift and the vertical length from the point of suspension to the water surface.

The correction will be $= K_a = (\sec\theta - 1) \times 100$

Thus, after applying the air-line correction, the length of the chain from the suspension point up to the water surface level will be

$$d_1 = Y_1 \times (1 + 0.01\, K_a)$$

where, Y_1 = Vertical height of chain in m
d_1 = Inclined length of chain in m
K_a = Air-line correction. It is dimensionless

Wet-line correction

- The chain under the water surface will not be having a straight profile but a curved one, depending upon the velocity of water, unit weight of water, unit weight of the chain material and also the weight of current meter and fish weight.
- The vertical depth of water from the water surface up to the current meter will be as follows:

$$d_2 = Y_2 \times (1 + 0.01\,K_w)$$

where, Y_2 = Vertical depth of water from the water surface up to the current meter
d_2 = Inclined length of the cable from the water surface up to the current meter
K_w = Wet-line correction. It is dimensionless.

The two correction factors, viz K_a and K_w will mainly depend on angle θ. The standard values in percentages are given in Table 9.1.

Table 9.1 Standard values of K_a and K_w for different values of θ

Serial no.	θ (degree)	K_a (%)	K_w (%)	Remarks
1	4	0.24	0.06	—
2	8	0.98	0.32	—
3	12	2.23	0.72	—
4	16	4.03	1.28	—
5	20	6.42	2.04	—
6	24	9.46	2.96	—
7	28	13.26	4.08	—

The angle θ should not be more than 30°.

Example 9.5

Observations are to be taken for the velocity of flow by lowering a current meter from bridge with the following details:

1. Depth of flow = 5.00 m
2. Distance between the index point and the water surface = 4.00 m
3. Angle made by the cable with the vertical = 24°

Find the length of the cable to reach the bed of the channel.

Solution:

Air-line correction $= 4 \times \frac{9.46^*}{100} = 0.378$ m

Wet-line correction $= 5 \times \frac{2.96^*}{100} = 0.148$ m

* Taken from the Table 9.1 for an angle of 24°

Therefore, length of the cable from the index point up to the bed of the channel

$$= 4 + .378 + 5 + .148$$
$$= 9.526 \text{ m.}$$

9.5.4.6 Discharge computations

Normally two methods are followed to calculate the discharge after calculating the cross-sectional area and the average velocity. These are as follows:

1. The area and the average velocity of each of the segments are evaluated and tabulated.

Then, $Q = Q_1 + Q_2 + Q_3 + \cdots$

$$= A_1 \times V_{a1} + A_2 \times V_{a2} + A_3 \times V_{a3} + \cdots$$

where, Q = Total discharge in the stream in m^3/s

A_1, A_2, A_3 = Area of each of the segment in m^2

V_{a1}, V_{a2}, V_{a3} = Average velocity in each of the segment in m/s

This method is known as *Mid-section Method.*

Example 9.6

Observations taken on a stream for the measurement of discharge by a current meter are as follows:

1. Width of the stream at the water surface = 19.70 m
2. Number of segments = 5
3. Width of each segment (m) = 3.1, 4.5, 5.0, 4.3, 2.8
4. Average depth of flow of each segment (m) = 0.75, 1.20, 1.75, 1.35, 0.60

The observed velocity of flow at various depths from the water surface are as shown in Fig. 9.15. Find the discharge.

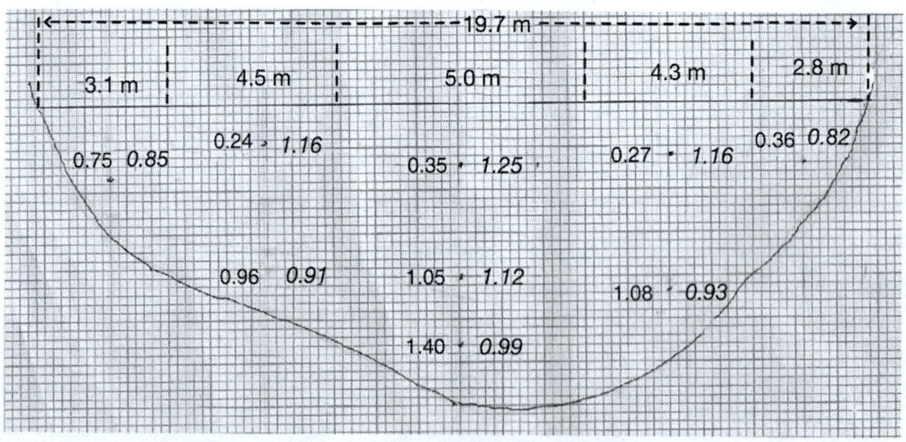

Italics indicate velocity of flow in m/s

Fig. 9.15 Velocity of flow observed at various depths

Solution:

The calculations done for the discharge are as under.

Serial no.	Area of flow (m²)	d (m)	d_0 (m)	d_0/d	Observed velocity (m/s)	V_A (m/s)	Discharge (m³/s)
1	3.1 × 0.75= 2.325	0.75	0.45	0.6	0.85	0.85	1.976
2	4.5 × 1.20= 5.400	1.20	0.24	0.2	1.16	1.03	5.562
			0.96	0.8	0.90		
3	5.0 × 1.75= 8.750	1.75	0.35	0.2	1.25	1.12	9.800
			1.05	0.6	1.12		
			1.40	0.8	0.99		
4	4.3 × 1.35= 5.805	1.35	0.27	0.2	1.16	1.05	6.096
			1.08	0.8	0.94		
5	2.8 × 0.60 = 1.680	0.60	0.36	0.6	0.82	0.82	1.377
						Total	24.811

d_0 is the depth where the velocity was observed.

$$\text{Here, } V_A = V_{0.6D} = \frac{V_{0.2D} + V_{0.8D}}{2} = \frac{V_{0.2D} + 2V_{0.6D} + V_{0.8D}}{4}$$

2. Based on the velocity observed at various depths in each segment isovels at equal interval are drawn on the cross section as shown in Fig. 9.16.

An Isovel May be Defined as a *Contour of Equal Velocity,* i.e., *a Line Joining Locations with Same Flow Velocity.*

The area between two consecutive isovels is measured accurately and tabulated. The discharge in the stream is calculated as

$$Q = Q_1 + Q_2 + Q_3 + \cdots$$
$$= A_1 \times V_{a1} + A_2 \times V_{a2} + A_3 \times V_{a3} + \cdots$$

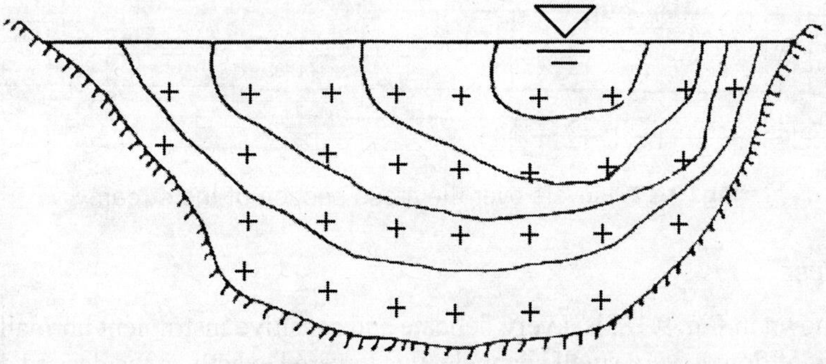

Fig. 9.16 Isovels

where, Q = Total discharge flowing in the stream in m^3/s

A_1, A_2, A_3 = Area between two consecutive isovels in m^2

V_{a1}, V_{a2}, V_{a3} = Average velocity of flow between two isovels in m/s

This method is more time taking, but it gives better results.

Example 9.7

Velocity observations taken over a cross section of a stream at various locations are as shown in Fig. 9.17. The width of the cross section at the surface and the maximum depth are 23.67 m and 5.5 m, respectively.

Draw the isovels and evaluate the discharge.

Solution:

The isovels were drawn taking into consideration the velocity and the location. The area between the consecutive isovels was measured accurately. The results are tabulated below.

Serial no.	Isovel range (m)	Average of isovel (m)	Area (m^2)	Discharge (m^3/s)
1	1.2 and above	1.20	1.41	1.692
2	1.2 and 1.1	1.15	6.21	7.142
3	1.1 and 1.0	1.05	8.75	9.188
4	1.0 and 0.9	0.95	5.45	5.177
5	0.9 and below	0.90	1.85	1.665
		Total	23.67	24.864

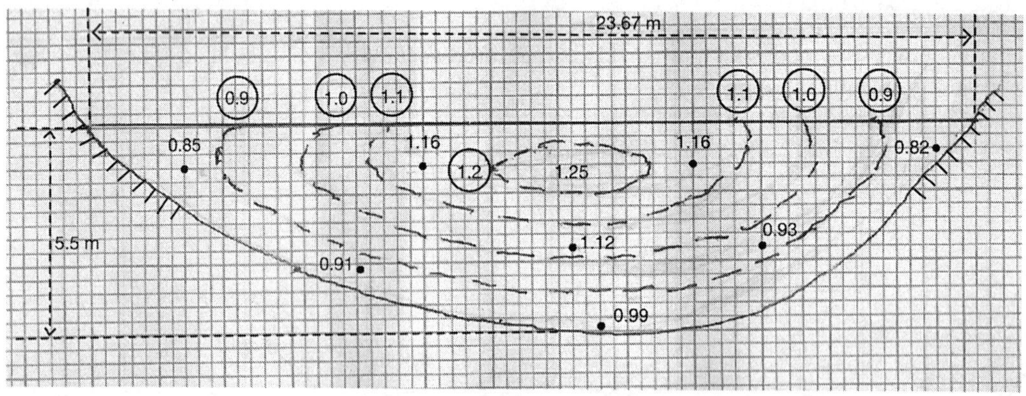

Fig. 9.17 Isovels over the cross section of the stream

9.5.5 Pitot Tube

Pitot tube, as shown in Fig. 9.18, is a very delicate and sensitive instrument normally used to measure the velocity of flow in very small channels. It is lowered exactly at the desired depth facing the flow and observations can be taken accurately.

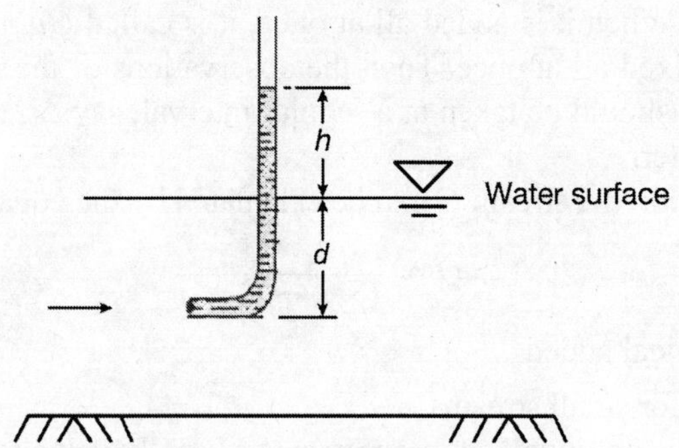

Fig. 9.18 Pitot tube

It is normally connected to a differential U-tube mercury manometer and the velocity head is measured in terms of the depth of the fluid. The velocity of flow at depth d can be equated as follows:

$$v = k\sqrt{(2gh)}$$

where, v = Velocity of flow in m/s
h = Rise of water column in the Pitot tube in m
k = Coefficient of Pitot tube to account for the loss of head. (This is dimensionless and always less than one.)

9.6 SALT TITRATION METHOD

In this method, a concentrated solution of a salt or a chemical in water is added to a stream at a constant rate q and is allowed to mix with the flow of the stream. A sample of the mixed flow is collected at a location such that thorough mixing is achieved. If necessary artificial means may be followed for thorough mixing. The sample thus collected is tested in the laboratory and the concentration is determined accurately.

The discharge Q in the stream can be calculated from the following equation:

$$n_1 \times q + n_0 \times Q = n_2 (Q + q)$$

where, Q = Discharge in the stream in m^3/s
q = Discharge of the concentrated salt/chemical in m^3/s
n_1 = Normality of the concentrated salt/chemical before mixing
n_2 = Normality of the sample collected after mixing
n_0 = Normality of water in the stream before mixing

$$\therefore \quad Q = q\,\frac{(n_1 - n_2)}{(n_2 - n_0)}$$

If n_0 is negligible, $Q = q\,\dfrac{(n_1 - n_2)}{n_2}$

The success of the method mainly depends on the thorough and complete mixing of the salt/chemical with the flowing water. When a salt or a chemical is added in the stream at a constant rate, it is called *Plateau Method*. When it is added all at once, it is called *Gulp Method*. In the case of gulp method, chemical is mixed all at once. Then the observations of the normality of the stream water after thorough mixing should be taken at a regular interval, say Δt, till no trace of chemical is observed in the stream water.

In this case, the discharge in the stream Q can be calculated by the equation:

$$Q = \frac{n_1 \times V}{(\Sigma n_2 - n_0)\,\Delta t}$$

where, V = Volume of chemical added

The method can be used for small streams.

It is simple because it does not involve measurements of depth, velocity, etc.

9.6.1 Properties of a Salt or Chemical

The salt/chemical to be used for this method should have the following properties:

- It should be easily soluble in water.
- It should not be harmful to plants and animals.
- It should not be present in the flowing water of the stream.
- It should be easily available and cheap.
- It can be easily detected after thorough mixing.

Example 9.8

20 mg/cc of salt at the rate of 10 cc/s was discharged in a stream having a concentration of the same salt of 0.1 parts per billion (ppb). The concentration after thorough mixing was found to be 5 ppb. Find discharge in the stream.

Solution:

$q = 10$ cc/s $= 1\times10^{-5}$ m^3/s

Concentration of salt added in the stream $= n_1 = \frac{20}{10^{-6}} \times \frac{1}{10^9} = 0.02$

Concentration of salt in the flowing water $= n_0 = \frac{0.1}{10^9} = 0.1\times10^{-9}$

Concentration after the thorough mixing $= n_2 = \frac{5}{10^9} = 5 \times 10^{-9}$

$$= \frac{q(n_1-n_2)}{(n_2-n_0)} = \frac{1\times10^{-5}(0.02-5\times10^{-9})}{5\times10^{-9}-0.1\times10^{-9}}$$

$= 40.81$ m^3/s

(1 mg/1 m^3 = 1 part = 1 per billion)

Example 9.9

30 litres of water with a salt concentration of 20 mg/cc was added to a flowing stream in gulp. Concentration of the stream water after thorough mixing was measured on the d/s every after 3 min and was noticed to be 0, 5, 10, 20, 15, 5, 0 ppb.

Find the discharge in the stream.

Solution:

Since the normality of the stream water n_0 before mixing is not given, it is assumed to be zero.

$$Q = \frac{\dfrac{20}{10^{-6}} \times \dfrac{30 \times 10^{-3}}{10^9}}{\dfrac{3 \times 60\,[(5-0)+(10-0)+(20-0)+(15-0)+(5-0)]}{10^9}}$$

$$= \frac{600 \times 10^{-6}}{\dfrac{3 \times 60 \times 55}{10^9}} = \frac{600 \times 10^3}{3 \times 60 \times 55} = 60.60 \text{ m}^3/\text{s}$$

9.7 DISCHARGE MEASUREMENT BY HYDRAULIC STRUCTURES

The following hydraulic structures can be used for the discharge measurement:
(1) notches (2) throated flumes and (3) weirs

9.7.1 Notches

Notches are used when the discharge is small. The different notches are as follows:

1. Triangular notch
2. Rectangular notch
3. Trapezoidal notch
4. Cippolette notch

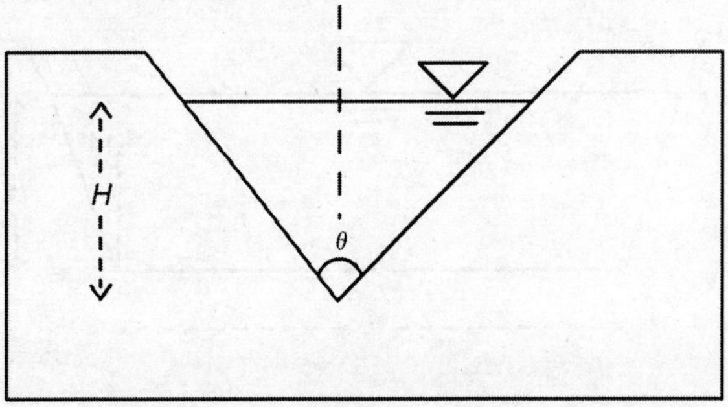

$$Q = \frac{8}{15} C_d \sqrt{2g} \tan\frac{\theta}{2} H^{5/2}$$

Fig. 9.19 Triangular notch

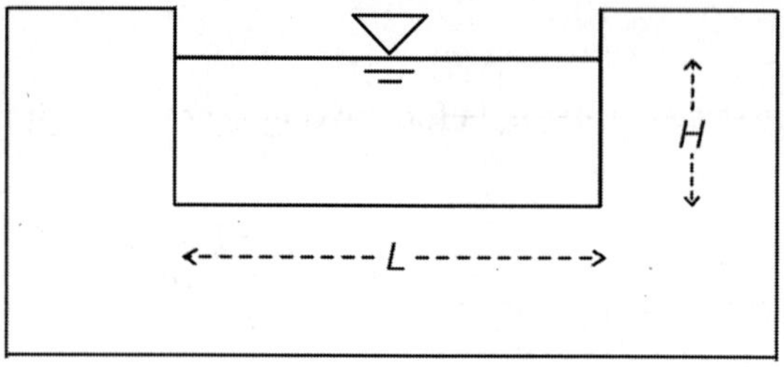

$$Q = \frac{2}{3} C_d \sqrt{2g}\, L H^{3/2}$$

Fig. 9.20 Rectangular notch

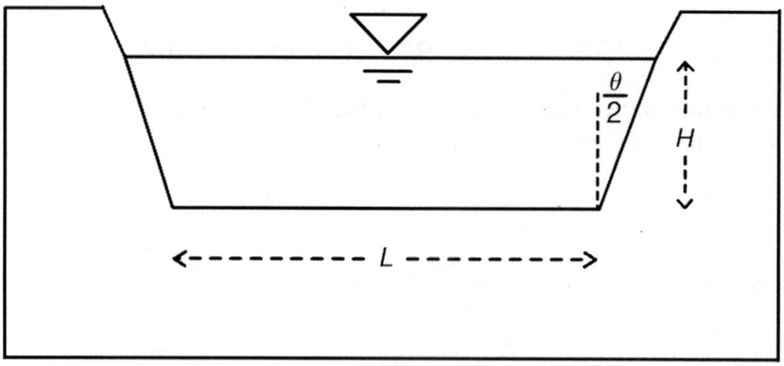

$$Q = \frac{2}{3} C_d \sqrt{2g} L\, H^{3/2} + \frac{8}{15} C_d \sqrt{2g} \tan\frac{\theta}{2}\, H^{5/2}$$

Fig. 9.21 Trapezoidal notch

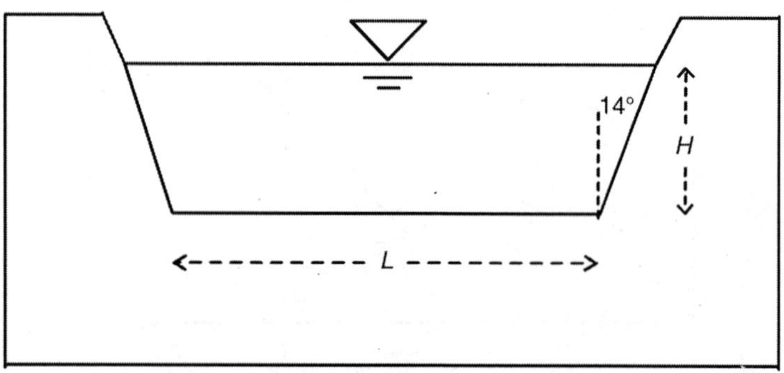

$$Q = 1.86\, L\, H^{3/2}$$

Fig. 9.22 Cippolette notch

Example 9.10

In a laboratory, following 4 notches were installed in parallel with the crest of all the notches at the same level:

The water level above the crest was 150 mm. Assume coefficient of discharge in all the cases to be 0.6.

1. Triangular notch with semi-vertical angle of 45°

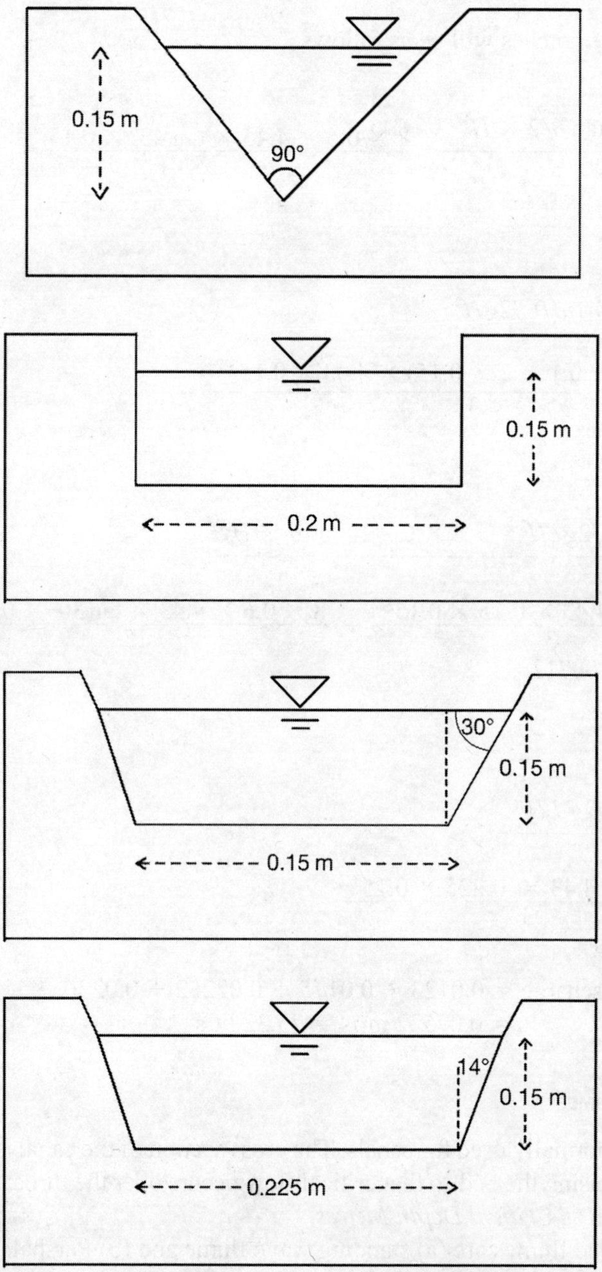

Fig. 9.23 Water levels on various notches in laboratory

2. Rectangular notch of crest length of 200 mm
3. Trapezoidal notch of crest length of 150 mm and side angle of 30°
4. Cippolette notch of crest length of 225 mm

The water level above the crest was 150 mm. Assume coefficient of discharge in all cases to be 0.6. Calculate the total discharge flowing over all the notches.

Solution:

The discharge over the notches will be as follows:

1. Triangular notch

$$Q = \frac{8 \times C_d \sqrt{2g} \tan \theta/2 \times H^{5/2}}{15} = \frac{8 \times 0.6 \times 4.43 \times \tan 45^\circ \times 0.15^{5/2}}{15}$$

$$= 0.0123 \text{ m}^3/\text{s}.$$

2. Rectangular notch

$$Q = \frac{2 \times C_d (L - 0.1nH) \sqrt{2g} H^{3/2}}{3}$$

$$= \frac{2 \times 0.6(0.2 - 0.1 \times 2 \times 0.15) \times 4.43 \times 0.15^{3/2}}{3}$$

$$= 0.0175 \text{ m}^3/\text{s}$$

3. Trapezoidal notch

$$Q = \frac{2 \times C_d \times \sqrt{2g} L H^{3/2}}{3} + \frac{8 \times C_d \sqrt{2g} \tan\theta \times H^{5/2}}{15}$$

$$= \frac{2 \times 0.6 \times 4.43 \times 0.15 \times 0.15^{3/2}}{3} + \frac{8 \times 0.6 \times 4.43 \times \tan 30^\circ \times 0.15^{5/2}}{15}$$

$$= 0.0154 + 0.00712$$

$$= 0.02252 \text{ m}^3/\text{s}$$

4. Cippolette notch

$$Q = \frac{2 \times C_d \times \sqrt{2g}\, L H^{3/2}}{3}$$

$$= \frac{2 \times 0.6 \times 4.43 \times 0.225 \times 0.15^{3/2}}{3}$$

$$= 0.0230 \text{ m}^3/\text{s}$$

Therefore, total discharge $= 0.0123 + 0.0175 + 0.02252 + 0.0230$
$= 0.07532 \text{ m}^3/\text{s} = 75.32 \text{ lit/s}$

9.7.2 Throttled Flumes

Throttled flumes are normally used for canals. The cross section of the canal is reduced by reducing the width as well as raising the bed so that critical flow occurs over the throat portion. Hence, these flumes are also known as *Critical Depth Metres*.

The normal throttled flumes are (a) standing wave flume and (b) Parshall flume

9.7.2.1 Standing wave flume

Standing wave flume (SWF) is also known as *Venturi-flume* (Fig. 9.24).

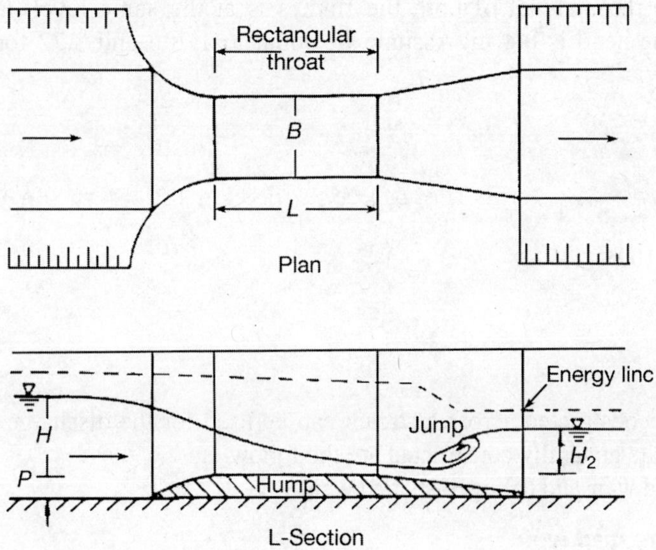

$$Q = 1.706\, B\, H^{3/2}$$

Fig. 9.24 Standing wave flume

9.7.2.2 Parshall flume

Parshall flume was developed by R. L. Parshall in 1910 and is widely used in the USA (Fig. 9.25).

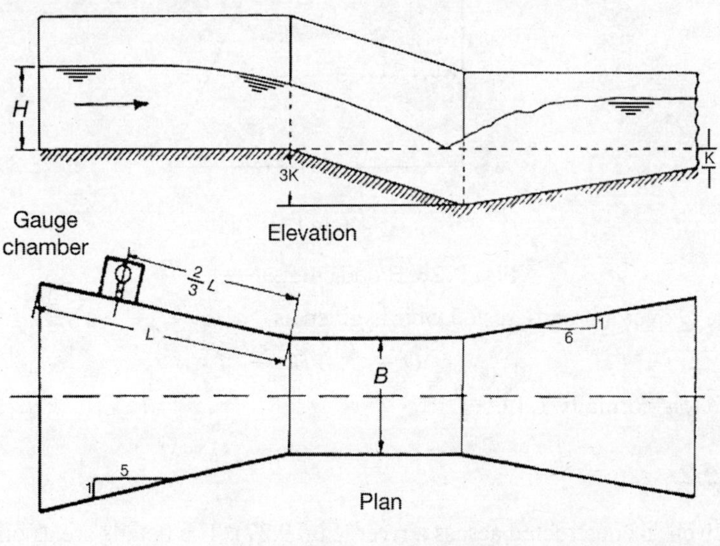

$$Q = 1.72\, B\, H^{1.49}$$

Fig. 9.25 Parshall flume

Example 9.11

A 0.5-m-wide SWF and a 0.25-m-wide Parshall flume are used to measure discharge in a branch canal. The throat level of both the flumes is at the same level. Find the combined discharge when the head is 0.4 m. Assume C_d equal to 1.706 and 1.72 for SWF and partial flume, respectively.

Solution:

$$Q = C_d \times L \times H^{3/2} + C_d \times L \times H^{1.49} = 1.706 \times 0.5 \times 0.4^{3/2} + 1.72 \times 0.25 \times 0.4^{1.49}$$

$$= 0.215 + 0.109$$

$$= 0.324 \text{ m}^3/\text{s}$$

9.7.3 Weirs

The existing weirs constructed across a stream can be used for the discharge measurement. The normal types of weirs generally constructed are the following:
(a) broad-crested weir and (b) ogee-shaped weir.

9.7.3.1 Broad-crested weir

Broad-crested weirs are normally of low heights, constructed over the entire length of the channel. It will act as a broad-crested weir only if $B > 2/3\ H$ (Fig. 9.26).

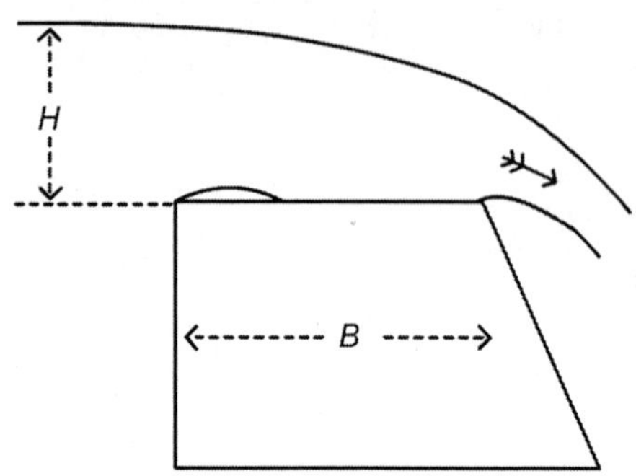

$Q = 1.49\ L\ H^{3/2}$

Fig. 9.26 Broad-crested weir

The discharge Q over a broad-crested weir is given as

$$Q = C_d L H^{3/2}$$

The value of C_d is normally 1.49.

Example 9.12

A broad-crested weir is constructed across a river (Fig. 9.27). The details are as under:

(1) Length of the weir = 10 m
(2) Width of the weir = 1.0 m

Find the discharge over the weir when the head over the weir is 0.5 m.

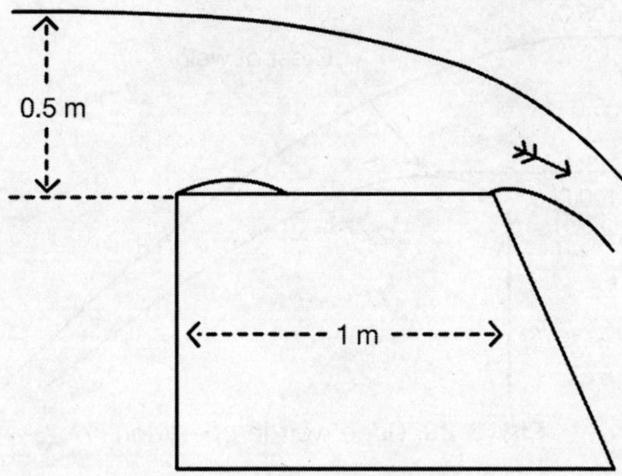

Fig. 9.27 Broad-crested weir in operation

Solution:

$$Q = 1.49 \times L \times H^{3/2} = 1.49 \times 10.0 \times 0.5^{3/2} = 5.26 \text{ m}^3/\text{s}.$$

9.7.3.2 Ogee-shaped weir

Normally, ogee-shaped weirs are provided for the spillway of a storage dam. The profile is designed to suit the lower nappe of the jet flowing over the weir (Fig. 9.28).

The discharge Q over an ogee weir is given as:

$$Q = C_d \times (L - nkH) \times H^{3/2}$$

where, Q, L and H carry the usual meaning. n is the number of end contractions caused due to the piers and k is a coefficient depending upon the shape of the pier. The value of C_d is normally 2.05.

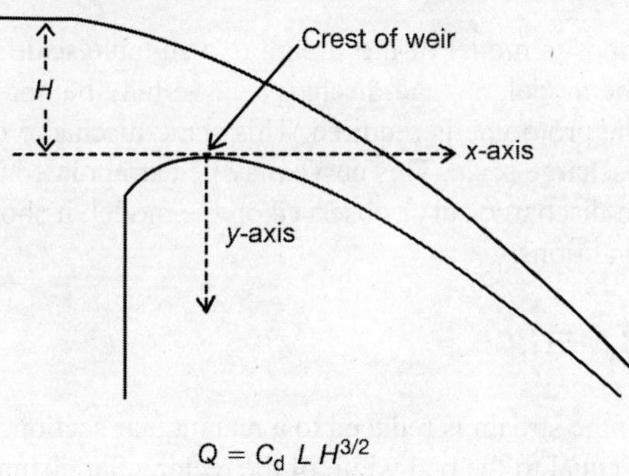

Fig. 9.28 Ogee weir

Example 9.13

A gated ogee spillway has been provided to a dam, the details are as under.

1. Spillway crest level = 100.0 m
2. Number and size of gates = 4, each 6 m × 9 m

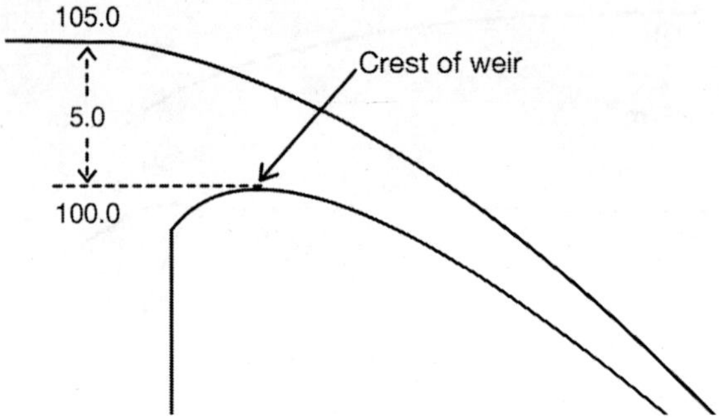

Fig. 9.29 Ogee weir in operation

3. Water level in the reservoir = 105.0 m
4. Coefficient of discharge = 2.05 (assumed)

Find the discharge over the spillway when the gates are fully open.

Solution:

$$Q = C_d \times (L - nk^* H) \times H^{3/2}$$

$$= 2.05\,(4 \times 9 - 0.1 \times 8 \times 5) \times (105 - 100)^{3/2}$$

$$= 2.05 \times 32 \times 5^{3/2}$$

$$= 733.4 \text{ m}^3/\text{s}$$

(k is assumed equal to 0.1.)

9.8 HYDRAULIC MODEL METHOD

A three-dimensional composite model of the stream to a suitable scale may be constructed in the laboratory. By running the model, a stage-discharge curve may be obtained for the cross section where the discharge in the prototype is required. This stage-discharge curve may be used for the prototype by using the discharge scale. This curve may be extrapolated if required.

Before using the stage-discharge curve observed on the model, it should be seen that the model is proved to prototype conditions.

9.9 ULTRASONIC METHOD

In this method, a reach of the stream is reduced to a rectangular section. This length of the reach L should be approximately equal to the bed width of the rectangular channel (Fig. 9.30).

Two transducers are located on each bank at a depth d. The line joining the transducers should make an angle of 45° with the axis of the stream. These two transducers should emit and also receive the ultrasonic signals. When an ultrasonic signal is emitted from A to B, it will travel with a velocity equal to $u + V_p$ and the time of travel T_1 to travel for this signal from A to B will be

$$T_1 = \frac{L}{(u + V_p)}$$

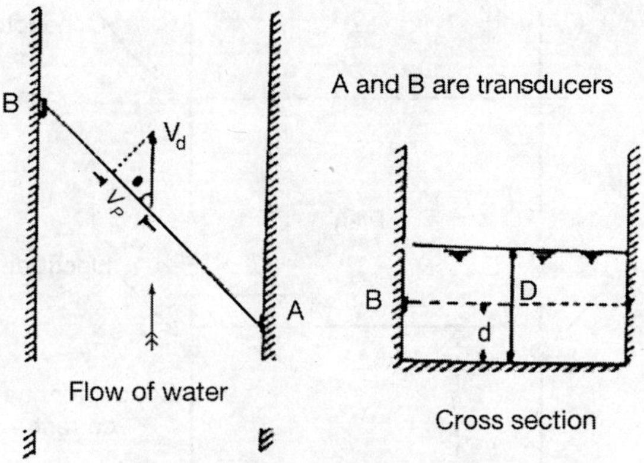

Fig. 9.30 Ultrasonic method

where, T_1 = Time of travel for the signal from A to B in seconds
L = Length between A and B in metres
u = Velocity of sound in water in m/s
V_d = Velocity of flow at a depth of d
V_p = Component of velocity of water V_d at a depth d along AB
= $V_d \cos 45°$

Similarly, T_2 will be the time of travel for the signal from B to A and can be mentioned as $T_2 = L/(u - V_p)$.

$$\text{Thus } V_d = \frac{L}{2\cos\theta}\left(\frac{1}{T_1} - \frac{1}{T_2}\right)$$

This method is known as *single path gauging method.*

Alternatively, a series of transducers may be installed at different depths and thus the vertical profile of velocity can be obtained to work out the average velocity and consequently the discharge, more accurately. This is known as *Multiple Path Gauging Method.*

The method can be used for river width up to 500 m and an accuracy up to 2% can be obtained.

With this method, immediate results are obtained. However, the observations may be affected due to weed growth, salinity of water, air entrainment, temperature of water, etc.

9.10 ELECTROMAGNETIC INDUCTION METHOD

This method is based on the Faraday's principle that *An Electromagnetic Force is induced in the conductor (water) when it passes through a normal magnetic field.*

A long conductor is buried at the bottom of the stream at right angles to the flow throughout the cross section. A current I is allowed to flow in this conductor. This will produce a controlled vertical magnetic field (Fig. 9.31) and this magnetic field will produce an electromagnetic force (e.m.f.) depending upon the following:

(a) Depth of flow (b) Discharge in the stream (c) Current I

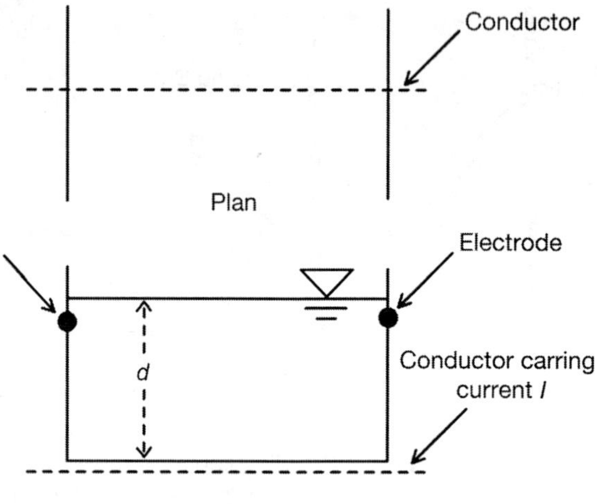

Fig. 9.31 Electromagnetic method

Two sensitive electrodes installed on the banks will measure the small voltage produced due to this e.m.f. The discharge in the stream can be calculated after the calibration as follows:

$$Q = K_1\left(\frac{E \times d}{I} + K_2\right)^n$$

where, Q = Discharge in the stream in m^3/s
E = Voltage produced due to the e.m.f. in millivolts
I = Current in main conductor in amperes
d = Depth of flow in metres
K_1, K_2, n = System constants to be obtained by calibration

The method is suitable for rivers

- Having a tendency to change its cross section
- Having weed growth
- Tidal channels where there is rapid change in direction and discharge

The method can be used up to a width of 100 m and has an accuracy of 3%. The discharge flowing can be found out immediately.

9.11 MOVING-BOAT TECHNIQUE

When a large river is in spate, it is very difficult to take discharge observations particularly because it is very difficult to keep the boat steady at a location, to lower the current meter and take the velocity observations at the same time. A technique called *Moving-boat Technique* for discharge measurement was developed by the US Geological Survey in 1969. In this method, the boat is kept moving and velocity observations are continuously taken. Hence, the method is named *moving-boat technique* (Fig. 9.32).

The boat is moved at an angle θ with the cross section P-P′ to Q-Q′, with a uniform velocity V_v. However, the flow having a velocity V_f will also act on the boat. Thus, the boat will actually move

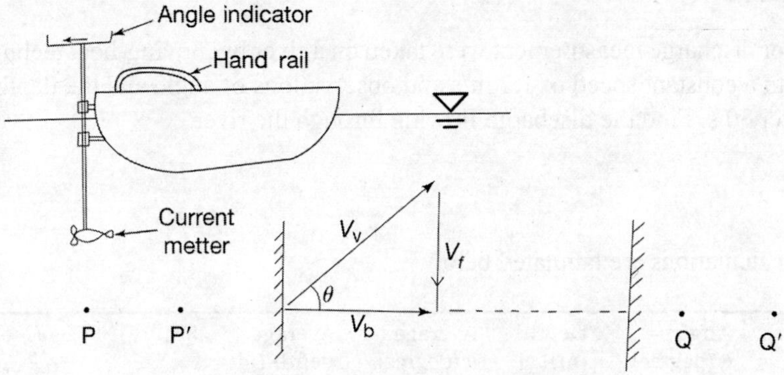

Fig. 9.32 Moving-boat technique

along the resultant of V_v and V_f, i.e. V_b. V_b must be along a well-defined cross section at right angles to the flow. Since the velocity of flow V_f goes on increasing from the bank towards the centre of the stream, the boatman has to change the θ so that V_b is along the fixed cross section.

Normally, a propeller-type current meter is used, as it is free to move about its vertical axis. The current meter is maintained at a depth of 0.5 m below the water surface. There is a vertical vane attached to the current meter so that it aligns itself in the direction of flow. An echo sounder is also fixed to the boat for recording the depth of flow or the bed profile.

The boatman starts the boat from the bank at a velocity V_v and keeps the boat always along the pre-determined cross section by changing the angle θ. To keep the boat along the cross section, he ranges the boat with the help of the posts P,P′ and Q,Q′ located on the banks. During the movement of the boat, observations of V_v, θ and the depths of flow at equal intervals of time for various segments are taken.

The discharge in each segment will be dQ

$$dQ = dA \times V_f$$

$$= \left[\frac{(d_1 + d_2)}{2}\right] \times W \times V_f$$

where, W = Width of segment $= V_b \times dt = V_v \cos\theta \times dt$

and, $V_f = V_v \sin\theta$

$$dQ = \left[\frac{(d_1 + d_2)}{2}\right] \times V_v \sin\theta \times V_v \cos\theta \times dt$$

$$= \left[\frac{(d_1 + d_2)}{2}\right] \times V_v^2 \sin\theta \times \cos\theta \times dt$$

Since all the values on the right hand side are observed during the ride of the boat, dQ can be evaluated.

$$\text{Total discharge, } Q = \sum dQ$$

The operation is repeated for the return journey of the boat. These operations are repeated and the average value of discharge is calculated. The method is used for streams having more than 3-m depth of flow. The method is strenuous as it is difficult to move the boat along a fixed cross section in a high velocity flow. However, there is no alternative but to follow this method for very large rivers.

Example 9.14

Observations for discharge measurement were taken on a river by moving-boat technique. The boat was moved with a constant speed of 1.2 m/s and observations of angle and the depth of flow were taken every after 30 s. Find the discharge flowing through the river.

Solution:

The discharge calculations are tabulated below.

Serial no.	Time (s)	Angle = θ (degree)	Depth (m)	Average θ (degree)	Average depth (m)	Sin θ	Cos θ	Discharge
1	0	0	0	—	—	—	—	—
2	30	20	2.6	10	1.3	0.1736	0.9848	9.601
3	30	24	3.2	22	2.9	0.3746	0.9271	43.508
4	30	28	4.8	26	4.0	0.4383	0.8987	68.065
5	30	36	6.2	32	5.5	0.5299	0.8480	106.766
6	30	40	8.0	38	7.1	0.6156	0.7880	148.787
7	30	32	6.2	36	7.1	0.5877	0.8090	145.829
8	30	26	4.4	29	5.3	0.4848	0.8746	97.080
9	30	20	2.0	23	3.2	0.3907	0.9205	49.716
10	30	00	0.0	10	1.0	0.1736	0.9848	7.385
							Total	676.837

Total discharge flowing in the river = 677 m^3/s

Q = Average depth $\times V^2 \sin\theta \times \cos\theta \times$ time

9.12 STAGE DISCHARGE RELATION

When a number of discharge observations along with the stage observations are taken at a gauging site, these observations are plotted on a simple graph, with discharge on the x-axis and stage on the y-axis. Once a relation of stage and discharge is established, it becomes easy to calculate the discharge flowing in the channel.

Such a graph is known as *Stage Discharge Relation* or *a Rating Curve*. Normally, this is parabolic in nature as shown in Fig. 9.33.

9.12.1 EXTRAPOLATION OF RATING CURVE

The rating curve will have to be extended for higher values of the stage. This can be done by the following methods:

(a) Simple judgement: The rating curve may be extended by simple judgement as shown in Fig. 9.33.
(b) Logarithmic method: Equation of the available curve may be assumed to be $Q = k\,d^n$

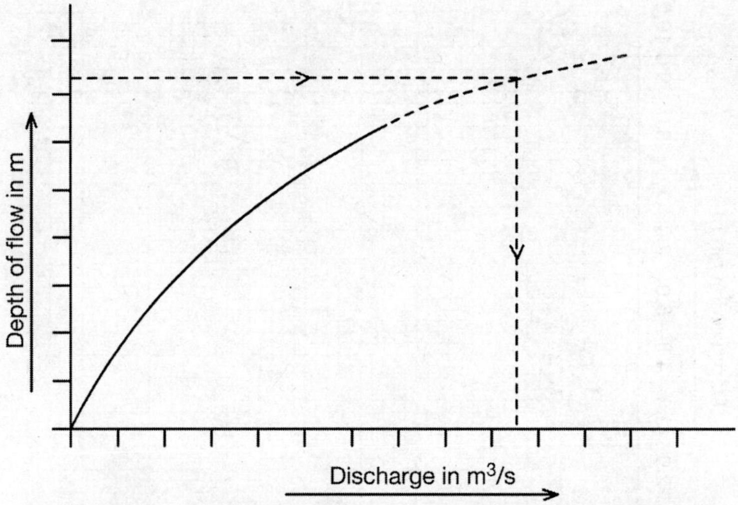

Fig. 9.33 Stage-discharge curve

where, Q = Discharge in m^3/s
d = Stage in metres
k, n = Some constants

Taking logarithms of both sides we get,

$$\text{Log } Q = \text{Log } k + n \log d$$

If the available curve is plotted on a log–log paper, then it will be a straight line. This straight line may be extended to determine the discharge corresponding to higher stage as shown in Figs. 9.34 and 9.35.

Example 9.15

The rating curve coordinates obtained at a gauging site are as follows:

Stage (m)	4.0	4.9	6.5	7.2	7.7
Discharge (m^3/s)	40	60	100	120	140

Find (i) Equation of the rating curve
(ii) Discharge when the stage would be 8.6 m
(iii) Stage corresponding to the discharge of 200 m^3/s

Solution:

(i) Let the equation of the rating curve be

$$Q = k\,d^n$$

Substituting the values of $Q = 40$ m^3/s and $Q = 100$ m^3/s in the assumed equation

We get, $40 = k \times 4^n$ and $100 = k \times 6.5^n$

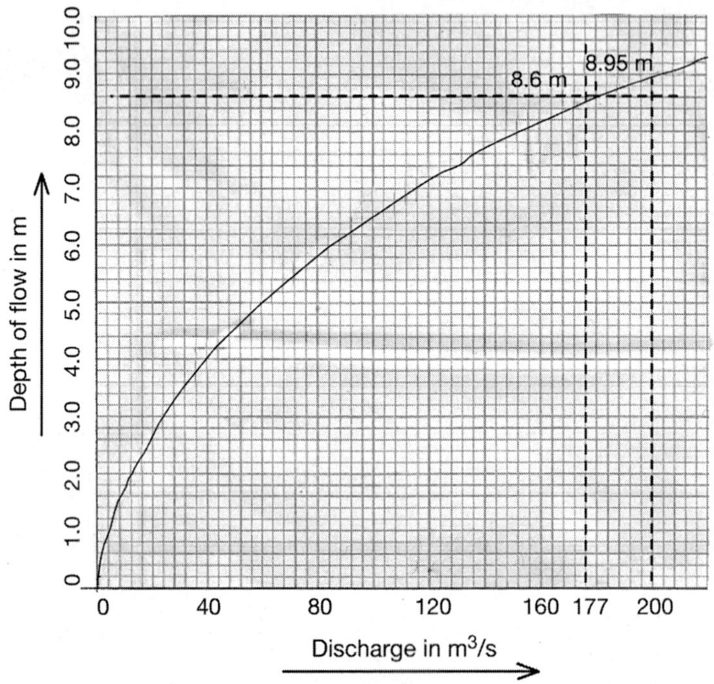

Fig. 9.34 A simple graph of stage discharge relation extended

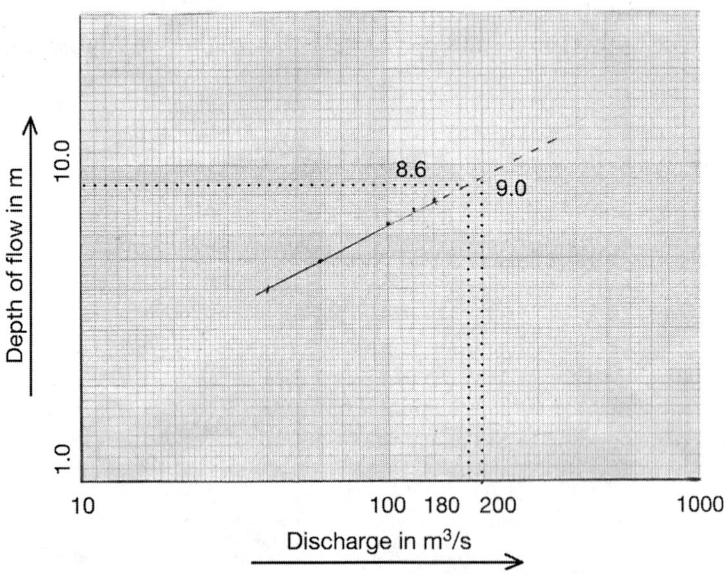

Fig. 9.35 A log–log graph of stage discharge relation extended

Dividing the second equation by the first

$$\frac{100}{40} = \left(\frac{6.5}{4.0}\right)^n = 2.5$$

Taking logarithms of both sides with base as 'e'

$$n \log 1.625 = \log 2.5$$
$$n\, 0.4855 = 0.9162$$

Therefore, $n = \dfrac{0.9162}{0.4855} = 1.887$

Substituting the value of n in one of the equations

We get, $40 = k \times 4^n = k \times 4^{1.887} = k \times 13.68$

Therefore $k = \dfrac{40}{13.68} = 2.92$

Thus, the equation of the rating curve will be $Q = 2.92 \times d^{1.887}$

The rating curve coordinates are plotted on a simple graph paper. The curve was extrapolated beyond and the details as required were observed as follows:

(i) Discharge when the stage is 8.6 m = 177 m³/s
(ii) Stage corresponding to a discharge of 200 m³/s = 8.95 m

The rating curve coordinates were also plotted on a log–log paper. It was noticed that the graph is a straight line. This straight line was extended beyond and the required details were observed as follows:

(i) Discharge when the stage is 8.6 m = 180 m³/s
(ii) Stage corresponding to a discharge of 200 m³/s = 9.0 m

9.13 STREAM GAUGING NETWORK

The stream flow data is the basic data. All further designs are based on these data and hence sufficient number of stream gauging stations should be established. The WMO has suggested the following criteria.

(a) Arid zones
    Minimum = 1 station per 5,000–20,000 km^2
(b) For temperate Mediterranean and tropical zones
    Minimum = 1 station per 1,000–2,500 km^2
(c) Mountainous areas of temperate Mediterranean and tropical zones
    Minimum = 1 station per 300–1,000 km^2

A stream gauging station should be established at the following locations:

(i) Downstream of a possible dam or a hydraulic structure
(ii) At the inlet or the outlet of a natural lake
(iii) Upstream of a city subject to flood damage
(iv) A point just above the tidal influence
(v) On a tributary just u/s of its confluence with the main river

REVIEW QUESTIONS

1. Define discharge. Discuss the various units used for it.
2. What are the main objectives of the measurement of discharge?
3. Discuss the importance of the measurement of discharge in the study of hydrology.
4. What is water potential? Discuss its units.
5. What is a stage? Discuss the different methods followed to measure the stage of a channel.
6. Explain with a neat sketch a water-stage recorder.
7. State the different methods adopted to measure the discharge in a channel and explain in details any one of them.
8. Discuss the criteria for the selection of a site for discharge measurement.
9. Discuss the procedure followed to evaluate the cross-sectional area and the hydraulic mean depth at a cross section of a channel.
10. Explain how the Manning's coefficient is used to find the discharge flowing in a channel.
11. Explain with a neat sketch the velocity profile along a cross section of a channel.
12. Explain why steady uniform flow is assumed in the case of area–slope method of discharge measurement.
13. Discuss the criteria for the selection of a float for discharge measurement.
14. Describe the procedure followed to evaluate the surface velocity.
15. Discuss the precautions to be taken while taking observations of velocity by floats.
16. Explain the construction of a cup-type current meter. Also explain how it is used to evaluate the velocity of flow.
17. Explain the construction of a propeller-type current meter. Also explain how it is used to evaluate the velocity of flow.
18. Discuss the various methods of lowering the current meter to the desired depth for taking the velocity observations.
19. Discuss the corrections due the flowing water to be applied, when a current meter is used to observe velocity of flow.
20. What is an isovel? Sketch a pattern of isovels along a cross section of a channel.
21. Explain with a neat sketch the working of a Pitot tube.
22. Explain the salt-titration method of discharge measurement. Also discuss the properties of the salt/chemical to be used in this method.
23. Explain with neat sketches the different notches that can be used for the measurement of discharge.
24. Explain with neat sketches the different types of flumes that can be used for discharge measurement.
25. Explain the hydraulic model method of discharge measurement.
26. Explain with a neat sketch the ultrasonic method of discharge measurement.
27. Explain with a neat sketch the electromagnetic induction method of discharge measurement.
28. Explain with a neat sketch the moving-boat technique for discharge measurement.
29. Discuss the advantages and the disadvantages of ultrasonic method and electromagnetic method relative to the method by current meter.
30. Write short notes on:
    a. Basic equation of discharge measurement
    b. Suspension weight gauge method of measuring the depth of flow in a channel
    c. Float type water-stage recorder

d. Area–slope method of discharge measurement
e. Echo sounder
f. Hydrographic survey
g. Merits and demerits of area–slope method of discharge measurement.
h. Types of floats
i. Rod float
j. Integrate method of velocity observation by a current meter
k. Mid-section method of evaluating the discharge in a channel
l. Crest gauge
m. Extension of the rating curve

31. Differentiate between:
   a. A channel and a canal
   b. Vertical gauges and inclined gauges
   c. Cup-type current meter and propeller-type current meter
   d. Single-point method, two-point method and three-point method of velocity observation by a current meter
   e. Air-line correction and wet-line correction
   f. Standing wave flume and Parshall flume

NUMERICAL QUESTIONS

1. A test was conducted for the measurement of discharge for a stream. The details are as follows:
   (i) Distance between two sections = 200 m
   (ii) Depth of flow at u/s Section A = 2.52 m
   (iii) Depth of flow at the d/s section = 2.61 m
   (iv) Water level at Section A = 256.45 m
   (v) Water level at Section B = 256.25 m
   (vi) Manning's coefficient (assumed) = 0.035
   Assume, Area of flow in $m^2 = 8 \times d^2$
   and wetted perimeter in m $= 12 \times d$
   Here d is in metres.
   Calculate the discharge flowing in the stream.

Ans: $Q = 68.13$ m^3/s

2. Observations were taken for the velocity of flow in a channel at various depths as follows:

Depth of flow (m)	0	0.5	1.0	1.5	2.0	2.5	3.0	3.5	4.0
Velocity of flow (m/s)	0	0.025	0.12	0.22	0.45	0.73	1.24	1.91	3.35

   Find the average velocity.

Ans: Average velocity = 0.69 m/s

3. Observations taken on a river by the float method for the measurement of discharge are as follows:
   (i) Number of segments = 6
   (ii) Width of stream at the water surface = 23.9 m
   (iii) Width of each segment in metres = 3.0, 4.6, 5.0 4.7, 3.9 and 2.7
   (iv) Depth of flow in metres at the centre of each segment = 0.75, 1.20, 1.78, 1.40, 1.10 and 0.60

(v) Distance between two cross sections = 150 m
(vi) Time taken by a surface float in each section is as follows:

Section	Minute	Second
1	2	16
2	1	46
3	1	23
4	1	20
5	1	40
6	2	15

Find the discharge flowing in the river.

Ans: Discharge = 39.9 m^3/s

4. A current meter was lowered from a bridge by means of a cable for the measurement of discharge in a stream. The observations are as follows:
   (i) Depth of flow in the stream = 4.0 m
   (ii) Vertical distance between the index point up to water surface = 3.0 m
   (iii) Angle made by the cable with the vertical = 28°

   Find the length of the cable from the index point up to the bed.

Ans: Length of the cable = 7.56 m

5. Isovels were drawn on the cross section of a channel covering an area of 25 m^2. The percentage of distribution of the area between the two consecutive isohyets is as follows:

Serial no.	Isovels (m/s)	% distribution
1	Less than 1.0	5
2	1.0–1.2	22
3	1.2–1.4	27
4	1.4–1.6	36
5	Above 1.6	10

   Find the discharge.

Ans: 33.57 m^3/s

6. In a stream, 15 mg/cc of salt at the rate of 15 cc/s was discharged. The concentration of the salt after thorough mixing at a location of 100 m on the d/s was found to be 5 ppb. Find the discharge.

Ans: 45 m^3/s

7. For a gauging site, the rating curve coordinates are as under:

Stage (m)	1.0	1.8	2.0	3.1	4.3	5.3	6.6
Discharge (m^3/s)	18	24	28	33	39	41	44

Find: (i) Rating curve equation
(ii) The discharge when the stage is 3.51 m
(iii) The stage when the discharge is 42 m^3/s

Ans: (i) $Q = 23.86\, d^{0.31}$
(ii) $Q = 35\ m^3/s$
(iii) $d = 6.3$ m

8. The discharge measurement observations by the moving-boat technique were taken on a river. The boat was kept moving at a constant speed of 1.0 m/s. Observations of angle and depth of flow were taken every after 30 s.

Time (s)	0	30	30	30	30	30	30	30	30
Angle (degree)	0	20	26	34	42	32	24	18	0
Depth (m)	0	2.0	3.0	4.2	5.8	6.6	5.0	3.2	0

Find the discharge.

Ans: $Q = 428.5\ m^3/s$

MULTIPLE CHOICE QUESTIONS

1. A water-stage recorder records:
(a) Maximum water level reached (b) Minimum water level reached
(c) Water level at that moment (d) Average water level

2. Echo sounder is an equipment that records:
(a) Velocity of flow (b) Discharge
(c) Depth of flow (d) Direction of flow

3. The velocity of flow at the surface is
(a) Equal to the average velocity (b) Two times the average velocity
(c) 0.85 times the average velocity (d) 1.250 times the average velocity

4. The average velocity of flow is observed at
(a) 0.2D (b) 0.4D
(c) 0.6D (d) 0.8D

5. The single-point method of measurement of velocity is to observe velocity at
(a) 0.2D (b) 0.4D
(c) 0.6D (d) 0.8D

6. The unit of discharge is
(a) m^3/s (b) Cumecs
(c) m/s per unit area of the catchment (d) Any of these three

7. The two-point method of measurement of velocity is the average velocity as
(a) $\dfrac{V_{0.2D}+V_{0.6D}}{2}$ (b) $\dfrac{V_{0.4D}+V_{0.8D}}{2}$

(c) $\frac{V_{0.2D}+V_{0.8D}}{2}$ (d) $\frac{V_{0.2D}+V_{0.4D}}{2}$

8. The three-point method of the measurement of velocity is the average velocity as

(a) $\frac{V_{0.2D}+2V_{0.6D}+V_{0.8D}}{4}$ (b) $\frac{V_{0.2D}+2V_{0.4D}+V_{0.8D}}{4}$

(c) $\frac{V_{0.2D}+2V_{0.4D}+V_{0.6D}}{4}$ (d) $\frac{V_{0.4D}+2V_{0.6D}+V_{0.8D}}{4}$

9. The unit of water potential is

(a) Million m^3 (b) ha m
(c) m per unit area of catchment (d) Any of these three

10. 1 million m^3 is equal to
(a) 1×10^1 ha m (b) 1×10^2 ha m
(c) 1×10^3 ha m (d) 1×10^4 ha m

11. Crest gauge is used to record:
(a) Maximum water level reached (b) Minimum water level reached
(c Average water level (d) Water level at the moment

12. A rating curve of a stream shows the variation of discharge w.r.t.
(a) Depth of flow (b) Velocity of flow
(c) Area of flow (d) All the above

13. The air-line and wet-line corrections depend on
(a) Velocity of flow (b) Unit weight of water
(c) Weight of chain and the current meter (d) All the above

14. The salt titration method is suitable for
(a) Small streams (b) Meandering rivers
(c) Very big streams (d) Rivers in flood

15. The science of discharge measurement is known as
(a) Hydrography (b) Hydrometry
(c) Hydrometeorology (d) None of the above

16. An observation well is necessary when the stage is to be measured by
(a) Vertical gauges (b) Inclined gauges
(c) Float gauges (d) None of the above

17. In the area–slope method, the flow is assumed to be
(a) Unsteady and non-uniform (b) Unsteady and uniform
(c) Steady and non-uniform (d) Steady and uniform

18. The method that is not a direct stream-flow measurement technique is
(a) Area–velocity method (b) Area–slope method
(c) Salt titration method (d) Ultrasonic method

19. In the moving-boat technique, the observations to be taken are
    (a) Depth of flow, velocity of the boat and current meter observations
    (b) Depth of flow, current meter observations and angle θ
    (c) Velocity of boat, angle θ and depth of flow
    (d) Angle θ, velocity of boat and depth of flow

20. The only information that is required to estimate the flood discharge from the rating curve is
    (a) Velocity of flow (b) Depth of flow
    (c) Maximum precipitation (d) Water surface slope

21. Isovel is a line joining locations having equal:
    (a) Depth of flow (b) Velocity of flow
    (c) Hydrostatic pressure (d) Velocity head

ANSWERS TO MULTIPLE CHOICE QUESTIONS

1. c	2. c	3. d	4. c	5. c	6. d	7. c	8. a	9. d
10. b	11. a	12. a	13. d	14. a	15. d	16. c	17. d	18. d
19. d	20. b	21. b						

*** ——— ***

10 Flood Routing

Chapter Outline

10.1 DEFINITION

Sometimes it becomes essential to know the depth of flow at a specific location of a channel. This can be done by following the flood-routing procedure.

When a river is in floods, the depth of the flow and the hydrograph at different locations along the river course are not the same, since the flow in the river is non-uniform and unsteady. In such a case, the depth of the flow and the hydrograph at each section depend on: (a) the depth of the flow and the hydrograph at a section u/s of it and (b) the hydraulic characteristics of the river in between the two sections such as the slope of the river, the length of the reach, the Manning's coefficient, the cross-sectional area and so on.

Thus, the depth of the flow and the hydrograph at the lower section can be calculated having the above information.

Similarly, the outflow from a reservoir will depend on the following factors:

1. The inflow hydrograph
2. The storage capacity of the reservoir
3. The discharging capacity of the spillway weir

Thus, the outflow hydrograph from a reservoir, whether gated or ungated, can be determined by knowing (1) the inflow hydrograph, (2) the hydraulic characteristics of the reservoir and (3) the discharging capacity of the spillway weir.

Flood routing can, therefore, be defined as *the procedure for calculating the depth of flow and the hydrograph at a section knowing the flood hydrograph at the u/s section and the hydraulic characteristics of the channel in between or the hydraulic characteristics of the reservoir*.

10.2 TYPES OF FLOOD ROUTING

Flood routing may be divided into two parts:

1. flood routing through a channel and
2. flood routing through a reservoir.

The flood-routing procedure may be in accordance with the following methods:

1. Hydraulic routing
2. Hydrologic routing
3. Routing machines

10.2.1 Hydraulic Flood Routing

Hydraulic flood routing through a river can be carried out by applying the following three basic equations:

1. Continuity equation
2. Energy equation
3. Momentum equation

When the flow phenomenon in a channel is non-uniform and unsteady, it is very difficult to obtain a true picture of the flow by this procedure since it is very complex. Also, the solution can be obtained only by assuming a uniform flow.

Therefore, the results obtained by this method are not accurate.

10.2.2. Hydrologic Flood Routing

In the hydrologic-routing procedure for a stretch of a river or for a reservoir, the following continuity equation is followed:

$$\text{Inflow} - \text{Outflow} = \text{Change of storage}$$

Over a time interval Δt, symbolically, it may be stated as:

$$I - Q = \Delta s \tag{10.1}$$

where, I = Inflow

Q = Outflow

Δs = Change in storage

The inflow-hydrograph ordinates at a time interval of Δt will be I_1, I_2 and I_3, and these will be known since the inflow hydrograph is known. The outflow ordinates may be Q_1, Q_2, Q_3 and so on. However, these are not known and are to be calculated.

Over a time interval Δt, the inflow volume will be $= \dfrac{(I_1 + I_2)}{2} \times \Delta t$.

And, over the same time interval Δt, the outflow volume will be $= \dfrac{(Q_1 + Q_2)}{2} \times \Delta t$.

The difference between the inflow volume and the outflow volume will be the change in the storage $= S_2 - S_1$. Here, neither the energy equation nor the momentum equation is applied, but a relation between inflow–outflow and storage is used.

Thus,

$$\frac{(I_1+I_2)}{2} \times \Delta t - \frac{(Q_1+Q_2)}{2} \times \Delta t = S_2 - S_1 \qquad (10.2)$$

10.2.3 Flood-Routing Machines

Mechanical flood routers have mechanical gears to represent the reservoir system (Fig. 10.1). These days, electrical voltage system and/or electronic systems are used to represent the fluctuations in a reservoir or in a channel (Fig. 10.2). The results obtained by this method are not accurate.

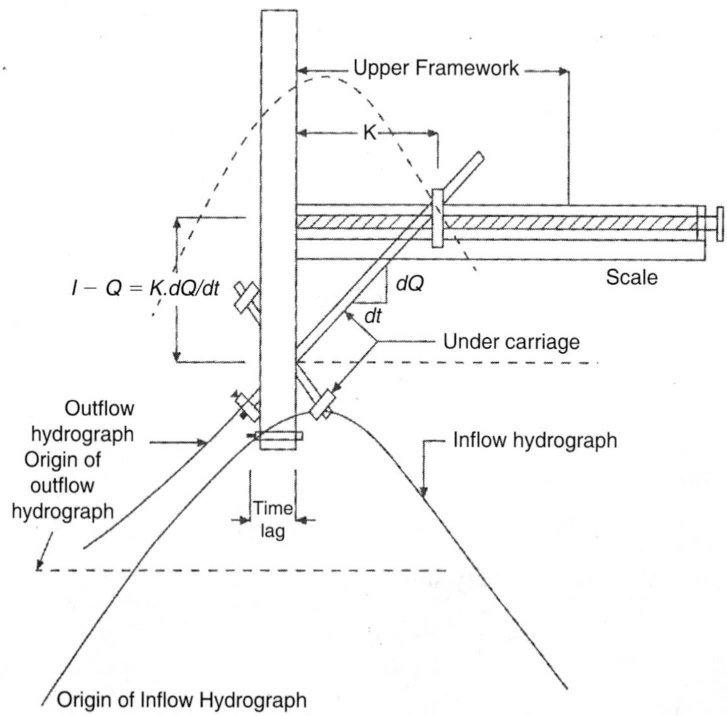

Fig. 10.1 Mechanical flow router

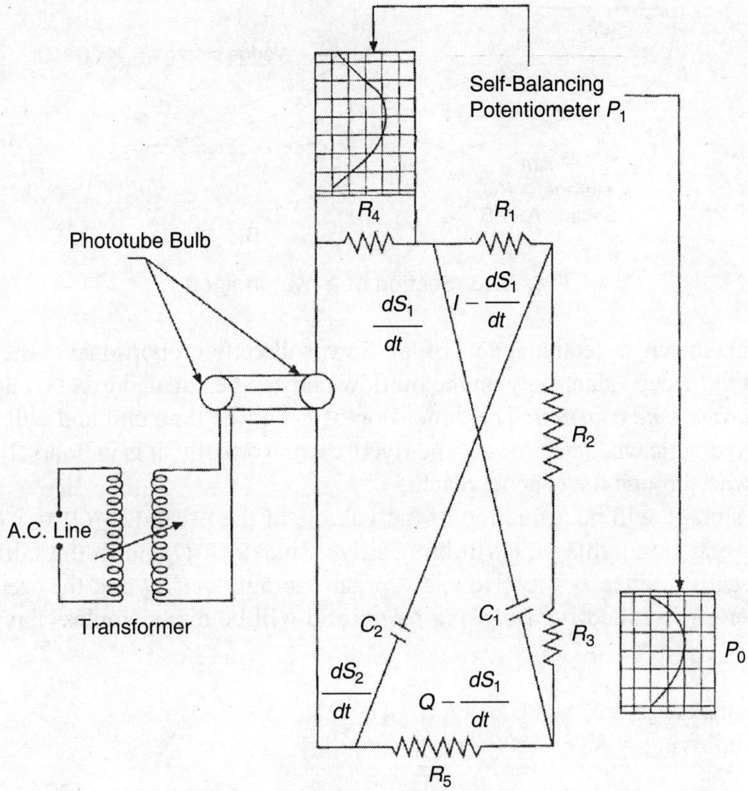

Fig. 10.2 Electric analogue routing machine

10.3 FLOOD ROUTING THROUGH CHANNELS

In a river during floods, the flow is non-uniform and unsteady. This type of flow is very difficult to solve. The hydraulic characteristics vary from stage to stage and also from channel to channel. There may be a lateral inflow or outflow, also.

All these conditions make the analysis of the problem increasingly difficult. However, neglecting all the above conditions and assuming changes occurring gradually with time, the problem can be simplified and solved. It is also known as the *stream-flow routing*.

10.3.1 Muskingum Method

This method was first followed by G. I. McCarthy for the Muskingum River in the USA. Hence, the method is known as the 'Muskingum method'.

Consider two sections A and B of a river at a distance L and assume that there is neither inflow nor outflow in this reach. During floods, the stages at the two cross sections will differ as shown in Fig. 10.3.

The storage in this river reach can be divided into two parts—*the prism storage* and the *wedge storage*. The prism storage is formed by a volume of constant cross section along the length of the river. The wedge storage is the volume between the top of the prism and the water surface.

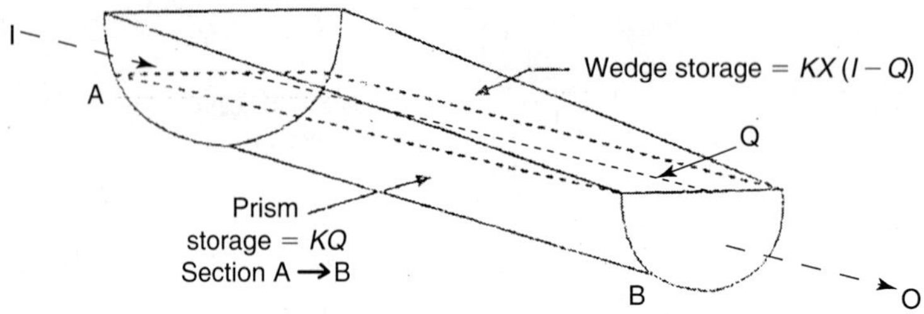

Fig. 10.3 Section of a river in flood

Assuming that the cross-sectional area of the flow is directly proportional to the discharge, the prism storage will be dependant only on the outflow and can be equated to $K \times$ outflow, where K is called the *storage time constant.* The dimension of K will be time unit and will depend on the length and the hydraulic characteristics of the river reach. Normally, it is in hours. It represents the *time of travel wave* through the channel reach.

The wedge storage will be a fraction of the volume of the prism $(I - Q) = x\,(I - Q)$, where $x < 1$. In the rising flood, this part will be positive, since $I > Q$ and in the falling flood, this part will be negative, since $I < Q$. Now, the prism storage $= K\,Q$ and the wedge storage $= x\,K\,(I - Q)$. Here, x is called *the weighing factor* and will be dimensionless having a range of $0 < x < 0.3$.

$$\text{The total storage} = S = KQ + xK(I - Q)$$

$$\text{or} \quad \text{after simplifying} = K[xI + (1 - x)Q] \tag{10.3}$$

Equation (10.3) is known as the ***Muskingum storage equation.***

Thus, at the beginning, $\quad S_1 = K[xI_1 + (1 - x)Q_1]$.

And after a time interval Δt, $\; S_2 = K[xI_2 + (1 - x)Q_2]$.

$$\text{Therefore, the change in storage} = S_2 - S_1 = K[x(I_2 - I_1) + (1 - x)\,(Q_2 - Q_1)] \tag{10.4}$$

When $x = 0$, there will not be any wedge storage. The inflow will be equal to the outflow and the water surface will be parallel to the bed.

While considering Eqs. (10.2) and (10.4), we see that I_1, I_2 and I_3 are the inflow-hydrograph coordinates (these are known) and Q_1, Q_2 and Q_3 the outflow-hydrograph coordinates (these are to be calculated).

By rearranging and combining the two equations, the following equation is derived:

$$Q_2 = C_0 I_2 + C_1 I_1 + C_2 Q1 \tag{10.5}$$

Equation (10.5) is popularly known as ***Muskingum routing equation.***

Such that

$$C_0 = \frac{\Delta t/K - 2x}{2(1 - x) + (\Delta t/K)}$$

$$C_1 = \frac{\Delta t/K + 2x}{2(1 - x) + (\Delta t/K)}$$

$$C_2 = \frac{2(1 - x) - (\Delta t/K)}{2(1 - x) + (\Delta t/K)}$$

C_0, C_1 and C_2 are called the *routing coefficients*. It may be checked that $C_0 + C_1 + C_2 = 1$.

C_0, C_1 and C_2 can be evaluated, if Δt, K and x are known and then from Eq. (10.5), Q_2 can be calculated. S_2 can be calculated from Eq. (10.4), once Q_2 is known.

Then, knowing C_0, C_1, C_2, Q_2 and S_2, we can calculate Q_3 and S_3.

Knowing C_0, C_1, C_2, Q_3 and S_3, we can calculate Q_4 and S_4. Thus, this forward substitution procedure can be followed to find the outflow-hydrograph ordinates, $Q_1, Q_2, Q_3, \ldots$.

This outflow hydrograph will be the inflow hydrograph for the next river reach. The inflow and the outflow hydrographs observed in a specific case are shown in Fig. 10.4.

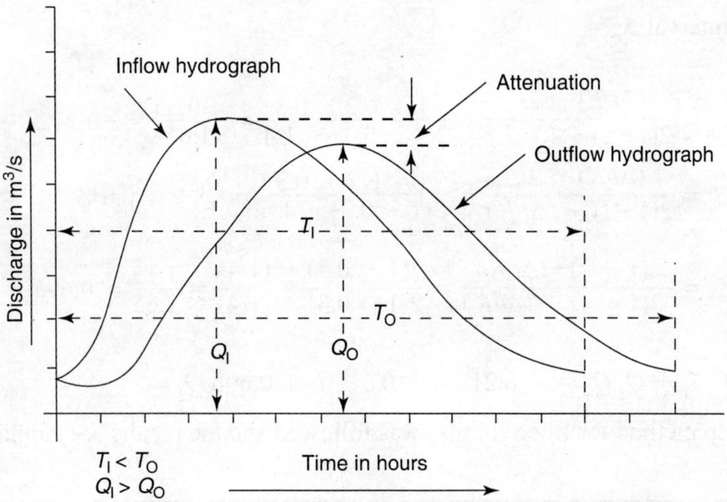

Fig. 10.4 Inflow and outflow hydrographs in a river

The following observations can be made:

1. The maximum outflow discharge is less than the maximum inflow discharge. This is known as *flood attenuation*.
2. The time of the maximum-outflow discharge is more than that of the maximum-inflow discharge. This difference is known as *translation or lag*.
3. The base period of the outflow hydrograph is more than that of the inflow hydrograph.
4. The inflow volume should equal the outflow volume.

10.3.2 Selection of Δt

Over a span of time Δt, the average inflow volume is calculated by assuming a straight-line variation between I_1 and I_2 and also the outflow volume, by assuming a straight-line relation between Q_1 and Q_2. The accuracy will, therefore, depend on Δt. If Δt is small, it will give more accurate results. Δt should also be less than the time of travel between the reach of the river, i.e sections A and B.

Normally, Δt is taken to be 1/2 or 1/3 time of the travel between the reaches of the river selected.

Care should be taken so that the peak of the inflow hydrograph is not missed while selecting Δt.

Example 10.1

In case of an open channel, the inflow hydrograph at a location is as follows:

Time (h)	0	1	2	3	4	5	6	7	8	9	10
Discharge (m^3/s)	10	16	28	45	37	30	24	19	15	12	10

The values of K and x are 4 h and 0.3, respectively. Route the hydrograph.

Solution:

Assume the time interval $\Delta t = 1$ h

Therefore,

$$C_0 = \frac{(\Delta t/K) - 2x}{2(1-x) + (\Delta t/K)} = \frac{1/4 - 2 \times 0.3}{2(1 - 0.3) + 1/4} = \frac{-0.35}{1.65} = -0.212$$

$$C_1 = \frac{(\Delta t/K) + 2x}{2(1-x) + (\Delta t/K)} = \frac{1/4 + 2 \times 0.3}{2(1 - 0.3) + 1/4} = \frac{0.85}{1.65} = 0.515$$

$$C_2 = \frac{2(1 - x)-(\Delta t/K)}{2(1 - x) + (\Delta t/K)} = \frac{2(1 - 0.3) - (1/4)}{2(1 - 0.3) + (1/4)} = \frac{1.15}{1.65} = 0.696$$

Now,

$$Q_2 = C_0 I_2 + C_1 I_1 + C_2 Q_1 = -0.212\, I_2 + 0.515\, I_1 + 0.696\, Q_1$$

The step-by-step method for flood routing was followed and the results are tabulated below.

Time (h)	Inflow (m^3/s)	$-0.212 I_2$ (m^3/s)	$0.515\, I_1$ (m^3/s)	$0.697\, Q_1$ (m^3/s)	Q_2 (m^3/s)
0	10	–	–	–	10
1	16	−3.392	5.150	6.970	8.728
2	28	−5.936	8.240	6.083	8.387
3	45	−9.540	14.420	5.846	10.726
4	37	−7.844	23.175	7.476	22.806
5	30	−6.360	19.055	15.896	28.591
6	24	−5.088	15.450	19.928	30.290
7	19	−4.028	12.360	21.112	29.444
8	15	−3.180	9.785	20.522	27.127
9	12	−2.544	7.725	18.907	24.088
10	10	−2.120	6.180	16.789	20.840
11	10	−2.210	5.150	14.530	17.560
12	10	−2.120	5.150	12.240	15.270
13	10	−2.120	5.150	10.640	13.660
14	10	−2.120	5.150	9.523	12.550
15	10	−2.120	5.150	8.750	11.780
16	10	−2.120	5.150	8.210	11.240
17	10	−2.120	5.150	7.830	10.860

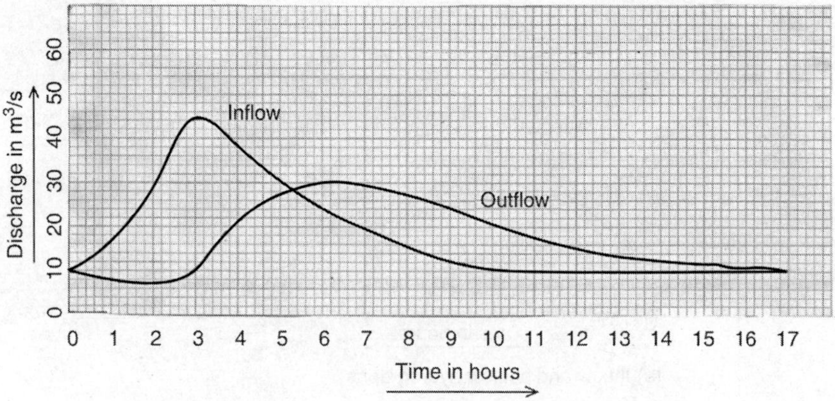

Fig. 10.5 Inflow and estimated outflow hydrographs

The inflow and outflow hydrographs are shown in Fig. 10.5.

Observations:

1. The maximum inflow and outflow discharges are 45 m^3/s and 30.29 m^3/s, respectively.
2. The inflow and the outflow base periods are 10.0 h and 17.0 h, respectively.

10.3.3. Determination of *K* and *x*

The values of K and x will have to be evaluated to find the outflow. These can be calculated from an observed set of inflow and outflow hydrographs in the reach of the river.

The observed inflow and outflow hydrographs in a specific case are shown in Fig. 10.6.

The values of K and x can be calculated by three methods.

First method

Now, $I - Q = \dfrac{dS}{dt}$

and $$S = K\,[x\,I + (1 - x)\,Q] \tag{10.3}$$

Therefore, $$\frac{dS}{dt} = K\left[x\frac{dI}{dt} + (1-x)\frac{dQ}{dt}\right]$$

Therefore, $$I - Q = K\left[x\frac{dI}{dt} + (1-x)\frac{dQ}{dt}\right] \tag{10.6}$$

When the two hydrographs cross each other, then $I = Q$.

Therefore, from Eq. (10.6) we get

$$x = \frac{dQ/dt}{(dQ/dt - dI/dt)}$$

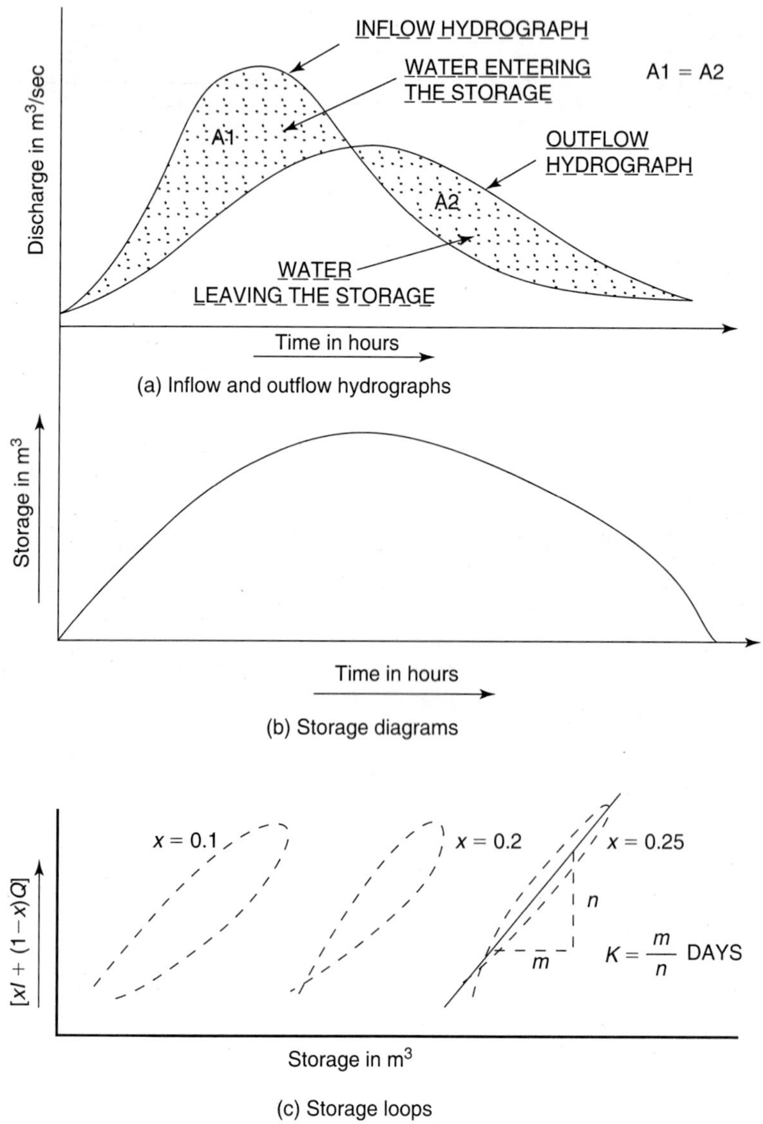

Fig. 10.6 Determination of *K* and *X*

Here,

$$\frac{dQ}{dt} = \text{Slope of the outflow hydrograph at the crossing}$$

$$\frac{dI}{dt} = \text{Slope of the inflow hydrograph at the crossing}$$

Once x is evaluated, K can be evaluated from Equation (10.6) by selecting two points on the two hydrographs for a specific value of T and substituting the values of I, Q, dI/dt, dQ/dt and x.

This method is not accurate since finding a slope to a curve is approximate.

Second method

The values of K and x can be evaluated from Eq. (10.6), by solving the two simultaneous equations. This can be done by selecting two points on each of the inflow and the outflow hydrographs for two specific times T_1 and T_2, so that I, Q, dI/dt and dQ/dt can be evaluated at these two points. Then the only two unknowns are K and x.

Third method

From the observed inflow and outflow hydrographs, it can be seen that initially the inflow will be more than the outflow. At one point, the inflow will be equal to the outflow and then the outflow will be more than the inflow.

When the inflow is more than the outflow, the difference in the volume will be temporarily stored in the river reach and when the outflow will be more than the inflow, this storage will be drained. The storage in the reach will be the shaded part, as shown in Fig. 10.6.

Then, some values of x are assumed and a graph of S versus $[x\,I + (1 - x)\,Q]$ is plotted for each value of x.

Normally, the graphs for the various values of x will represent loops. But for a certain value of x, the graph will be a straight line. This value of x is finalized when the loop is straight or almost straight. The slope of this line will be K. Thus, x and K can be evaluated.

Example 10.2

The observed values of the inflow and the outflow at two cross sections of a stream are as follows:

Time (h)	0	1	2	3	4	5	6	7	8	9	10	11	12	13	14	15	16	17
Inflow (m^3/s)	10	16	28	45	37	30	24	19	15	12	10	10	10	10	10	10	10	10
Outflow (m^3/s)	10	9	8	11	23	29	30	29	27	24	21	18	15	14	13	12	11	10

Calculate the values of K and x, for the Muskingum method of flood routing through channel. The inflow and the outflow hydrographs are as shown in Fig. 10.7.

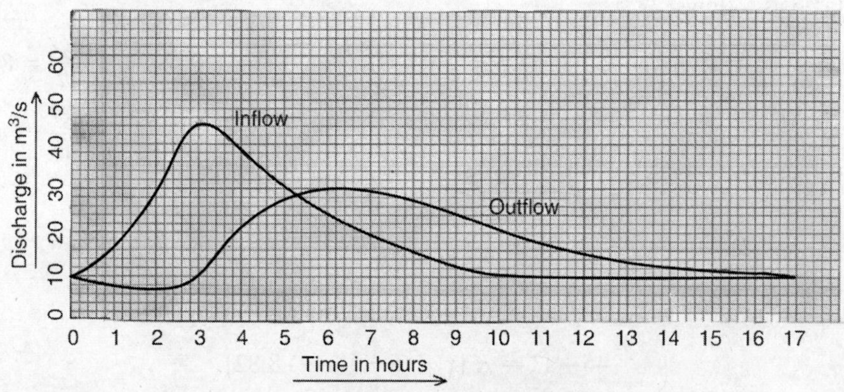

Fig. 10.7 Inflow and outflow hydrographs

Solution:

It can be seen that the two hydrographs cross each other at 5.1 h, when the discharge is 29.3 m^3/s. The values of K and x can be evaluated by the three different methods as follows:

First method

$$x = \frac{dQ/dt}{(dQ/dt - dI/dt)}$$

The values of dQ/dt and dI/dt, where the two hydrographs cross each other, were calculated and were 2.250 and − 6.071, respectively.

Therefore,
$$x = \frac{2.250}{2.250-(-6.071)} = \frac{2.250}{8.321} = 0.27$$

Now,
$$I - Q = K\left[x\frac{dI}{dt} + (1-x)\frac{dQ}{dt}\right] = 0$$

Two points on the two hydrographs for the same value of time are selected as follows:

At $t = 3$ h,

$$I = 45 \text{ m}^3/\text{s},\ Q = 11 \text{ m}^3/\text{s},\ \frac{dI}{dt} = 0,\ \frac{dQ}{dt} = 8.62 \quad \text{and} \quad x = 0.27$$

Substituting these values in the equation, we get

$$45 - 11 = K(0.27 \times 0 + 0.73 \times 8.62)$$

$$K = \frac{34}{6.2} = 5.4 \text{ h}$$

Second method

Two points are selected from each of the two hydrographs, and the slopes at these four points are also calculated as follows:

T	I	Q	$\frac{dI}{dt}$	$\frac{dQ}{dt}$
$T = 3$ h	$I = 45$ m^3/s	$Q = 11$ m^3/s	$\frac{dI}{dt} = 0.0$	$\frac{dQ}{dt} = 8.82$
$T = 2$ h	$I = 28$ m^3/s	$Q = 8$ m^3/s	$\frac{dI}{dt} = 13.38$	$\frac{dQ}{dt} = -0.1$

Now we have,

$$I - Q = K\left[x\frac{dI}{dt} + (1-x)\frac{dQ}{dt}\right] = 0$$

Substituting these values in this equation, we get the following two equations:

$$45 - 11 = K[x \times 0 + (1-x)\,8.82] \tag{i}$$

and

$$28 - 8 = K[x \times 13.38 + (1-x)(-1)] \tag{ii}$$

Dividing (i) by (ii), we get

$$\frac{34}{20} = \frac{(-8.82x+8.82)}{(13.48x-0.1)}$$

Therefore, $$x = \frac{179.8}{634.72} = 0.283$$

Substituting the value of x in (i), we get

$$45 - 11 = K\,(0.283 \times 0 + 0.717 \times 8.62)$$

Therefore, $K = \dfrac{34}{6.18} = 5.5$ h

Third method

From the plot of the two hydrographs, the additional storage in the channel is calculated for every hour. A graph of S vs $[x\,I + (1-x)\,Q]$, for some assumed values of x, is plotted. For this, $x = 0.10$, 0.30 and 0.50 are assumed. The calculations are tabulated below.

(A) $x = 0.10$

Serial no.	S (m³/h)	I (m³/s)	Q (m³/s)	$x\,I$	$(1-x)\,Q$	$x\,I + (1-x)\,Q$
1	3.5	16	9	1.60	8.10	9.70
2	17	28	8	2.80	7.20	10.00
3	44	45	11	4.50	9.90	14.40
4	68	37	23	3.70	19.70	23.40
5	76	29.3	29.3	2.93	26.37	29.30
6	73	24	30	2.40	27.00	29.40
7	65	19	29	1.90	26.10	28.00
8	54	15	27	1.50	24.30	25.80
9	42	12	24	1.20	21.60	22.80
10	30.5	10	21	1.00	18.90	19.90
11	20	10	18	1.00	16.20	17.20
12	12.5	10	15	1.00	13.50	14.50
13	8	10	14	1.00	12.60	13.60
14	4.5	10	13	1.00	11.70	12.70
15	2	10	12	1.00	10.80	11.80
16	0.5	10	11	1.00	9.90	10.80
17	0	10	10	1.00	9.00	10.00

(B) $x = 0.30$

Serial no.	S (m^3/h)	I (m^3/s)	Q (m^3/s)	$x\,I$	$(1 - x)\,Q$	$x\,I + (1 - x)\,Q$
1	3.5	16	9	4.80	6.30	11.10
2	17	28	8	8.40	5.60	14.00
3	44	45	11	13.50	7.70	21.20
4	68	37	23	11.50	16.10	27.20
5	76	29.3	29.3	8.79	20.51	29.30
6	73	24	30	7.20	21.00	28.20
7	65	19	29	5.70	20.30	26.00
8	54	15	27	4.50	18.90	23.40
9	42	12	24	3.60	16.80	20.40
10	30.5	10	21	3.00	14.70	17.70
11	20	10	18	3.00	12.60	15.60
12	12.5	10	15	3.00	10.50	13.50
13	8	10	14	3.00	9.80	12.80
14	4.5	10	13	3.00	9.10	12.10
15	2	10	12	3.00	8.40	11.40
16	0.5	10	11	3.00	7.70	10.70
17	0	10	10	3.00	7.00	10.00

(C) $x = 0.50$

Serial no.	S (m^3/h)	I (m^3/s)	Q (m^3/s)	$x\,I$	$(1 - x)\,Q$	$x\,I + (1 - x)\,Q$
1	3.5	16	9	8.00	4.50	12.50
2	17	28	8	14.00	4.00	18.00
3	44	45	11	22.50	5.50	28.00
4	68	37	23	18.50	11.50	30.00
5	76	29.3	29.3	14.66	14.66	29.32
6	73	24	30	12.00	15.00	27.00
7	65	19	29	9.50	14.50	24.00
8	54	15	27	7.50	13.50	21.00
9	42	12	24	6.00	12.00	18.00
10	30.5	10	21	5.00	10.50	15.50
11	20	10	18	5.00	9.00	14.00
12	12.5	10	15	5.00	7.50	12.50
13	8	10	14	5.00	7.00	12.00
14	4.5	10	13	5.00	6.50	11.50
15	2	10	12	5.00	6.00	11.00
16	0.5	10	11	5.00	5.50	10.50
17	0	10	10	5.00	5.00	10.00

Three graphs of S vs $[x\,I + (1 - x)\,Q]$ were plotted for $x = 0.10$, 0.30 and 0.50. These are shown in Fig. 10.8.

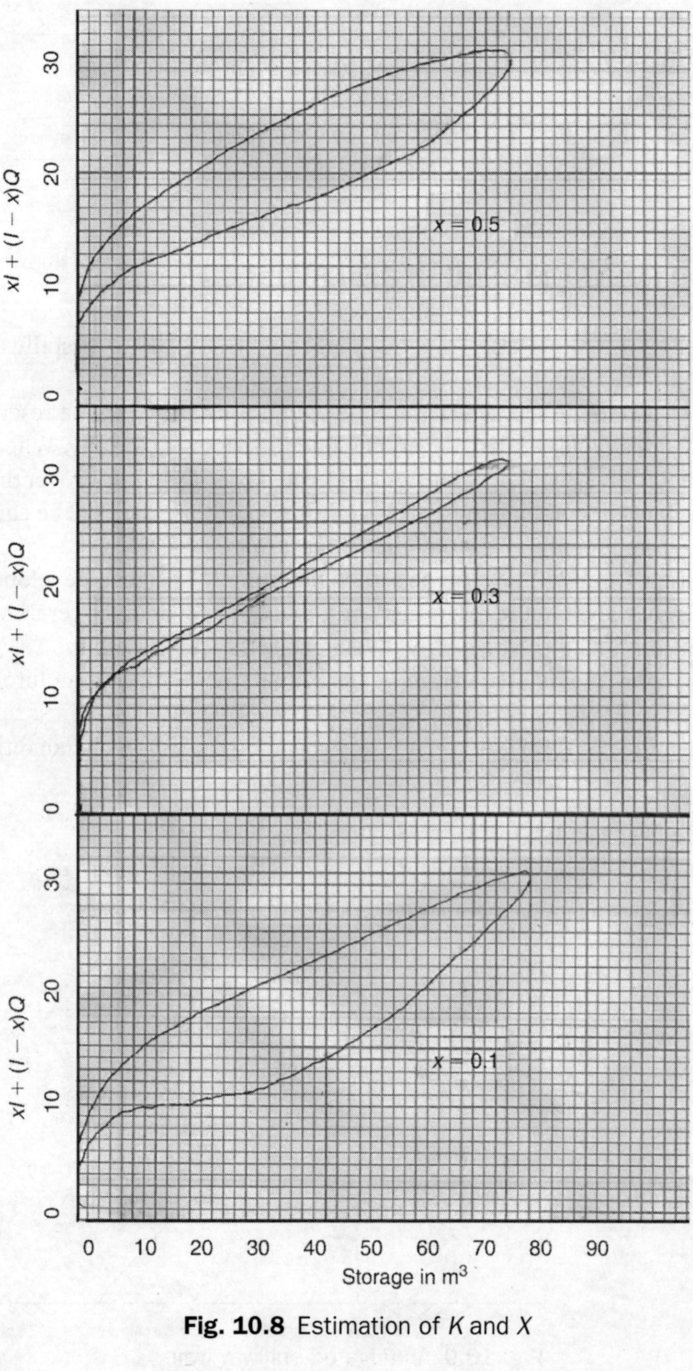

Fig. 10.8 Estimation of K and X

It can be seen that for $x = 0.3$, the graph is practically a line. Hence, this value of x is accepted. The slope of the straight line $= K = 53/13 = 4.07$ h. The values of x and K, evaluated by the three methods, are tabulated below:

	x	K (h)
First method	0.270	5.40
Second method	0.283	5.50
Third method	0.300	4.07

10.4. FLOOD ROUTING THROUGH RESERVOIR

Consider an ungated spillway of a reservoir, vide Fig. 10.9. It is full up to its spillway crest weir and flood water enters into the reservoir.

As the inflow starts into the reservoir, it will be temporarily stored in the reservoir and, hence, the water level in the reservoir will rise depending upon its storage capacity. When the water level rises, a head over the spillway weir develops. As the head develops, outflow over the weir starts. All these phenomena will occur simultaneously. This outflow, therefore, will not be equal to the inflow, since some volume from the inflow is stored in the reservoir.

The reservoir water level will be assumed to be horizontal. The head developed over the weir will depend on the water level in the reservoir. Thus, the water level in the reservoir, the storage in the reservoir and the outflow discharge, all will be dependent on each other.

Since reservoir water level is assumed to be horizontal, the flood routing through reservoir is, therefore, also called *level pool routing*.

The outflow from the reservoir will, therefore, depend on the following conditions:

1. Inflow hydrograph
2. The absorption capacity of reservoir
3. The discharging capacity of the spillway weir

The outflow hydrograph can be evaluated by the **Puls method**.

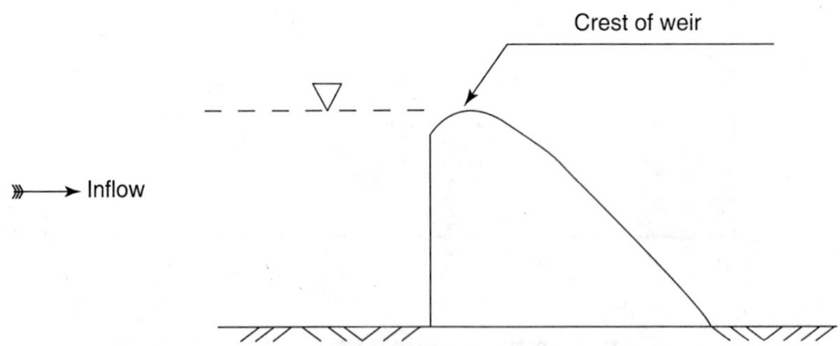

Fig. 10.9 An ungated spillway weir

10.4.1. Puls Method

This method was first suggested by L. G. Puls of the US Army Corps of Engineers and hence is known as the Puls method. It is also known as the *inflow–storage–discharge method (ISD method)*. It is a step-by-step procedure.

For a reservoir, the continuity equation can be written as:

Inflow − Outflow = Change in storage

i.e., $I - Q = \Delta s$

This identity can be rearranged over a span of time Δt as:

$$\frac{(I_1+I_2)}{2} \times \Delta t - \frac{(Q_1+Q_2)}{2} \times \Delta t = S_2 - S_1 \tag{10.2}$$

In this method, Δt has to be assumed suitably,

where, Δt = Time interval in seconds as assumed
I_1 = Inflow in m^3/s at the beginning of time
I_2 = Inflow in m^3/s at the end of time interval
Q_1 = Outflow in m^3/s at the beginning of time interval
Q_2 = Outflow in m^3/s at the end of time interval
S_1 = Storage in reservoir in m^{3}at the beginning of time interval
S_2 = Storage in reservoir in m^3 at the end of time interval

Δt should be selected such that it will be approximately equal to $t_p/5$, where t_p is the time of maximum inflow discharge.

Care should also be taken that the peak inflow discharge is not missed.

Thus,

$$\frac{(I_1+I_2)}{2} \times \Delta t = \text{Average inflow volume in m}^3\text{in time interval } \Delta t$$

And $\dfrac{(Q_1+Q_2)}{2} \times \Delta t$ = Average outflow volume in m^3 in time interval Δt

$$S_2 - S_1 = \text{Change in storage in the reservoir in m}^3 \text{ in time interval } \Delta t$$

Here I_1 and I_2 are known, since these are coordinates of the inflow hydrograph. Q_1 and S_1 are also known at the beginning of the time.

Equation (10.2) can be rearranged as follows:

$$\frac{(I_1 + I_2)}{2} \times \Delta t + S_1 - Q_1 \frac{\Delta t}{2} = S_2 + Q_2 \frac{\Delta t}{2} \tag{10.7}$$

In this equation, all the terms in the LHS are known. And in the RHS, S_2 and Q_2 are not known. However, a relation between the storage and the outflow can be established and Q_2 can be calculated.

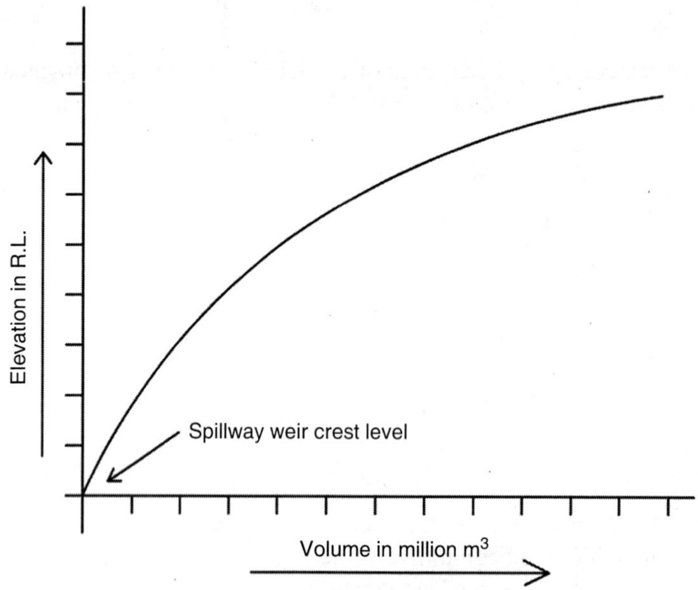

Fig. 10.10 Reservoir storage capacity

The storage in a reservoir and the outflow from a reservoir are related to the water level in the reservoir. These are known as *hydraulic characteristics of the reservoir*.

These relations are as follows:

1. The storage capacity curve of the reservoir (Fig. 10.10).
2. The discharging capacity of the spillway weir (Fig. 10.11).

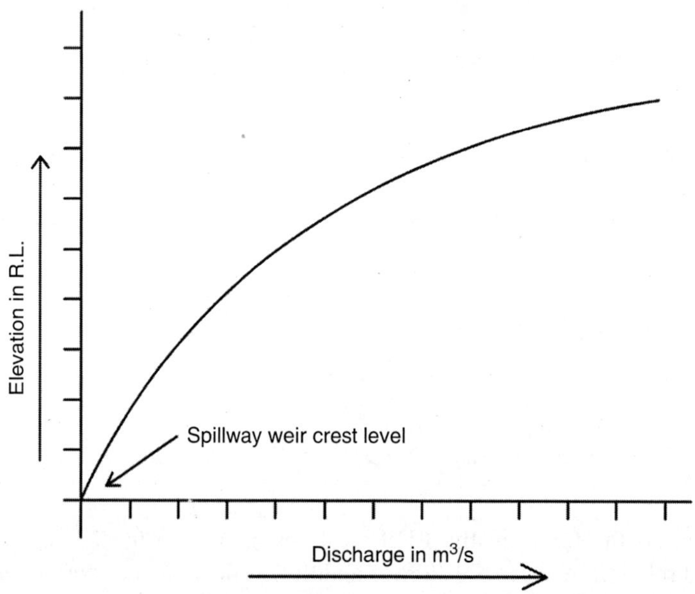

Fig. 10.11 Discharging capacity of a weir

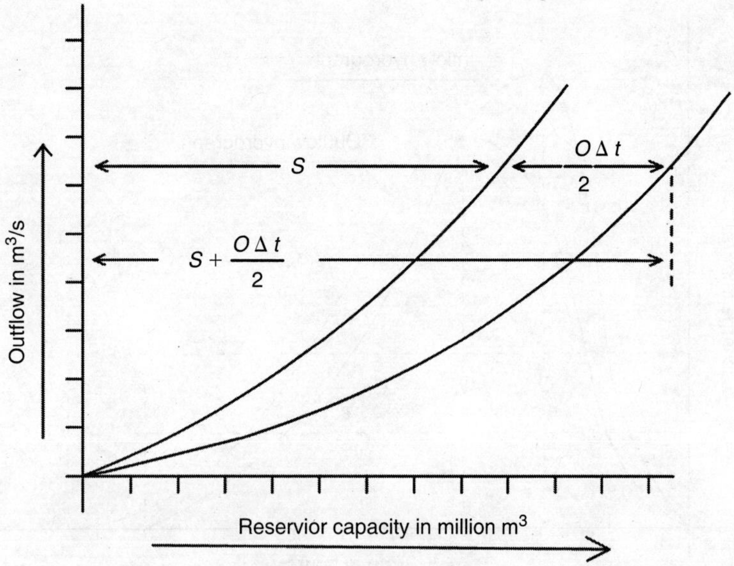

Fig. 10.12 Outflow vs storage capacity of weir

From these two curves, for a specific value of water level, S and Q can be calculated. This means that S and Q are related to each other, and their values can be calculated for different water levels. Thus, a relation of S vs Q can be derived as shown in Fig. 10.12.

On the same graph, in addition, a relation Q vs $(S + \frac{Q}{2}\,\Delta t)$ (Fig. 10.12) can also be established. Here, $\frac{O}{2}\,\Delta t$ can be calculated, since Δt is known and O can be read on the y-axis.

The LHS can be calculated from Eq. (10.7) and then the RHS, i.e. $(S_2 + Q_2/2\,\Delta t)$, can be known. Q_2 and S_2 can be calculated from the graphs in Fig. 10.12. From the value of S_2, in the graph in Fig.10.10, the water level in the reservoir can also be calculated. Thus Q_2, S_2 and the water level in the reservoir are known.

The next step will be,

$$\frac{(I_2+I_3)}{2} \times \Delta t - Q_2\frac{\Delta t}{2} = S_3 + Q_3\frac{\Delta t}{2}$$

From this equation, S_3, Q_3 and the water level in the reservoir can be calculated.

Then the next step will be,

$$\frac{(I_3+I_4)}{2} \times \Delta t - Q_3\frac{\Delta t}{2} = S_4 + Q_4\frac{\Delta t}{2}$$

From this equation, S_4, Q_4 and the water level in the reservoir can be calculated.

Thus, following this step-by-step procedure, Q_1, Q_2, Q_3 and so on, can be evaluated, which is the outflow hydrograph.

Care should be taken, so that the peak of the inflow hydrograph is not missed while selecting Δt. The inflow and the outflow hydrographs observed in a specific case are shown in Fig. 10.13. The following observations can be made:

- The maximum outflow discharge is less than the maximum inflow discharge. This is known as *flood attenuation*.

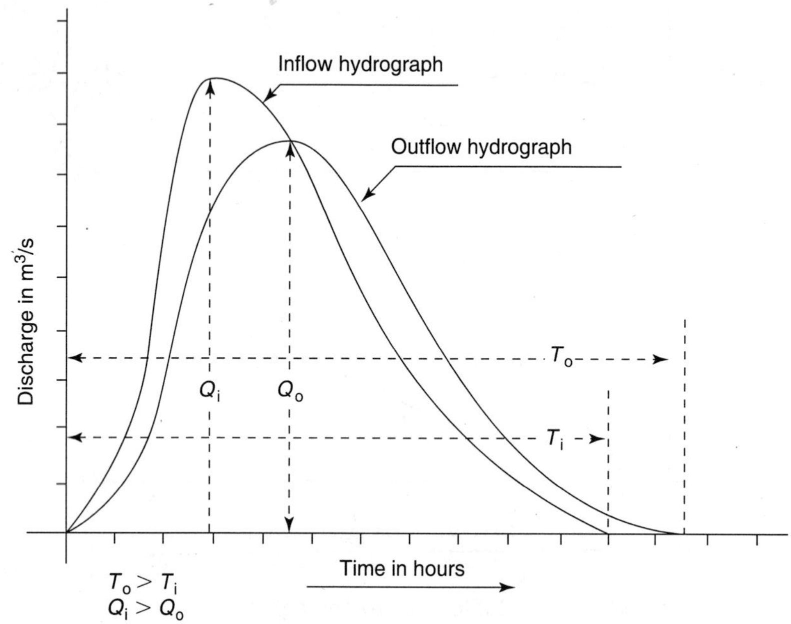

Fig. 10.13 Inflow and outflow hydrographs in a reservoir

- The time of maximum outflow discharge is more than that of the maximum inflow discharge. This difference is known as *translation or lag.*
- The base period of the outflow hydrograph is more than that of the inflow hydrograph.
- The outflow discharge is the maximum when the outflow hydrograph crosses the inflow hydrograph.
- The inflow volume is equal to the outflow volume.

10.4.2. Practical Cases

The Puls method can be applied in the following five practical cases.

1. The spillway weir is constructed and the gates are not installed, or the spillway weir is without gates. When the water level in the reservoir is up to the crest level and flood occurs as per the inflow-flood hydrograph.
2. This case is the same as mentioned above, in the previous point, only that here the water level in the reservoir is below the crest level and flood occurs as per the inflow-flood hydrograph.
3. The spillway weir is a gated one, without any flood lift. The water level in the reservoir is upto the full reservoir level (FRL), and flood occurs as per the inflow-flood hydrograph.
4. The spillway weir is a gated one having a flood lift. The water level in the reservoir is upto the FRL and flood occurs as per the inflow-flood hydrograph.
5. The spillway weir is a gated one, having a flood lift .The water level in the reservoir is below the FRL, and flood occurs as per the inflow-flood hydrograph.

Case no. 1

In this case, as flood enters the reservoir, the water level in the reservoir will start rising. As the water level rises, the water will simultaneously start overflowing the spillway weir. Initially, the outflow will be less than the inflow and the balance will be absorbed temporarily in the reservoir, resulting in the rise of water level in the reservoir. This will cause an increase in the outflow.

As the inflow discharge reduces and is less than the outflow, the water level in the reservoir will start receding. This will reduce the outflow. Finally, first the inflow and then the outflow will reduce to zero.

As there is no change in the water level in the reservoir, there will not be any change in the storage. The inflow volume and the outflow volume will be equal.

It should be noted that the outflow will start to recede when it crosses the inflow hydrograph. The inflow hydrograph, the outflow hydrograph and the water level in the reservoir are shown in Fig. 10.14.

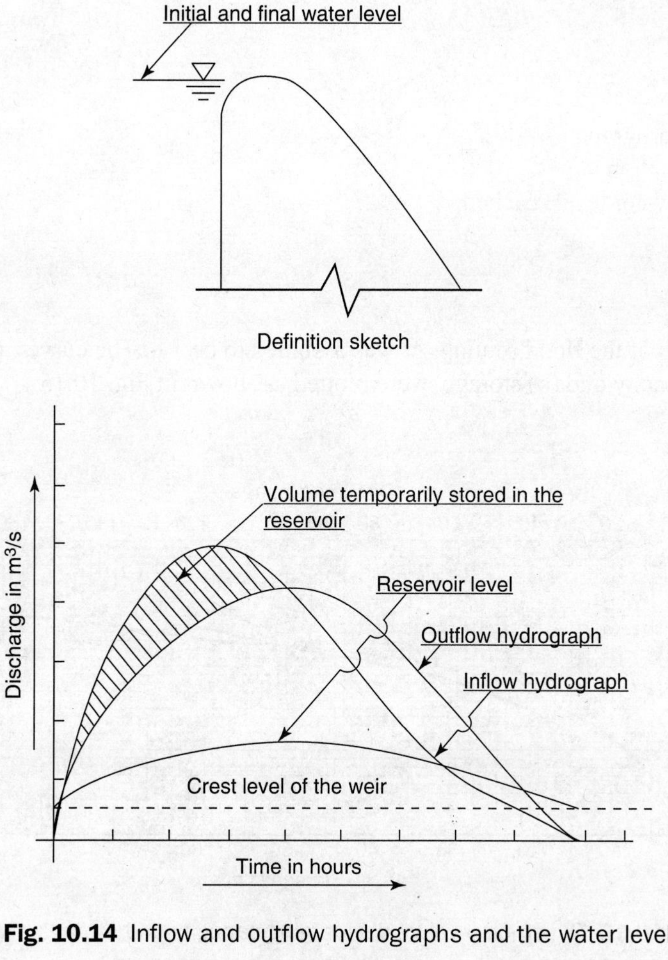

Fig. 10.14 Inflow and outflow hydrographs and the water level

Example 10.3

An ungated spillway has its crest at 550 m. The elevation, storage and discharging capacity of the spillway are as follows:

Elevation (m)	550	551	557	562	565	568	570	572	574	575	576	577
Storage (ha m)	0	3	8	13.6	20.8	31.0	40.0	52.0	66.0	70.0	76.0	80.0
Discharge (m^3/s)	0	4	12	20	32	45	62	92	115	140	160	200

The inflow into the reservoir, when the water level in the reservoir is at 550 m, is as follows:

Time (h)	0	1	2	3	4	5	6	7	8	9	10	11	12	13	14
Discharge (m^3/s)	0	12	50	160	270	240	205	150	95	68	36	23	14	7	0

Find the following:

1. The outflow hydrograph
2. The flood attenuation
3. The maximum water level reached

Solution:

For the calculations of the flood routing, Δt was assumed to be 1 h. The curves, reservoir level vs (a) discharging capacity and (b) storage, were plotted as shown in Fig. 10.15.

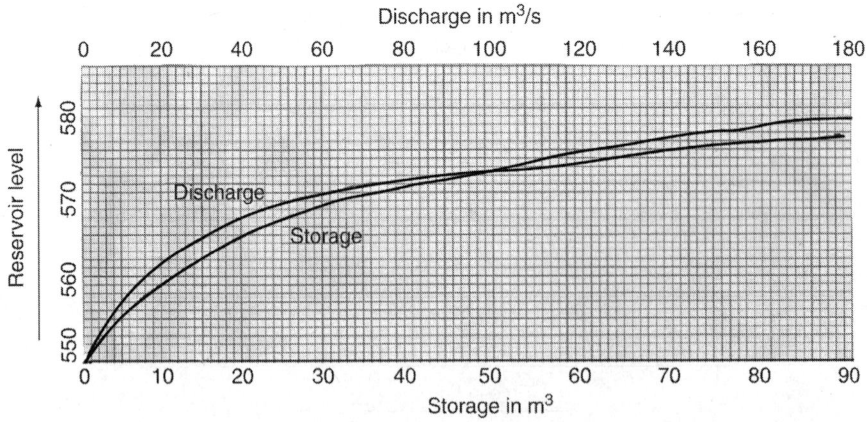

Fig. 10.15 Reservoir level vs discharging capacity and storage of weir

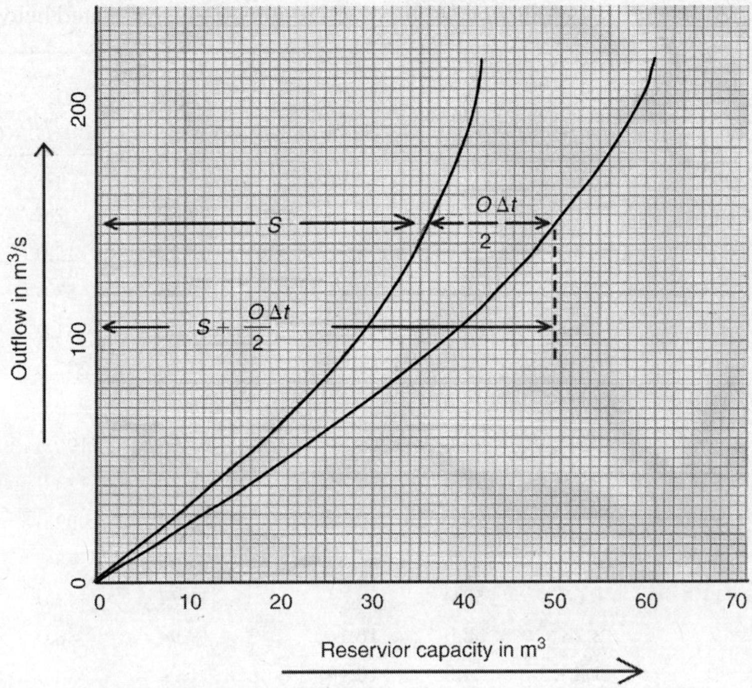

Fig. 10.16 Outflow discharge vs storage capacity of the reservoir

From Fig. 10.15, the two relations—discharge vs (a) storage and (b) storage + $O/2 \times \Delta t$—were established and plotted in Fig. 10.16.

Serial no.	Discharge (m^3/s)	Storage (ha m)	$Q/2 \times \Delta t$ (ha m)	Storage + $Q/2 \times \Delta t$ (ha m)
1	5.0	2.0	0.90	2.90
2	7.5	5.0	1.35	6.35
3	12.0	8.0	2.16	10.16
4	15.0	10.0	2.50	12.50
5	32.0	20.0	5.76	25.76
6	46.0	30.0	8.28	38.28
7	64.0	40.0	11.52	51.52
8	84.0	50.0	15.12	65.12
9	108.0	60.0	19.44	79.44
10	137.0	70.0	24.66	94.66
11	200.0	80.0	36.00	116.00

The step-by-step procedure was followed for the flood routing and is tabulated below.

Time (h)	I_1 (m³/s)	I_2 (m³/s)	$\left(\frac{I_1+I_2}{2}\right)\times\Delta t$ ha m	Q_1 (m³/s)	$\left(S_1-\frac{Q_1}{2}\right)\times\Delta t$ ha m	$\left(S_2+\frac{Q_2}{2}\right)\times\Delta t$ ha m	Q_2 (m³/s)	S_2 (ha m)	Water level (m)
1	0	12	2.16	0.0	0.0	2.16	–	–	–
2	12	50	11.16	2.6	1.64	12.80	2.6	2.1	551.0
3	50	160	37.80	15.0	5.32	43.12	15.0	10.0	559.0
4	160	270	77.40	53.0	22.46	99.86	53.0	32.0	568.8
5	270	240	91.80	151.0	45.82	137.62	151.0	73.0	575.6
6	240	205	80.10	237.0	52.14	132.24	237.0	94.8	577.5
7	205	150	63.90	230.0	50.60	114.50	230.0	92.0	577.4
8	150	95	44.10	190.0	43.80	87.90	190.0	78.0	576.8
9	95	68	29.34	123.0	45.86	75.20	123.0	68.0	574.3
10	68	36	18.72	99.0	40.18	58.90	99.0	58.0	573.0
11	36	23	10.62	75.0	31.50	42.12	75.0	45.0	571.2
12	23	14	6.66	52.0	24.64	31.24	52.0	34.0	568.7
13	14	7	3.78	38.0	16.16	19.94	38.0	23.0	566.6
14	7	0	1.26	24.0	10.68	11.94	24.0	15.0	563.5
15	0	0	0.00	14.0	6.48	6.48	14.0	9.0	558.9
16	0	0	0.00	7.5	3.65	3.65	7.5	5.0	558.0
17	0	0	0.00	3.0	1.30	1.30	3.0	1.8	552.0

The inflow and the outflow hydrographs are as shown in Fig. 10.17.
Flood attenuation = 270 – 237 = 33 m³/s
Maximum water level reached = 577.5 m

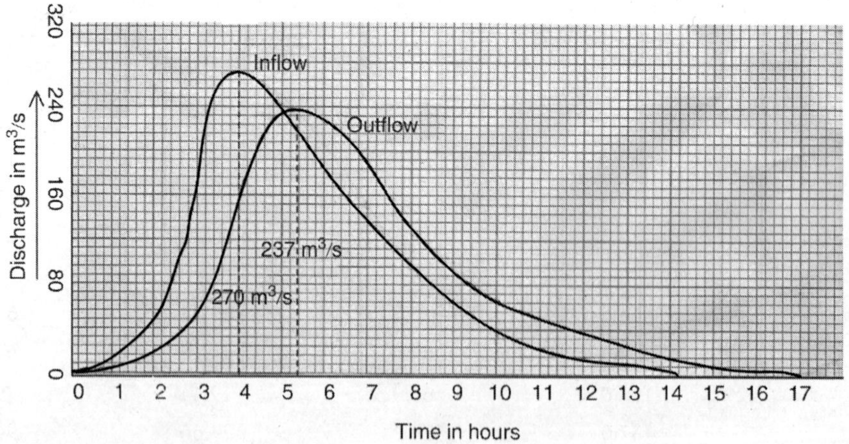

Fig. 10.17 Inflow and estimated outflow hydrographs

Case no. 2

In this case, when the flood occurs, the reservoir level is lower than the crest level of the spillway. The spillway is ungated or the gates are yet to be installed.

As the inflow discharge starts flowing into the reservoir, it will be absorbed in the reservoir. The water level in the reservoir will go on rising and will continue to rise until it reaches the spillway crest level. Upto this stage, there will not be any outflow and all the inflow will be absorbed in the reservoir.

Similar to Case no. 1, as the inflow increases further, the water level in the reservoir will increase above the crest level and will develop a head over the spillway weir and the weir will start functioning. The outflow will start flowing over the weir. The inflow and the outflow will not be equal. Initially, the outflow will be less than the inflow and the difference will be temporarily absorbed in the reservoir resulting in an increase in the reservoir water level. This will increase the outflow.

As the inflow reduces and is less than the outflow, the water level in the reservoir will also reduce. A stage will be reached when the outflow and the inflow will reduce to zero and the water level in the reservoir will be up to crest level.

In such a case, the inflow volume will be the summation of the outflow volume and the volume of water absorbed into the reservoir initially from the original water level up to the crest of the spillway weir.

In this case, also, the outflow will start receding when the inflow and the outflow hydrographs cross each other.

The inflow hydrograph, the outflow hydrograph, the water absorbed in the reservoir and the water level in the reservoir are shown in Fig. 10.18.

Case no. 3

In this case, the spillway gates are installed. There is no flood lift and the water level is up to the FRL or the HFL. (Normally, the water level in the reservoir is kept as high as possible, so that the storage in the reservoir is the maximum.)

As the inflow occurs, the spillway gates will be operated partially and the outflow will be kept equal to the inflow, thus maintaining the water level constant at the FRL. As the inflow increases, the gates will be operated and opened more and more, keeping the water level at FRL. Thus the outflow will be equal to the inflow. For the maximum discharge, the gates will be fully opened and there will be no change in the water level. It will be at FRL only.

As the inflow decreases, the gates will be lowered keeping the water level constant at FRL, and the inflow will be equal to the outflow. When the inflow stops, the gates will be closed completely and the outflow will also stop. In such a case, since there is no flood lift, the outflow will always be equal to the inflow and the water level will be constant at FRL.

Thus, there is no flood routing in this case and the inflow and outflow hydrographs will be the same as shown in Fig. 10.19.

Case no. 4

In this case, there is a flood lift provided in the design. HFL is higher than the FRL by an amount equal to the flood lift. The water level is at the FRL, with gates completely closed.

As the inflow occurs, the gates will be operated partially to maintain the water level at FRL so that the inflow is equal to the outflow. As inflow increases, the gates will be operated such that the water level is maintained constant at the FRL. Thus the outflow will be equal to the inflow.

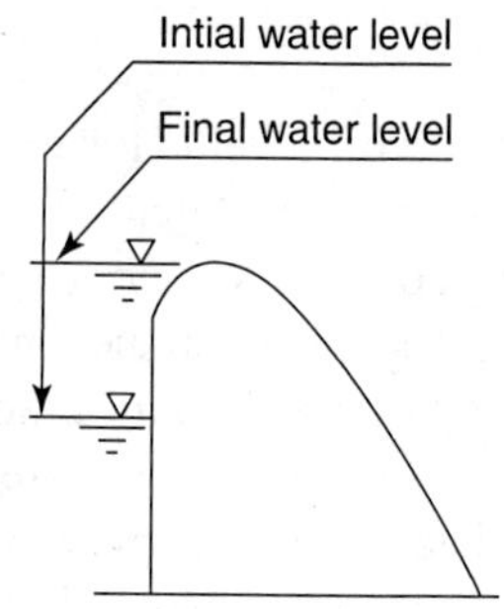

Definition sketch

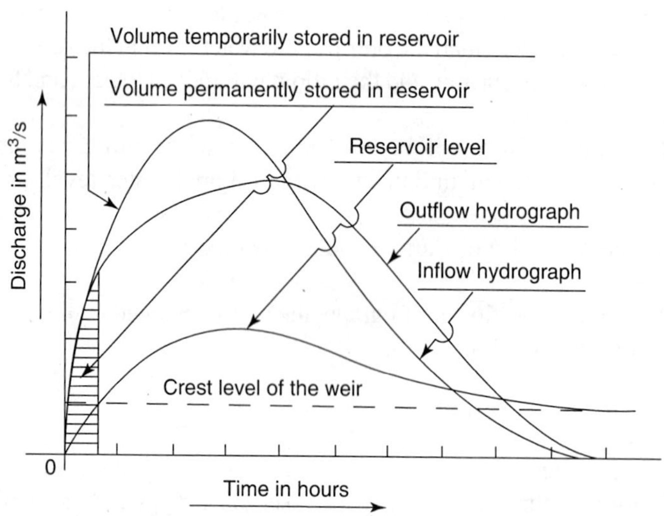

Fig. 10.18 Inflow and outflow hydrographs with the reservoir and crest levels

A stage will be reached when the gates will be completely opened with the water level is at the FRL, and the outflow will be equal to the inflow. As inflow increases still further, the water level in the reservoir will increase and the outflow will also increase, but will be less than the inflow. The water level will rise above the FRL and will reach HFL. when the outflow will be the maximum. This point will be reached when the inflow and outflow hydrographs cross each other.

As the inflow reduces further, a stage will be reached such that the outflow is equal to the inflow. From this point, the outflow and the water level will start receding, but the inflow will be less than the outflow. When the water level reaches the FRL, the gates will have to be operated to maintain the water level in the reservoir constant at FRL. In this case, the inflow will be equal to the outflow.

Finally, the gates will be closed completely; the inflow and the outflow will stop simultaneously, with the water level in the reservoir at the FRL.

Since the water level remains unchanged, the inflow volume and the outflow volume will be equal. The inflow and the outflow hydrographs and the water level in the reservoir are shown in Fig. 10.20.

Case no. 5

In this case, the spillway is gated with a flood lift provision in the design. The HFL is above the FRL, by a margin equal to the flood lift. The water level in the reservoir is below the FRL.

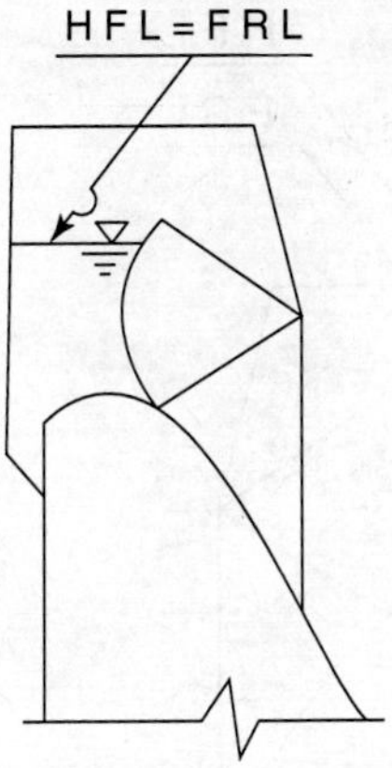

Definition sketch

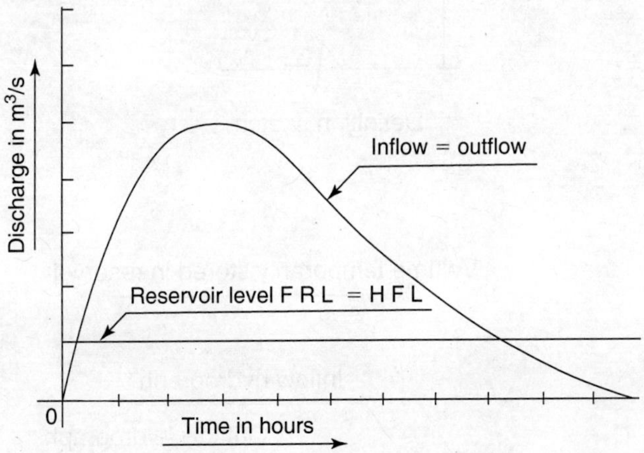

Fig. 10.19 Inflow and outflow hydrographs when there is no flood routing

As the inflow starts into the reservoir, the water level will go on increasing, absorbing the flood in the reservoir. The gates will remain in the closed position and there will be no outflow. When the water level reaches the FRL, the spillway gates will be operated partially, maintaining the water level at the FRL. Here the inflow will be equal to the outflow. A stage will be reached when the gates are fully open and the water level is at the FRL. This is the limit when the inflow is equal to the outflow.

For the higher inflow, the water level will go on increasing absorbing water in the reservoir above the FRL, with a result that both the water level in the reservoir and the outflow will increase. The inflow after reaching the maximum will start receding. A stage will be reached when the outflow will be equal to the inflow. From this point, the water level and hence the outflow will start receding. When the water level reaches the FRL, the gates will be operated partially so that the water level will not drop below the FRL. In this case, the inflow and the outflow will be equal.

Finally the inflow will stop. The gates will be closed the outflow will also stop. The inflow volume will be equal to the summation of the outflow volume and the volume absorbed in the reservoir from the initial water level up to the FRL.

Figure 10.21 shows the inflow and the outflow hydrographs, the volume absorbed by the reservoir and the water level in the reservoir.

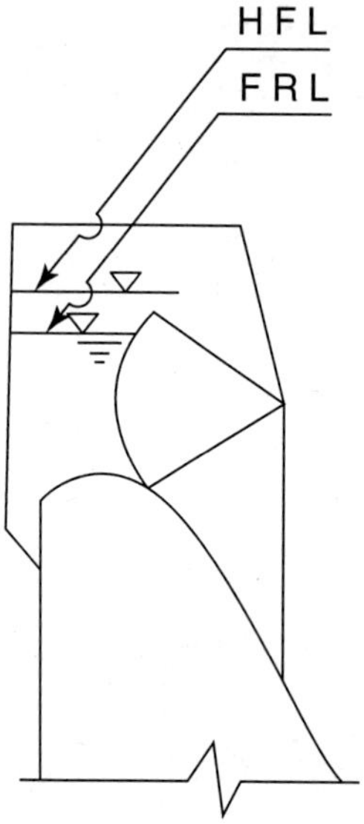

Definition sketch

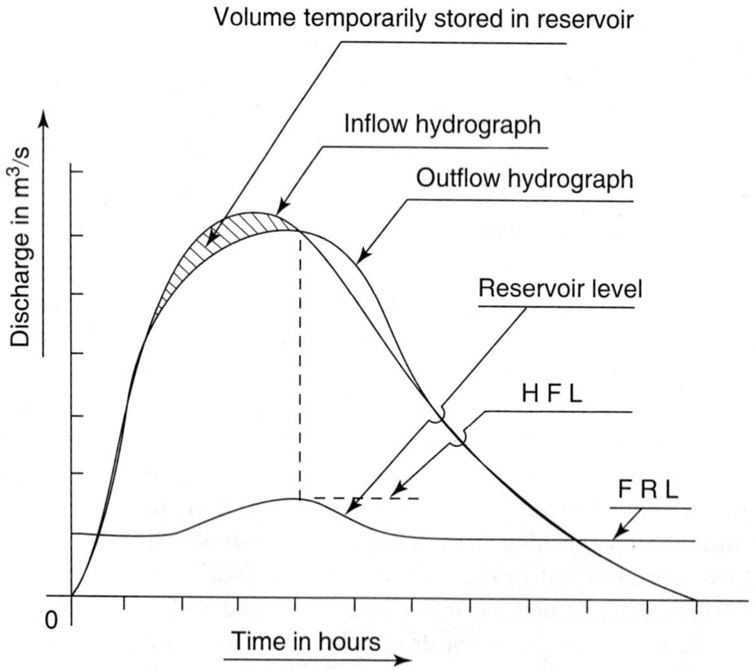

Fig. 10.20 Inflow and outflow hydrographs and water level in the reservoir

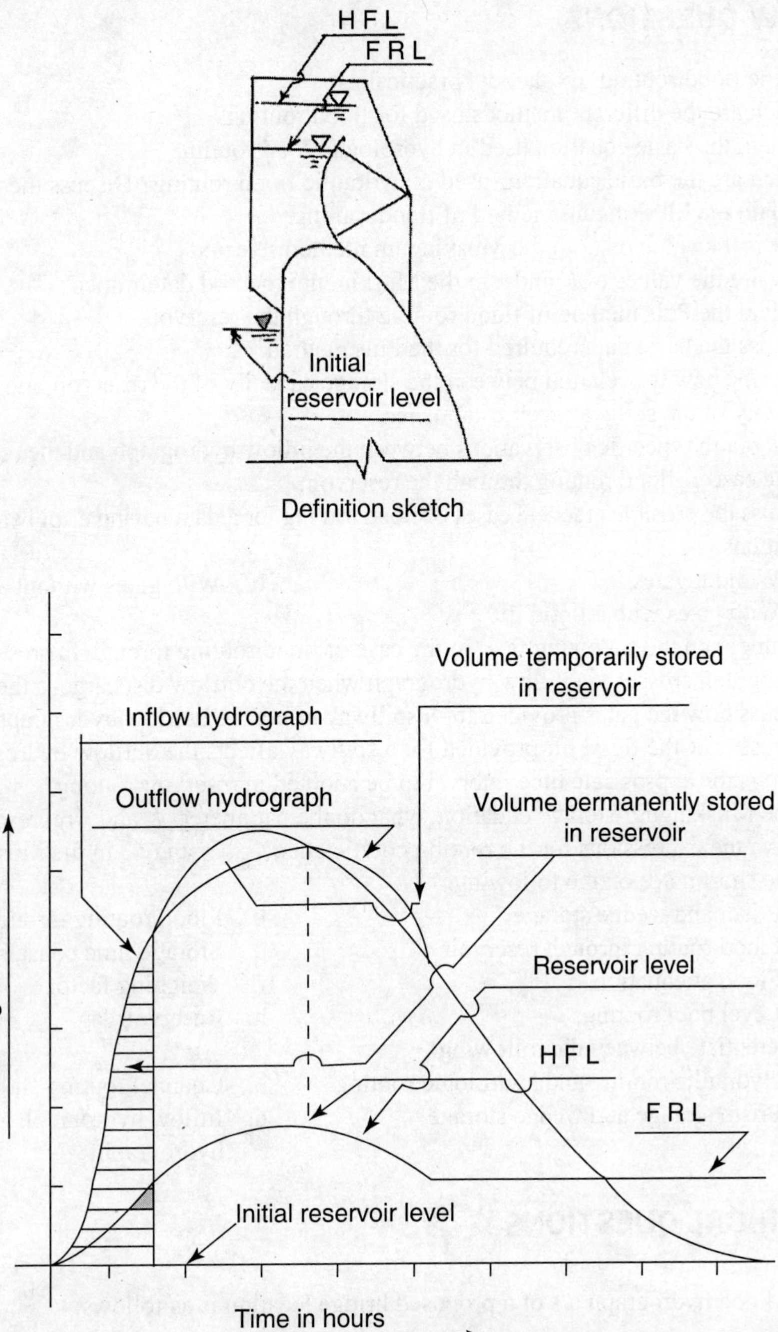

Fig. 10.21 Inflow and outflow hydrographs and the water level in the reservoir

REVIEW QUESTIONS

1. Define flood routing. Explain its practical use.
2. Which are the different methods used for flood routing?
3. Explain the basic equation used in hydrologic flood routing.
4. Which are the basic equations used in hydraulic flood routing? Discuss them in brief.
5. Explain the Muskingum method of flood routing.
6. How is the value of Δt in the Muskingum method selected?
7. How are the values of K and x in the Muskingum method determined?
8. Explain the Puls method of flood routing through the reservoir.
9. Discuss the basic data required for the Puls method.
10. Describe how the relation between the storage capacity of the reservoir and the discharging capacity of the spillway weir established.
11. Point out the specific observations between the inflow hydrograph and the outflow hydrograph in the case of flood routing through the reservoir.
12. Discuss the possible practical cases of flood routing for a dam having a spillway in the following situations:
    a. Without gates
    b. With gates without a flood lift
    c. With gates with a flood lift
13. Comment on the following fact: In the case of flood routing through the reservoir, the outflow hydrograph crosses the inflow hydrograph when the outflow discharge is the maximum.
14. Discuss how the gates provided for a spillway affect the outflow hydrograph.
15. Discuss how the flood lift provided for a spillway affects the outflow hydrograph.
16. What is the appropriate time interval to be adopted in reservoir routing?
17. In the Muskingum-storage equation, what do the parameters K and x represent?
18. Derive the expressions for the routing coefficients, C_0, C_1 and C_2 in the Muskingum method.
19. Write short notes on the following:
    a. Prism and wedge storage
    b. Flood routing through channel
    c. Flood routing through reservoir
    d. Storage time constant
    e. Flood attenuation
    f. Weighing factor
    g. Level pool routing
    h. Reservoir lag
20. Differentiate between the following:
    a. Hydraulic routing and hydrologic routing
    b. Channel routing and reservoir routing
    c. Prism storage and wedge storage
    d. Inflow hydrograph and the outflow hydrograph

NUMERICAL QUESTIONS

1. The flood hydrograph u/s of a proposed bridge location is as follows:

Time (h)	0	1	2	3	4	5	6	7	8	9
Discharge (m^3/s)	8.0	14.0	26.0	43.0	35.0	25.0	18.0	15.0	12.0	8.0

The values of K and x for this specific channel section are 3 h and 0.3, respectively. Find the hydrograph at the bridge location and the flood attenuation.

Ans: The flood hydrograph at the bridge location is as follows:

Time (h)	0	1	2	3	4	5	6	7	8	9	10	11	12	13	14	15	16	17	18
Discharge (m^3/s)	8.0	7.06	7.83	12.10	25.13	30.44	29.44	25.53	21.96	18.77	14.65	12.11	10.53	9.56	8.96	8.59	8.36	8.21	8.13

Flood attenuation = 43.0 − 30.44 = 12.56 m^3/s

MULTIPLE CHOICE QUESTIONS

1. The flow in a river during the flood is
   (a) Steady uniform (b) Unsteady uniform
   (c) Unsteady non-uniform (d) Steady non-uniform
2. The outflow from a reservoir will depend on
   (a) The inflow hydrograph (b) The storage capacity of the reservoir
   (c) The discharging capacity of the spillway weir (d) All the three above
3. The hydraulic flood routing through a river can be followed by applying
   (a) Continuity equation (b) Energy equation
   (c) Momentum equation (d) All the three above
4. The hydrologic flood routing through a reservoir can be followed by applying
   (a) Continuity equation (b) Energy equation
   (c) Momentum equation (d) All the three above
5. The storage time factor K has the dimensions of
   (a) Dimensionless (b) m^2
   (c) m (d) h
6. The prism storage is a function of
   (a) Inflow (b) Outflow
   (c) Inflow − outflow (d) Inflow + outflow
7. The wedge storage is a function of
   (a) Inflow (b) Outflow
   (c) Inflow − outflow (d) Inflow + outflow
8. The weighing factor x has the dimensions of
   (a) Dimensionless (b) m^2
   (c) m (d) h

9. The condition satisfied by the three routing coefficients in the Muskingum method is
   (a) $C_0 + C_1 + C_2 = 0$ (b) $C_0 + C_1 + C_2 = 1$
   (c) $C_0 + C_1 + C_2 = 10$ (d) $C_0 + C_1 + C_2 = 100$

10. Out of the following which is the Muskingum-storage equation?
   (a) $S = [x \times Q + (1 - x) \times I]$ (b) $S = [x \times Q - (1 - x) \times I]$
   (c) $S = [x \times I + (1 - x) \times Q]$ (d) $S = [x \times I - (1 - x) \times Q]$

11 . The Muskingum-routing equation is
   (a) $Q_2 = C_0 \times Q_1 + C_1 \times I_2 + C_2 \times I_1$ (b) $Q_2 = C_0 \times I_2 + C_1 \times I_1 + C_2 \times Q_1$
   (c) $Q_2 = C_0 \times I_1 + C_1 \times Q_1 + C_2 \times I_2$ (d) $Q_2 = C_0 \times Q_1 + C_1 \times I_1 + C_2 \times I_2$

ANSWERS TO MULTIPLE CHOICE QUESTIONS

1. c	2. d	3. d	4. a	5. d	6. b	7. c	8. a
9. b	10. c	11. b					

*** ———— ***

11 Groundwater

Chapter Outline

11.1 DEFINITION

Groundwater may be defined as *water that occurs below the surface of the earth*. It is also known as *sub-surface water*.

11.1.1 Sources of Groundwater

The main source of groundwater is precipitation. The water that infiltrates, after meeting the requirement of the soil, percolates and then becomes groundwater. It is also known as *meteoric water*.

Water present in the rock at the time of its formation is known as *connate water*. It is highly saline. It is also known as *fossil water* or *interstitial water*.

Juvenile water is formed chemically within the earth. It is in a very small quantity and is also known as *primitive water*.

11.1.2 Peculiarities of Groundwater

The peculiarities of groundwater are as follows:

- It is normally free from pollution but vulnerable to the pollution entering from the surface sources with infiltration.
- It can be made available on the surface at a small capital cost and in a short duration, but involves repetitive cost of energy for lifting it to the surface.
- It is very useful for domestic purposes, for small towns and isolated places since it requires minimum treatment.
- It is available up to a depth of 3 km.
- It is not an unlimited source and hence has to be managed against excessive exploitation and contamination.
- It exhibits less fluctuations in alternate wet and dry periods as compared to the surface water.
- It is free from weeds, plant organisms, turbidity and bacterial pollution.
- It is uniform in quality, temperature, chemical composition and soluble mineral contents, compared to the surface water.

11.2 FORMATION OF GROUNDWATER

The level below which the soil is saturated with water is called *groundwater table*. The groundwater is divided into two zones (1) unsaturated zone and (2) saturated zone.

These two zones are separated by water table where the pressure is atmospheric, as shown in Fig. 11.1.

11.2.1 Unsaturated Zone

The zone above the water table is known as *unsaturated zone*. It is also known as *vadose zone* or *aeration zone*. In this zone, the pores in the soil may contain water or air, and the pressure is atmospheric. This zone may be divided into three subzones:

(1) soil–water zone, (2) intermediate zone and (3) capillary zone.

11.2.1.1 Soil–water zone

It lies close to the ground surface in the major root zone of vegetation. There may be loss of water from this zone by evapotranspiration. It may extend from a few metres to 15 m, depending upon the nature of soil and vegetation.

11.2.1.2 Intermediate zone

This subzone is between the soil–water zone and the capillary zone. Water from the soil zone may flow downwards due to gravity. The thickness of this zone may vary from zero to several metres.

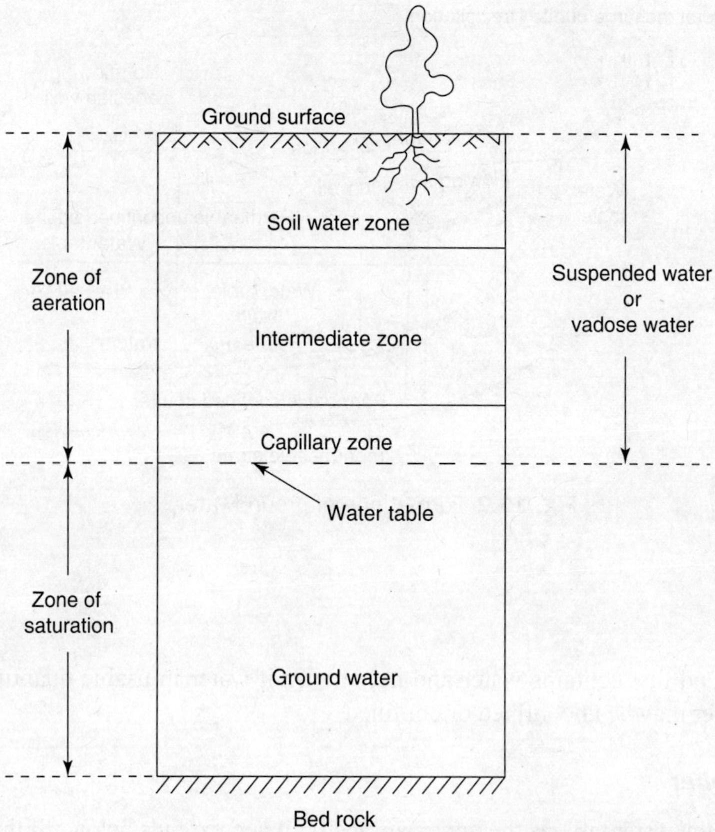

Fig. 11.1 Occurrence of groundwater

11.2.1.3 Capillary zone

From the saturated zone, i.e. from the water table, water may rise above due to capillarity, hence this zone is known as the *capillary zone* or *the capillary fringe*. The thickness may extend from a few centimetres to a few metres, depending on the porosity and structure of the soil. Water from this subzone may not move freely.

11.2.2 Saturated Zone

Below the water table, all the pores in the soil are filled with water. Hence, it is known as the *saturated zone*. It is also known as the *phreatic zone*.

Water in this zone moves freely and may extend till the impermeable rock below. The pressure in this zone is more than atmospheric and increases as the depth increases.

11.3 OCCURRENCE OF GROUNDWATER

Groundwater occurs at various locations below the earth surface, depending on the formation of ground. The various formations are shown in Fig. 11.2.

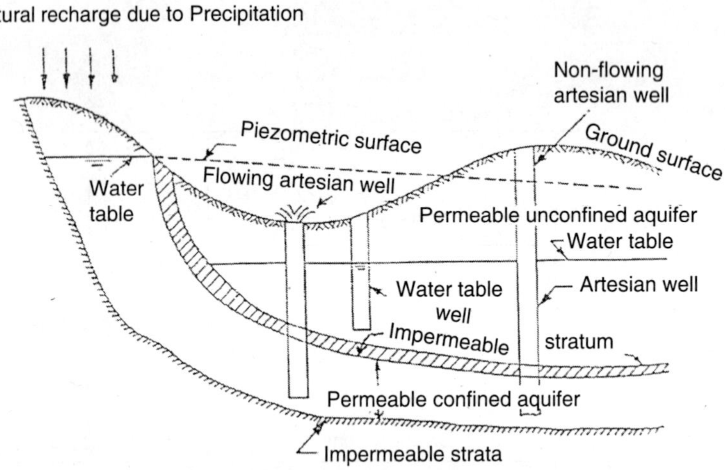

Fig. 11.2 Formations of groundwater

11.3.1 AQUIFER

Formation of ground that contains water and may transmit water in usable quantity is known as *aquifer*. The aquifer may be unconfined or confined.

Unconfined aquifer

An aquifer where the water table is the upper surface limit and extends below till the impermeable rock strata is called the *unconfined aquifer*, as shown in Fig. 11.3. It is also known as *free aquifer*, *phreatic aquifer*, *water table aquifer* and *non-artesian aquifer*.

The water level in a well is an unconfined aquifer will be upto the watertable.

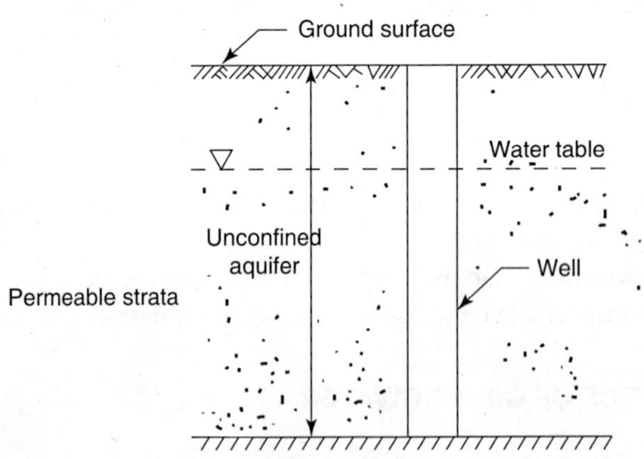

Fig. 11.3 Unconfined aquifer

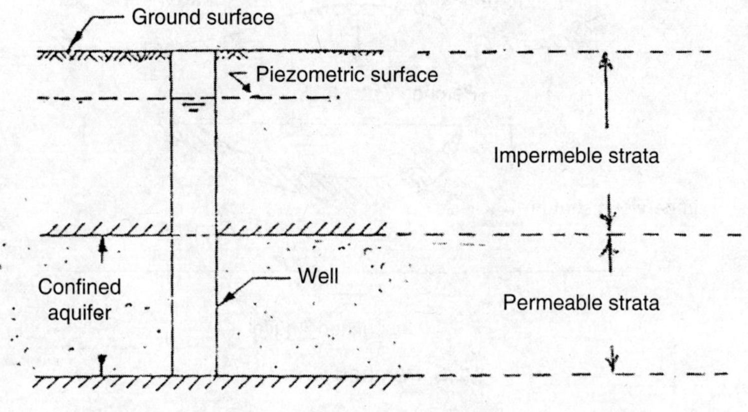

Fig. 11.4 Confined aquifer

Confined aquifer

When an aquifer is sandwiched between two impermeable layers, it is known as a *confined aquifer*. It will not have a free water table, and the aquifer will be under pressure as shown in Fig. 11.4.

If there is a well in this layer, the water in the well will rise up to the piezometric head. If the piezometric head is above the ground, then the water from the well in this layer will flow over the ground. Then the well is called as *free-flowing well* or *flowing well*.

If the water level in a well in this layer is above the upper-confining layer level, but below the ground level, then such a well is called as *artesian well*.

Groundwater may move to the ground surface at a very small rate through faults, permeable material in joints, discontinuities, and so on. It is then called a *spring*.

11.3.2 Aquiclude

A geological formation that may contain water because of high porosity but cannot transmit it is called an *aquiclude*.

11.3.3 Aquitard

A geological formation that has poor permeability, but through which seepage is possible, and is insignificant as compared to an aquifer is known as an *aquitard*.

11.3.4 Aquifuge

A geological formation that neither contains nor transmits water is called an aquifuge. Solid rock is an *aquifuge*.

11.3.5 Leaky Aquifer

An aquifer bound by two aquitards is known as a *leaky aquifer*. It is also known as *semi-confined aquifer*.

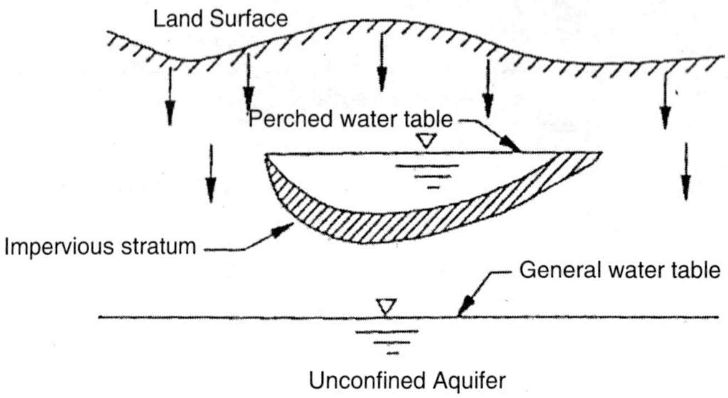

Fig. 11.5 Perched water table

11.3.6 Perched Groundwater

An impermeable saucer-shaped stratum of a small aerial extent occurring in the zone of aeration may retain and hold some amount of water. It is called *perched groundwater*, as shown in Fig. 11.5. It yields a limited quantity of water.

11.4 PARAMETERS OF AN AQUIFER

The various parameters of an aquifer are as follows:

(1) porosity, (2) permeability, (3) specific yield, (4) specific retention, (5) storage coefficient, (6) safe yield and (7) transmissibility.

11.4.1 Porosity

Any soil sample contains some pores or voids. The porosity of the sample is the ratio of the volume of voids and the volume of soil sample. It is a dimensionless number and is normally denoted by n and is expressed as percentage.

Thus, porosity $= n = \frac{V_v}{V} \times 100$

where, V_v = Volume of pores or voids
V = Volume of sample

Porosity is classified as follows:

20% or more—large, 5–20%—medium and less than 5%—small.

The *primary porosity* of a material is one that existed when the material was formed. The *secondary porosity* results from fractures, joints, solution channels and so on.

11.4.2 Permeability

Permeability is the property of an aquifer to transmit water through its pores.

The horizontal permeability and the vertical permeability may differ.

11.4.3 Darcy's Law

The velocity of the flow of water in an aquifer is very low and hence the flow is laminar. Darcy's law is applicable here. The law states, *The rate of flow per unit area of an aquifer is proportional to the gradient of potential flow in the direction of flow.*

i.e., $v = K \times i = -K\frac{dh}{dl}$

where, v = Velocity of flow in m/s
K = Coefficient of permeability
i = Hydraulic gradient
$= -\frac{dh}{dl}$

The negative sign indicates that the flow is in the falling head.

Coefficient of permeability, K, is also known as *hydraulic conductivity* and has the dimensions as those of the velocity of flow. It depends on the fluid property as well as the property of the aquifer soil sample. K is given by the following identity:

$$K = \frac{cd^2\rho g}{\mu}$$

where, K = Hydraulic conductivity
c = A constant factor
d = Average pore size of the material
ρ = Density of material
g = Gravitational acceleration
μ = Coefficient of dynamic viscosity of water

The velocity of flow through the soil sample v is also known as *Darcy's velocity, discharge velocity* and *apparent velocity*. Darcy's law can be verified in a laboratory.

11.4.4 Storage Coefficient

Storage coefficient of an aquifer is the volume of water received or discharged per unit head and per unit surface area. It is also known as *storativity*. It is normally denoted as 'S'.

The unit head in case of an unconfined aquifer will be the unit drop in water table, and the unit head in case of a confined aquifer will be the unit drop in the piezometric head.

The storage coefficient of a confined aquifer is very small as compared to that of an unconfined one.

11.4.5 Specific Yield and Specific Retention

If water is allowed to drain from a saturated sample of an aquifer, some water will drain freely under gravity.

The specific yield is the ratio of the volume of water drained divided by the volume of soil sample.

That is, $S_f = \frac{V_w}{V}$

where, S_f = Specific yield
V_w = Volume of water drained
V = Volume of soil sample

S_f is dimensionless. It is also known as *effective porosity* or *practical porosity*.

Some quantity of water will be retained by the soil sample and will not be drained under gravity. This is due to molecular attraction and surface tension. *Specific retention* is defined as the *ratio of the volume of water retained divided by the volume of sample*. It is denoted by S_r.

Thus, $S_r = \frac{W_f}{V}$

where, S_r = Specific retention
W_f = Volume of water retained by the soil sample
V = Volume of soil sample

S_r is dimensionless. It is also known as *field capacity*.

Naturally, $n = S_f + S_r$

where, n = Porosity
S_f = Specific yield
S_r = Specific retention

11.4.6 Transmissibility

It is the product of the coefficient of permeability and the thickness of the aquifer.

Thus, $T = K \times b$

It has the dimensions of m^2/s. It is also known as *transmissivity*.

For a unit width

$$Q = \text{area} \times \text{velocity} = b \times 1 \times K \times \left(-\frac{di}{dh}\right) = T\left(-\frac{di}{dh}\right)$$

K = Hydraulic conductivity of the soil
T = Transmissibility of the soil

11.5 STEADY ONE-DIMENSIONAL GROUNDWATER FLOW

Consider two rivers, A and B, flowing in parallel at a distance L from each other with a water level difference of h, as shown in Fig. 11.6.

Water will flow from river A to river B in only one direction, i.e. 'x' direction, and hence it is also known as *unidirectional flow*.

The velocity of flow will depend on the following factors:

1. h = Difference in the water levels in the two rivers
2. K = Hydraulic conductivity of soil

For the analysis of the flow, the following assumptions are made:

1. The flow is steady
2. The flow is horizontal and uniform in vertical direction

The characteristics of the flow will vary depending on the aquifers in between the two rivers as mentioned below:

1. Unconfined aquifer
2. Confined aquifer

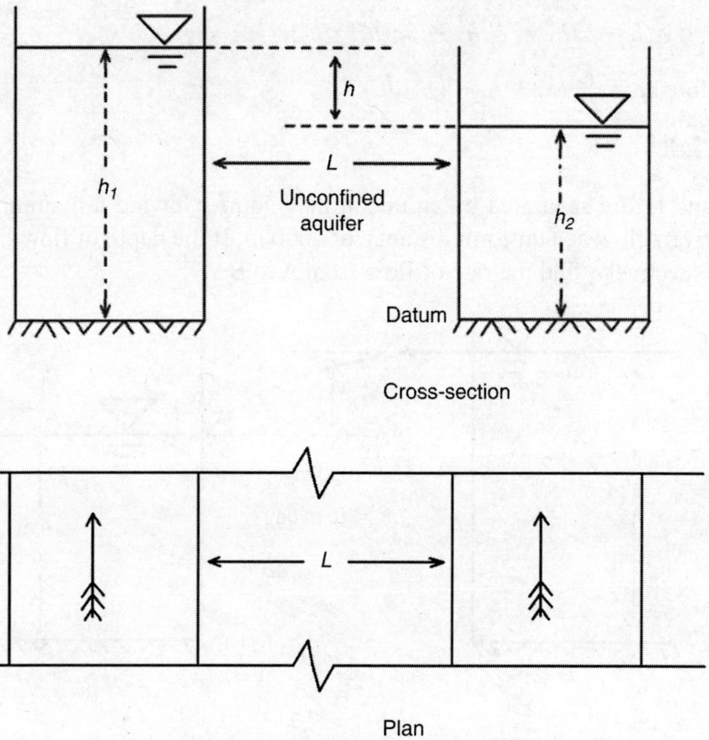

Fig. 11.6 Unidirectional flow

11.5.1 Unconfined Aquifer

Consider two rivers, A and B, with water depths of flow h_1 and h_2 at a distance L, separated by an unconfined aquifer having a hydraulic conductivity K as shown in Fig. 11.6.

The horizontal velocity of flow v at a distance x will be as follows:

$$v = -K \times \left(\frac{dh}{dx}\right)$$

where v = Velocity of flow

K = Hydraulic conductivity of the aquifer

$\left(\frac{dh}{dx}\right)$ = Hydraulic gradient at a distance x

The negative sign indicates that the flow is in the falling direction.

The intensity of discharge q per unit width will be

$$q = -K \times h \times \left(\frac{dh}{dx}\right)$$

Therefore, $q \times dx = -K \times h \times dh$

integrating $q \times x = -K \times \frac{h^2}{2} + C$

When, $x = 0, h = h_1,$

$$C = \frac{Kh_1^2}{2}$$

and when $x = L, h = h_2$

$$\therefore \quad q \times L = \frac{K}{2}h_1^2 - \frac{K}{2}h_2^2 = \frac{K}{2}(h_1^2 - h_2^2)$$

Therefore, $q = \frac{K}{2L}(h_1 + h_2)(h_1 - h_2)$

Example 11.1

Two rivers, A and B, are separated by an unconfined aquifer having a hydraulic conductivity of 10 m/day. The rivers flow at a uniform distance of 1000 m. If the depth of flow in both the rivers is 15 and 10 m, respectively, find the rate of flow from A to B.

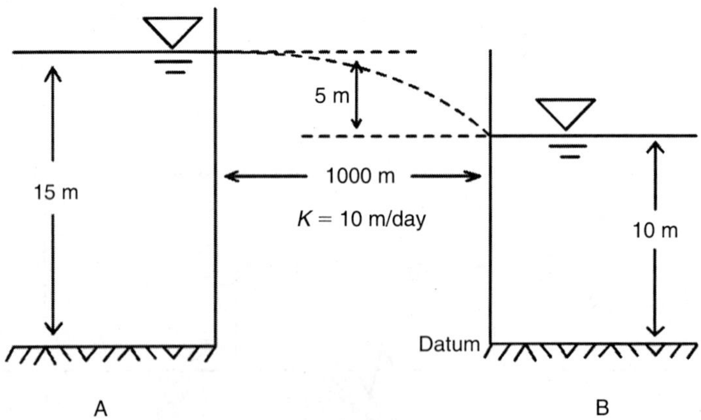

Fig. 11.7 Unidirectional flow in unconfined aquifer

Solution:

The flow between the two rivers is shown in Fig. 11.7.

$$\text{Therefore, } q = \frac{K}{2L}(h_1^2 - h_2^2) = \frac{10}{2 \times 1000}(15^2 - 10^2) = 0.625 \text{ m}^3/\text{m/day}$$

11.5.2 Confined Aquifer

Consider two rivers, A and B, with water depths of flow h_1 and h_2 at a distance L, separated by a confined aquifer having a hydraulic conductivity K, as shown in Fig. 11.8.

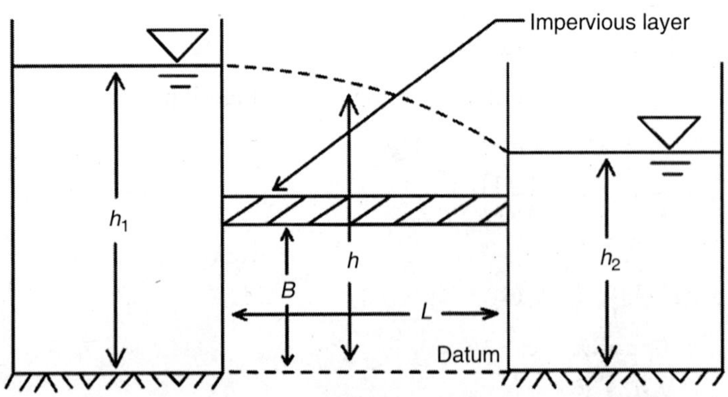

Fig. 11.8 Unidirectional flow in a confined aquifer

The velocity of flow $v = K \times \dfrac{(h_1 - h_2)}{L}$

Therefore, $q = B \times K \times \dfrac{(h_1 - h_2)}{L} = T \times \dfrac{(h_1 - h_2)}{L}$

since T = Transmissibility = $B \times K$

Example 11.2

Two rivers, P and Q, are separated by a confined aquifer of thickness 2 m. The hydraulic conductivity of the confined aquifer is 15 m/day. If the depth of flow in both the rivers is 15 and 10 m, respectively, find the rate of flow from P to Q.

Solution:

The flow between the two rivers is shown in Fig. 11.9.

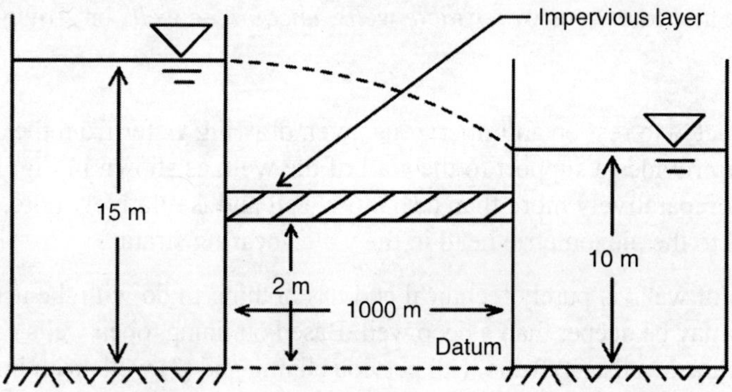

Fig. 11.9 Unidirectional flow in a confined aquifer

$$Q = \frac{B \times K}{L}(h_1 - h_2) = \frac{2 \times 15}{1000}(15 - 10) = 0.15 \text{ m}^3/\text{m/day}$$

11.6 WELL HYDRAULICS

A well is a hole or a shaft, normally vertical, excavated in the ground for bringing water to the surface. Wells are generally classified as: (a) open wells and (b) tube wells.

11.6.1 Open Well

An open well is a well 2–13 m in diameter, dug into the ground to tap water from the top pervious stratum. The depth is normally limited to 30 m. Large quantity of water is stored in the open wells.

Normally, these wells are either square or circular of uniform cross section. Open well is also known as a *dug well*. Open wells are further classified as: (a) shallow well and (b) deep well.

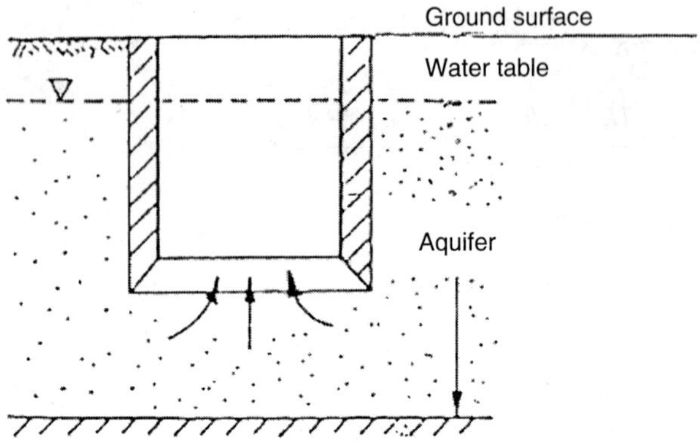

Fig. 11.10 Shallow well

Shallow well

An open well that is constructed to tap water from the topmost water-bearing stratum, i.e. from the unconfined aquifer, is called *shallow well*. The bottom of such a well will not rest on an impervious layer as shown in Fig. 11.10. The water level in such a well will be equal to the level of the water table. These wells are also known as *water table wells, unconfined wells* or *gravity wells*.

Deep well

A deep well is constructed to rest on an impervious layer, drawing water from the aquifer below it. The impervious layer provides a support to the wall of the well, as shown in Fig. 11.11. The yield from a deep well is comparatively more than a shallow well and is relatively pure. The water level in such wells is equal to the piezometric head in the water-bearing strata.

This classification of wells is purely technical and has nothing to do with the actual depth of the wells. A shallow well may be deeper than a deep well. Based on lining, open wells can also be classified as follows: (1) unlined wells, (2) wells with pervious lining and (3) wells with impervious lining.

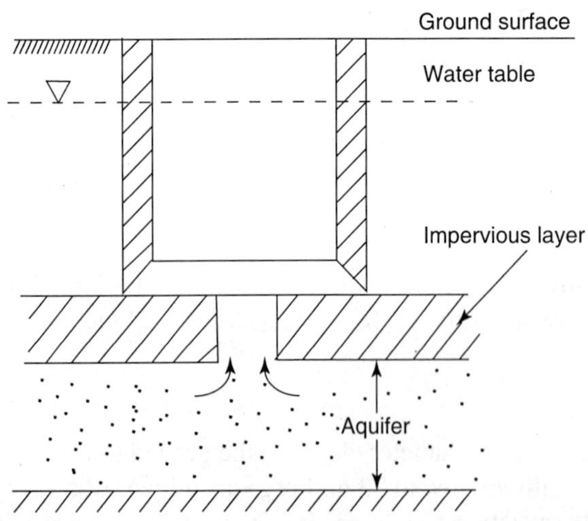

Fig. 11.11 Deep well

11.6.2 Tube Well

A tube well is a long pipe sunk deep into the ground, intercepting more than one stratum. Its diameter varies from 80 mm to 600 mm. The depth may vary from 50 to 150 m. Blind pipes are used for the impervious strata, and strainers are used against the pervious strata. These strainers allow passage of water through them, but prevent sand from coming in with the help of designed filters. Tube wells are classified as: (1) strainer well, (2) cavity well and (3) slotted well.

Strainer well

This is the most common and widely used type. The term 'tube well' is generally referred to this type of well. In this type, a special designed wire mesh with filter pack covering the water-bearing strata is used to draw water free from silt and sand. The pipe in this portion is perforated. The pipe is generally plugged at the bottom, to permit settlement of the sand passing through the strainers. In such type of wells, the inflow is radial. Figure 11.12 shows a strainer well.

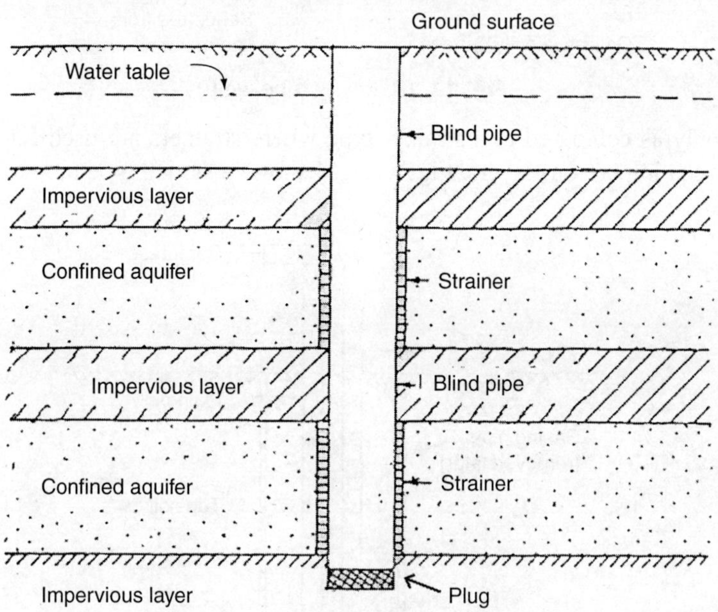

Fig. 11.12 Strainer tube well

Cavity well

This is a special type of tube well in which water is not drawn through the sides but is drawn through the bottom of the well where a cavity is formed as shown in Fig. 11.13. Normally this type is used when a clay layer is met with. The zone of the flow in this case is spherical.

Slotted tube well

This type of well is used when a strainer type or a cavity type cannot be used. In this case, a slotted tube is installed to penetrate into a water-bearing-confined stratum. A slotted tube is used at the

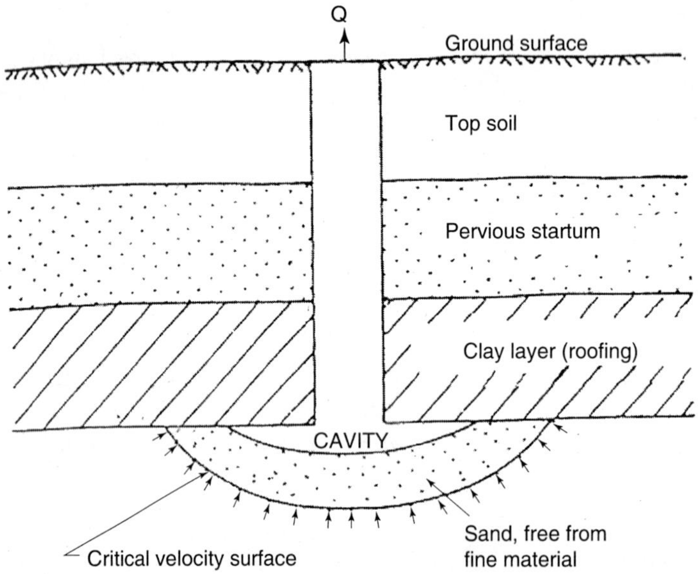

Fig. 11.13 Cavity tube well

bottom portion only as compared to a strainer type where strainers are used for several aquifers sandwiched between impervious layers. Figure 11.14 shows a slotted type tube well.

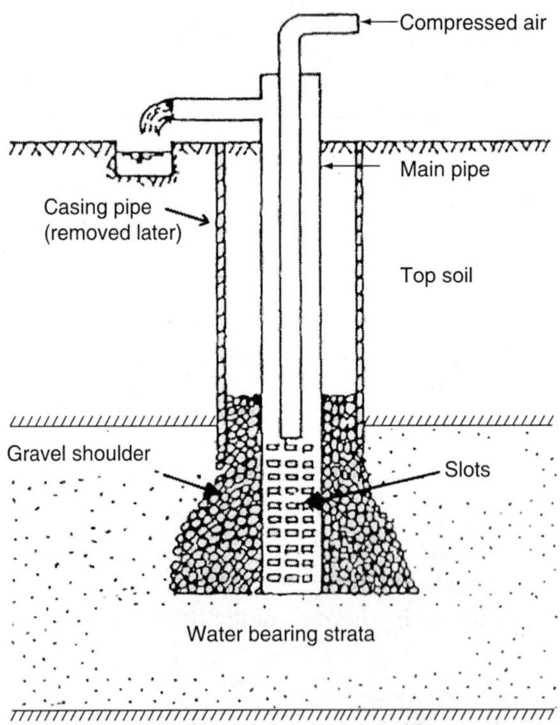

Fig. 11.14 Slotted tube well

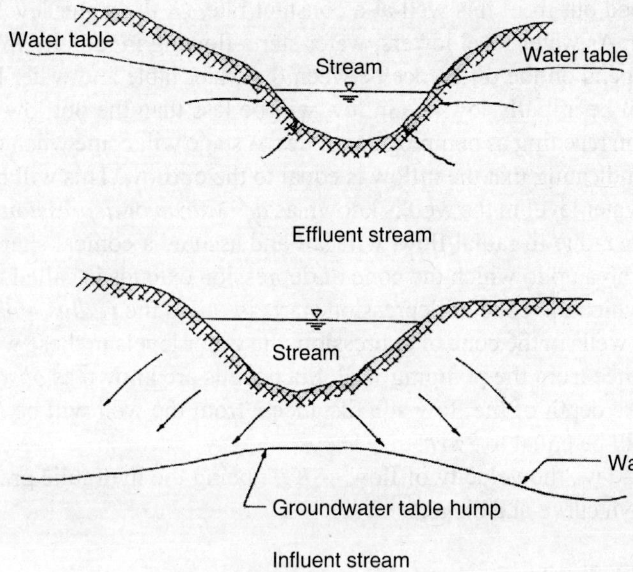

Fig. 11.15 Effluent and influent streams

11.7 GROUNDWATER MOVEMENT

The theory of groundwater movement into a well is commonly known as well hydraulics. The movement of groundwater is in accordance with the Darcy's law, coupled with the hydraulic principles. When the water table is higher than the bed of the stream, groundwater will flow to the stream. Such a stream is known as *effluent stream*. On the other hand, when the water table is lower than the bed of the stream, water will flow from the stream to the groundwater. Such a stream is called *influent stream*. Both types are shown in Fig. 11.15.

In both the cases, the flow will depend on the soil characteristics. Similar is the case of a well. When water flows into a well, the well is known as *discharging well*; when water flows from the well to the adjacent ground, it is called *recharging well*.

The groundwater movement into a well is classified as under:

1. Steady radial flow in an unconfined aquifer.
2. Steady radial flow in a confined aquifer.

In both the cases, the following assumptions are made:

- The flow is horizontal and uniformly distributed in a vertical section.
- The velocity of flow is proportional to the tangent of the hydraulic gradient.
- The well fully penetrates the aquifer.
- The discharge taken from the well is constant.
- The aquifer is homogeneous and isotropic.

11.7.1 Steady Radial Flow in an Unconfined Aquifer

Consider a well in an unconfined aquifer as shown in Fig. 11.16.

When water is pumped out from this well at a constant rate, Q, the water level in the well will naturally go on receding. As water level lowers, water starts flowing in radially from the adjoining area. The inflow will depend on the difference between the water table and water level in the well. Since this difference will be initially low, the inflow will be less than the outflow. Thus, the water level in the well will go on receding as pumping continues. A stage will come when the water level in the well will be steady, indicating that the inflow is equal to the outflow. This will be a *steady state*.

The decrease in the water level in the well is known as *drawdown* or *depression head*. The water table in the adjoining area, due to radial flow, will fall and assume a conical shape. This is called *cone of depression*. The area up to which the cone of depression extends is called the *area of influence*, and the radius at which the cone of depression starts is called the *radius of influence*.

If there are any other wells in the cone of depression, the water levels in these wells will also fall depending on their distance from the pumping well. Such wells are known as *observation wells*.

In the steady state, the depth of the flow at a distance r from the well will be h. Being a radial flow, the area of flow will be equal to $2\pi rh$.

According to Darcy's law, the velocity of flow $= Ki$, i being the hydraulic gradient dh/dr, i.e., the slope of the drawdown curve at a distance of h.

Therefore, the velocity of flow $= Ki = K\dfrac{dh}{dr}$

Now,
$$Q = \text{area} \times \text{velocity} = 2\pi rh \times K\frac{dh}{dr}$$

or
$$hdh = \frac{Q}{2\pi K} \times \frac{dr}{r}$$

The depths of the flow are h_1 and h_2 in the two observation wells at distances r_1 and r_2. Integration of the above equation in these limits yields

$$\int_{h_2}^{h_1} hdh = \frac{Q}{2\pi K}\int_{r_2}^{r_1}\frac{dr}{r}$$

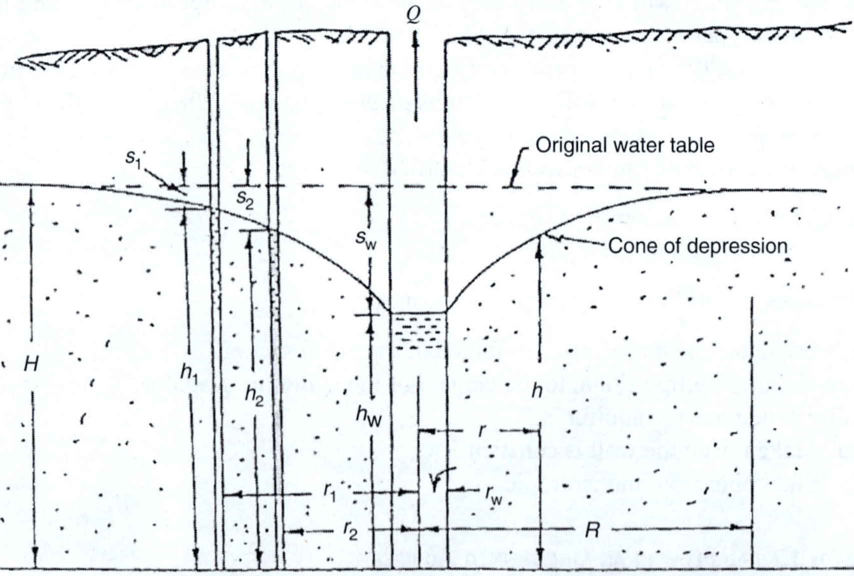

Fig. 11.16 Steady radial flow in an unconfined aquifer

Therefore, $\dfrac{h_1^2 - h_2^2}{2} = \dfrac{Q}{2\pi K} \log_e \dfrac{r_1}{r_2}$

The constant discharge Q can be equated as

$$Q = \frac{\pi K(h_1^2 - h_2^2)}{\log_e (r_1/r_2)}$$

where, Q = Steady discharge in m^3/s
K = Hydraulic conductivity in m/s
h_1 and h_2 = Depths of flow in the two observation wells in metres
r_1 and r_2 = Radial distances of the two observation wells from the pumping well in metres

If the radius of influence is known, then the equation can be integrated in the limits of R and r_w as well as H and h_w. This will yield

Now $$hdh = \frac{Q}{2\pi K} \times \frac{dr}{r}$$

Integrating the above equation in these limits yields

$$\int_{h_w}^{H} hdh = \int_{r_w}^{R} \frac{Q}{2\pi K} \times \frac{dr}{r}$$

Therefore, $(H^2 - h_w^2)/2 = \dfrac{Q}{2\pi K} \log_e \dfrac{R}{r_w}$

$$Q = \frac{\pi K(H^2 - h_w^2)}{\log_e (R/r_w)}$$

where, H = Thickness of the aquifer in m
h_w = Depth of flow at the pumping well in m
R = Radius of influence in m
r_w = Radius of the well in m

The numerator $\pi K (H^2 - h_w^2)$, in this equation, can be written as $\pi K (H + h_w) \times (H - h_w)$.

Here, $(H - h_w) = s_w$, will be the drawdown in the well, and $(H + h_w)/2$ will be the average thickness of the aquifer. Thus, $(H + h_w)/2 \times K = T$.

Therefore, $\pi K(H^2 - h_w^2)$ will be equal to $2\pi T \times s_w$, where T is the transmissibility of the aquifer.

The formula will reduce to

$$Q = \frac{2\pi T s_w}{\log_e (R/r_w)}$$

These equations were first derived by Dupuit in 1863, and then modified by Thiem in 1906. An empirical formula may be used to evaluate R as follows:

$$R = 3000\, s_w \sqrt{K}$$

where, R = Radius of influence in m
s_w = Drawdown in the pumping well in m
K = Coefficient of permeability in m/s

These equations have the following limitations:

1. The drawdown curve near the pumping well is assumed vertical, which may not be correct.
2. The steady state of the equilibrium occurs after long pumping periods, since the velocity of the groundwater flow is very low.

Example 11.3

Water is pumped out at the rate of 2400 lit/min, from a well of 0.25 m diameter, penetrating fully in an aquifer of 32 m thickness. The drawdown observed in two adjoining wells at 15 m and 110 m from the pumping well is 7.0 m and 0.5 m, respectively. Determine the average hydraulic conductivity.

Solution:

$$Q = \frac{\pi K(h_1^2 - h_2^2)}{\log_e (r_1/r_2)}$$

here, $Q = 2400 \text{ lit/min} = 2400/(1000 \times 60) = 0.04 \text{ m}^3/\text{s}$
$h_2 = 32 - 7 = 25$ m and $h_1 = 32 - 0.5 = 31.5$ m
$r_1 = 110$ m and $r_2 = 15$ m

$$\log_e\left(\frac{r_1}{r_2}\right) = \log_e\left(\frac{110}{15}\right) = 1.99$$

Substituting the above values, we get

$$0.04 = \frac{3.14K(31.5^2 - 25^2)}{1.99}$$

Therefore, $K = 6.89 \times 10^{-5}$ m/s = 5.952 m/day.

Example 11.4

Water was pumped out from a well of diameter 1.2 m, from an unconfined aquifer of 50 m thickness. The well penetrates fully in the aquifer and the depth of water in the well is 40 m. The effect of pumping was reaching up to a distance of 450 m. Find the discharge if $K = 10$ m/day.

Solution:

$$Q = \frac{\pi K(H^2 - h_w^2)}{\log_e (R/r_w)}$$

Here, $K = 10$ m/day $= 10/(24 \times 3600) = 1.157 \times 10^{-4}$ m/s
$H = 50$ m and $h_w = 40$ m
$R = 450$ m and $r_w = 1.2/2 = 0.6$ m
$\log_e (R/r_w) = \log_e (450/0.6) = 6.62$

Substituting the above values, we get

$$Q = \frac{\pi 1.157 \times 10^{-4} (50^2 - 40^2)}{6.62}$$
$$= 0.0494 \text{ m}^3/\text{s} = 49.4 \text{ lit/s}$$

11.7.2. Steady Radial Flow in a Confined Aquifer

Consider a well in a confined aquifer as shown in the Fig. 11.17.

Initially, the water level in the well will be the same as that of the piezometric head. When pumping is started from the pumping well at a constant discharge Q, the water in the well will start receding and will reach a constant level after some time. This means that the inflow is equal to the outflow. This is the steady state.

h_1 and h_2 are the piezometric heads in the two observation wells located at r_1 and r_2 from the observation well, s_1 and s_2 are the drawdowns in these two wells, and h_w and s_w are the piezometric head and the drawdown in the pumping well.

$$\text{The velocity of flow} = Ki = K\,dh/dr.$$

Being radial flow, the area of flow will be $= 2\pi rb$

$$Q = \text{area} \times \text{velocity} = 2\pi rb \times K\,dh/dr$$

or
$$dh = \frac{Q}{2\pi bK} \times \frac{dr}{r}$$

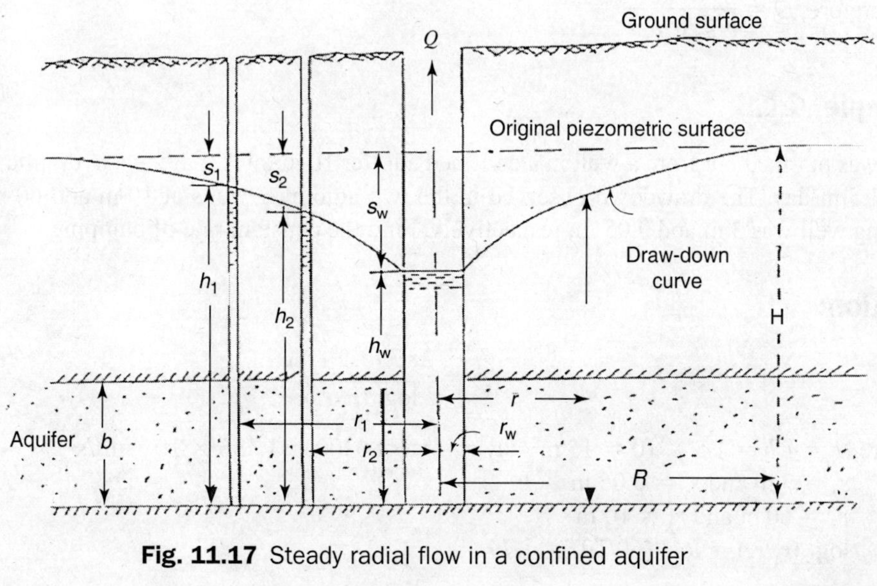

Fig. 11.17 Steady radial flow in a confined aquifer

Integrating the limits yields

$$h_1 - h_2 = \frac{Q}{2\pi bk} \times \log_e \frac{r_1}{r_2}$$

or

$$Q = \frac{2\pi bK (h_1 - h_2)}{\log_e (r_1/r_2)}$$

As $Kb = T$ (the transmissibility of the aquifer), the discharge can be expressed as follows:

$$Q = \frac{2\pi T (h_1 - h_2)}{\log_e (r_1/r_2)}$$

Now, $h_1 + s_1 = h_2 + s_2$

Therefore, $h_1 - h_2 = s_2 - s_1$

$$\therefore \quad Q = \frac{2\pi T (s_2 - s_1)}{\log_e (r_1/r_2)}$$

where, Q = Steady discharge in m^3/s

If observation wells are not available, then the following relation may be used for applying the boundary conditions to the pumping well.

Now, $dh = \frac{Q}{2\pi bK} \times \frac{dr}{r}$

Integrating the equation in the above limits yields

$$\int_{h_w}^{H} dh = \int_{r_w}^{R} \frac{Q}{2\pi bK} \times \frac{dr}{r}$$

$$H - h_w = \frac{Q}{2\pi T} \log_e \frac{R}{r_w}$$

Now, $H - h_w = s_w$

Therefore, $Q = \frac{2\pi T s_w}{\log_e (r/r_w)}$

Example 11.5

Water was pumped out from a well in a confined aquifer 10 m thick, having a hydraulic conductivity of 1.5 m/day. The drawdown observed in the two adjoining wells at 10 m and 60 m from the pumping well was 3 m and 0.05 m, respectively. Find the constant rate of pumping.

Solution:

$$Q = \frac{2\pi T (s_2 - s_1)}{\log_e (r_1/r_2)}$$

Here, $T = Kb = 1.5 \times 10 = 15$ m^2/day $= 15 \div 86400 = 1.736 \times 10^{-4}$ m^2/s

$s_2 = 3$ m and $s_1 = 0.05$ m

$r_1 = 60$ m and $r_2 = 10$ m

$\log_e (r_1/r_2) = \log_e (60/10) = 1.79$

Substituting the above values, we get

$$Q = \frac{2\pi \times 1.736 \times 10^{-4}\,(3.0-0.05)}{1.79}$$

$$= 0.001798\ \text{m}^3/\text{s} = 107\ \text{lit/min}$$

Example 11.6

A well 0.5 m in diameter penetrates fully a confined aquifer of thickness 25 m, having a conductivity of 1.6/day. It is expected that the drawdown in the well be limited to 2.5 m. Assuming that the radius of influence is 300 m, find the maximum discharge that can be pumped out of the well.

Solution:

$$Q = \frac{2\pi T s_w}{\log_e (R/r_w)}$$

Here, $T = K \times b = 1.6 \times 25\ \text{m}^2/\text{day} = 1.6 \times 25 \div 86400 = 0.00462\ \text{m}^2/\text{s}$.
$R = 300$ m and $r_w = 0.5 \div 2 = 0.25$ m and $s_w = 2.5$ m
$\log_e (R/r_w) = \log_e (300/0.25) = 7.090$

Substituting the above values, we get

$$Q = \frac{2\pi \times 0.00462 \times 2.5}{7.09}$$

$$= 0.01026\ \text{m}^3/\text{s} = 613.8\ \text{lit/min}$$

11.8 GROUNDWATER EXPLORATION

Assessment of groundwater, qualitatively as well as quantitatively, is necessary for the proper planning of groundwater development. It is done to determine whether water occurs under the conditions prevailing at the site for economical utilization in quality and in quantity.

Occurrence and availability of groundwater is very much dependent on the geology of the area and hence its thorough knowledge is necessary. Groundwater exploration is done by various methods. The methods are discussed in the following sections.

11.8.1 Electric-Resistivity Method

In this method, a current is applied to the ground and the measurement of resistivity is done from the ground surface by inserting electrodes at different locations and observing drop in potential. After scanning the data of electrical potential applied, and its drop measured at different locations, the possibility of groundwater occurrence can be assessed.

11.8.2 Seismic-Refraction Method

A very heavy hammer, when dropped from a distance, produces a seismic shock on striking the ground and the velocity of shock wave is recorded. The velocity of travel of the shock wave in different media, with or without water, is different. By analysing and properly interpreting the collected data of velocities of the shock waves, the possibility of groundwater can be assessed.

11.8.3 Remote-Sensing Method

This is the modern method followed. Photographs of the earth taken by an aircraft or from a satellite, with electromagnetic waves, indicate with a sufficient accuracy the possibility of groundwater.

11.8.4 Radio-Isotope Method

Use of radio-isotopes helps in determining various parameters of groundwater. Normally *Tritium isotopes* are used. The radioactive material is injected and the activity of the isotopes is studied. After analysing the data thus collected, the possibility of groundwater can be assessed.

11.8.5 Test-Drilling Method

This is the surest way of studying the possibility of the availability of groundwater. However, it is costly and time-taking. Drilling is done up to a depth of about 100 m and samples are collected to record the log of the hole. The test drill reveals the character, depth, and thickness of the various strata, water table and water quality. This method is normally used in a flat country. The test holes also serve as observation wells during pumping tests.

11.8.6 Water Divining

In ancient Sanskrit literature, a number of clues have been suggested by Varah Mihir, which indicate the possibility of the occurrence of groundwater at a station.

Some people claim that they have a God-given gift by which they can confidently say where the groundwater is available at a station.

According to this method, some use a stick of a specific tree and go round the area with that stick in their hands held in a specific way. There are so many similar procedure/approaches followed. Then they advise the possibility of groundwater at a station. There are cases where these people have successfully indicated the presence of groundwater. However, they do not have any scientific basis. Such water diviners are found in almost all countries.

11.9 YIELD FROM A BASIN

Safe yield from a groundwater basin is the amount of water, which can be annually withdrawn from a basin without producing any undesirable effect. This safe yield is the amount of water that enters into the basin annually. Any withdrawal more than the safe yield is called *overdraft*.

When water more than the safe yield is withdrawn from a basin, the excess comes from the storage within the aquifer. Such permanent depletion of storage within the aquifer is known as *mining of groundwater*.

11.10 YIELD FROM A WELL

When pumping is started from a well, the water level in the well goes on receding. This drop in water level in the well is known as *drawdown* or *depression head.*

When the depression is more than a critical limit, dislodging of soil particles from the adjoining strata is noticed due to high velocity of water entering into the well. The depression head, when such dislodging is noticed, is known as *critical depression head* and corresponding to this head, the yield is known as *maximum yield* or *critical yield.*

Safe depression head, also known as *working head,* is the head when there is no possibility of any dislodging of soil particles. Normally, it is taken as one-third of the critical depression head, and the yield corresponding to this working head is known as *maximum safe yield.*

The rate of inflow in a well Q is proportional to the depression head H.

$$\text{i.e., } Q \propto H \text{ or } Q = k \times H \text{ or } Q = \frac{k}{A} \times A \times H \text{ or } Q = C \times A \times H$$

where, k is the constant of proportionality having dimensions of m^2/s and A is the area of cross section of the well. Normally, A is taken as 4/3 of the cross-sectional area of the well, since additional cavity is formed at the bottom of the well.

Here, $C = k/A$, is known as *specific yield* or *specific capacity*. Its dimensions are T^{-1}.

11.10.1 Safe Yield from a Well

The safe yield from an open well can be estimated provided its size and the working head is known. The safe yield can be evaluated by conducting a pumping test.

The specific yield or specific capacity, i.e. C of a well, is estimated by a recuperation test, so that the discharge for any depression head can be evaluated.

11.10.2 Pumping Test

In this test, a constant discharge is pumped out from the test well. The inflow into the well will depend upon the drawdown. Initially, the drawdown will be less and hence inflow will be less. The drawdown will go on increasing with the inflow into the well. At a certain stage, the drawdown will reach the working head. This stage will be maintained steady by regulating the pump discharge. The constant discharge for the drawdown equal to working head is the *safe yield from the well*, and normally it is expressed as m^3/h.

Example 11.7

A pumping test was conducted for an open well of diameter 3.5 m. The water was pumped out at a constant rate of 300 lit/min. Find the specific yield.

Solution:

$$Q = C A H$$

Here, $Q = 300 \text{ lit/min} = 300 \times 10^{-3} \div 60 = 0.005 \text{ m}^3/\text{s}$

$D = 3.5$ m

$$A = \frac{\pi D^2}{4} = \frac{\pi \times 3.5^2}{4} = 9.6211 \text{ m}^2$$

Substituting the above values, we get

$$0.005 = C \times 9.6211 \times 3.5$$

$$\therefore \quad C = 1.48 \times 10^{-4}/\text{s} = 0.5345/\text{h}.$$

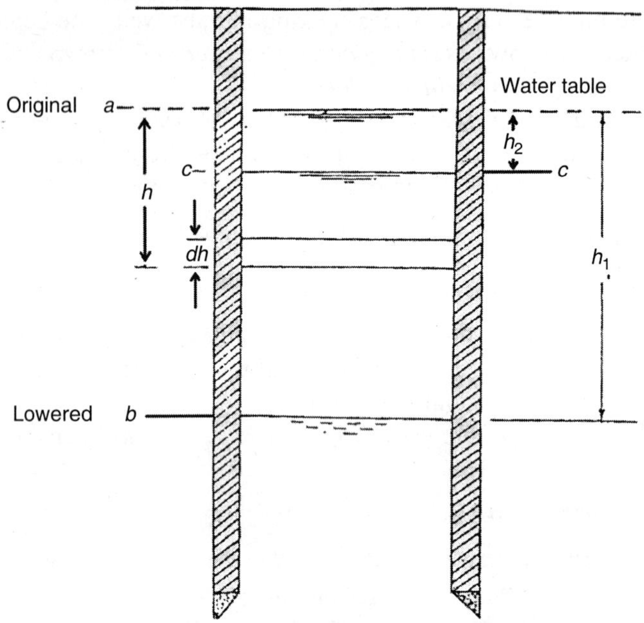

Fig. 11.18 Recuperation test

11.10.3 Recuperation Test

From the test well, a constant discharge is pumped out of the well. Naturally, the water level in the well starts receding. When the water level in the well goes below the normal water level, pumping is stopped. The water level in the well will go on rising. The time required for the water level to reach some specific higher level is noted (vide Fig. 11.18).

Where, A = Cross-sectional area of the well in m^2
aa = Static water level in the well before pumping has been started
bb = Water level in the well when the pumping is stopped
h_1 = Depression head in the well in metres, when the pumping is stopped
T = Time in hours for the rise in water level to reach a specific level, after the pump is stopped (from bb to cc)
cc = Water level in the well at time T, after the pumping is stopped
h_2 = Depression head in the well in metres, at time T after the pumping is stopped
h = Depression head in the well in metres, at time t after the pumping is stopped (a general term)

Now, in time dt the volume incoming into the well = $Qdt = -A\,dh$ ($-$ve sign indicates that when T increases h decreases), so also $Q = K \times h$.

$\therefore$ $K \times h \times dt = -A \times dh$ and hence

$$K \times dt = -A \times dh/h$$

Integrating between the limits: when $t = 0$, $h = h_1$ and $t = T$, $h = h_2$

We get $\frac{K}{A}\int_{0}^{T} = \int_{h1}^{h2} - \frac{dh}{h}$

Or $\frac{K}{A}\,[t]_0^T = -\log_e\,[h]_{h_1}^{h_2} = \log_e\,[h]_{h_2}^{h_1}$

Therefore, $\frac{K}{A}T = \log_e \frac{h_1}{h_2}$

Since $\frac{K}{A} = C$

$$C = \frac{1}{T}\log_e \frac{h_1}{h_2} = \frac{2.303}{T}\log_{10}\frac{h_1}{h_2}$$

where, C = Specific yield of the well per hour
h_1 = Depression head in metres, when pumping is stopped
h_2 = Depression head in metres, after time T
T = Time required in hours for depression head to rise from h_1 to h_2.

Knowing the values of h_1, h_2 and T, from the recuperation test, the value of C can be calculated. As C is known, the safe yield can be found out for any depression head as $Q = C\,A\,H$

where, Q = Safe yield in m^3/h
C = Specific yield per hour
A = Area of cross section of the well in m^2
H = Depression head in metres

In absence of the recuperation test results, the approximate values of C as given in Table 11.1 may be used.

Table 11.1 Approximate values of C

Serial no.	Type of soil	C (h^{-1})
1	Clay	0.25
2	Fine sand	0.50
3	Coarse sand	1.00

Example 11.8

A recuperation test was conducted on an open well 5.0 m in diameter. The water levels observed during the test were as follows:

1. Groundwater table level = 250.0 m
2. Water level when the pumping was stopped = 243.0 m
3. Water level in the well 2 h after pumping was stopped = 245.0 m

Find the safe yield of the well if the working head is 3.0 m.

Solution:

$$Q = C\,A\,H$$

Here, $h_1 = 250 - 243 = 7.0$ m and $h_2 = 250 - 245 = 5.0$ m
$T = 2.0$ h
$H = 3.0$ m

$$D = 5.0 \text{ m}$$

$$A = \frac{\pi \times D^2}{4} = \frac{\pi \times 5^2}{4} = 19.64 \text{ m}^2$$

$$C = \frac{2.303}{T} \log_{10} \frac{h_1}{h_2} = \frac{2.303}{2} \log_{10} \frac{7}{5}$$

$$= 1.152 \times 0.1461 = 0.1683/\text{h}$$

Substituting the above values, we get

$$Q = 0.1683 \times 19.64 \times 3.0$$

$$= 9.918 \text{ m}^3/\text{h} = 2.755 \text{ lit/s}$$

11.11 WELL INTERFERENCE

When two wells are located close to each other, and when pumping is started in both the wells, a cone of depression will be developed for each well. If these cones of depression overlap each other, the yield from each of the wells will naturally be reduced. When two or more cones of depression of the pumping wells overlap, affecting their yield, the phenomenon is known as *well interference* and is shown in Fig. 11.19.

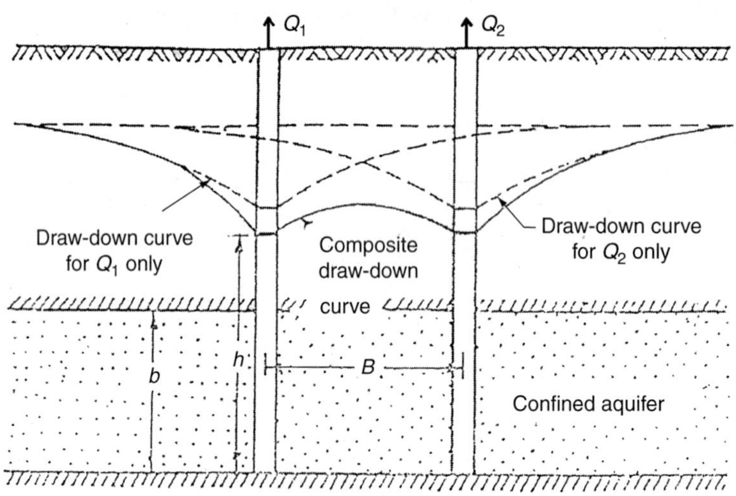

Fig. 11.19 Well interference

11.12 SEA WATER INTRUSION

If pumping of fresh groundwater is done near the sea, the sea water being heavier tries to occupy the lower level space vacated by the fresh water and causes serious water quality problems. This is known as *sea water intrusion*. There is a plane that separates the fresh water from the sea water, and it is known as *interface*. Normally, it is parabolic in shape as shown in Fig. 11.20.

Sea water is slightly heavier than the fresh water due to the dissolved salts. Specific gravity of sea water is 1.025 and that of fresh water is 1.000. As the fresh groundwater table increases and

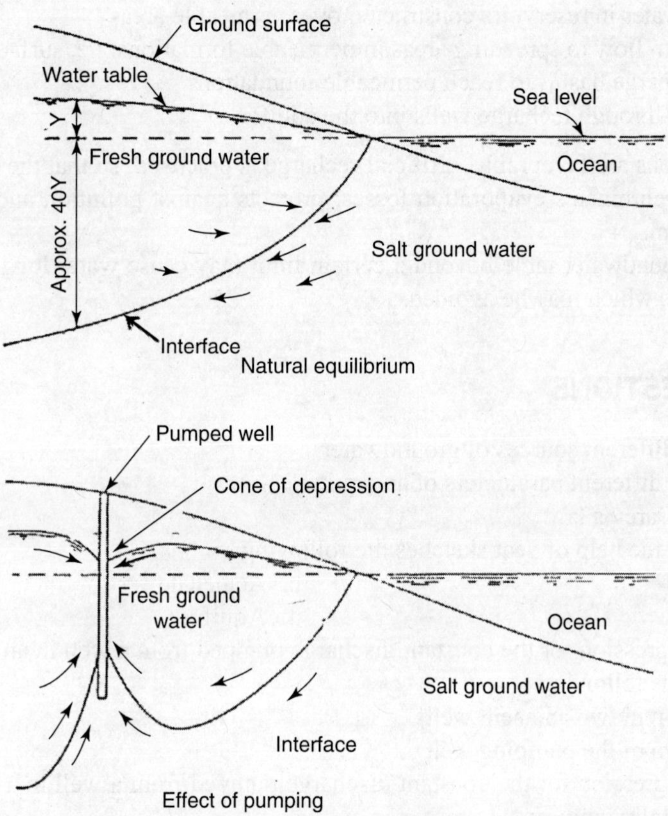

Fig. 11.20 Sea water intrusion

approaches the earth surface by 1.00 m, the sea water interface is pushed back by 40.00 m and vice versa. When the sea water intrusion is noticed, it can be pushed back by recharging the dug wells near the seashore by fresh water.

11.13 RECHARGING OF GROUNDWATER

Water may be withdrawn from an aquifer by (1) pumps, (2) discharge to surface water systems or by (3) evaporation and transpiration (to a minor extent). If this water is not replenished, the aquifer storage will reduce resulting in depletion of groundwater table.

Normally, this replenishment occurs naturally by the following ways:

- Infiltration from precipitation
- Surface bodies like reservoirs, streams, and so on

Artificial recharge is adhered to meet the deficit of withdrawal and replenishment or even to increase the aquifer yield. It is also followed to control the saltwater intrusion along the sea coast.

The following recharge methods are employed, taking into consideration the geological situation and economic consideration.

- Storing flood water in reservoirs constructed over permeable areas
- Diverting stream flow to spreading areas in permeable formations, i.e. surface spreading
- Excavating recharge basins to reach permeable foundations
- Pumping water through recharge wells into the aquifer

When conditions are favourable, artificial recharge is practised, so that the aquifer can be used as a reservoir. It eliminates evaporation losses, protects against pollution and provides low-cost distribution system.

Raising of groundwater table beyond a certain limit may cause water logging and consequent soil efflorescence, which may be avoided.

REVIEW QUESTIONS

1. Discuss the different sources of groundwater.
2. What are the different parameters of an aquifer?
3. Explain the Darcy's law.
4. Explain with the help of neat sketches the following:
   a. Aquifer
   b. Aquiclude
   c. Aquitard
   d. Aquifuse
5. Derive an expression for the constant discharge pumped from a well in an unconfined aquifer in terms of the following:
   a. Drawdown in two adjacent wells
   b. Drawdown of the pumping well
6. Derive an expression for the constant discharge pumped from a well in a confined aquifer in terms of the following:
   a. Drawdown in two adjacent wells
   b. Drawdown of the pumping well
7. What is salt water intrusion? How can it be controlled?
8. Discuss the different methods of groundwater exploration.
9. Why is recharging of wells necessary? Discuss the different methods.
10. Explain the pumping test to estimate the safe yield from an open well.
11. Explain the recuperation test to estimate the safe yield of an open well.
12. What are the peculiarities of groundwater?
13. Discuss the assumptions made in the analysis of steady radial flow into a well.
14. Write short notes on the following:
    a. Perched water table
    b. Cone of depression
    c. Well interference
    d. Salt water intrusion
    e. Safe yield from a well
    f. Hydraulic conductivity
    g. Storage coefficient
    h. Porosity of an aquifer
    i. Overdraft from a basin
    j. Mining of groundwater
    k. Capillary zone in an aquifer
    l. Area of influence
    m. Critical depression head
    n. Specific capacity
15. Differentiate between the following:
    a. Confined aquifer and unconfined aquifer
    b. Zone of saturation and zone of aeration
    c. Specific yield and specific retention
    d. Open well and tube well
    e. Shallow well and deep well
    f. Strainer well, cavity well and slotted well

g. Pumping test and recuperation test
h. Discharging well and charging well
i. Pervious layer and impervious layer
j. Groundwater table and piezometric level
k. Primary porosity and secondary porosity
l. Artesian well and flowing well
m. Safe depression head and critical depression head
n. Cone of depression and depression head

NUMERICAL QUESTIONS

1. Water was pumped at a constant discharge from a well of diameter 0.5 m located in an 35-m thick unconfined aquifer having a hydraulic conductivity of 6 m/day. The drawdown observed in two observation wells located at 16 m and 115 m from the pumping well were 5.0 and 0.5 m respectively. Find the constant discharge.

   Ans: $Q = 0.032\ m^3/s$

2. Water at the rate of 100 lit/min from a well of diameter 1.0 m, located in a confined aquifer 8.0-m thick, was pumped. The drawdown observed in two observation wells at 10.0 m and 60.0 m from the pumping well were 3.5 m and 0.1 m respectively. Find the hydraulic conductivity of the aquifer.

   Ans: $K = 1.5$ m/day

3. A recuperation test was conducted on a well of 4.0-m diameter. The water levels observed were as follows:

   (i) Ground water table level = 300 m

   (ii) Water level when the pumping was stopped = 292 m

   (iii) Water levelling the well 2 h after the pumping was stopped = 295 m

   Find the safe yield of the well if the working head is 3.5 m.

   Ans: $Q = 10.23\ m^3/h$

MULTIPLE CHOICE QUESTIONS

1. Which of the following formations contains water and also transmits it?
   (a) Aquifuse (b) Aquifer
   (c) Aquitard (d) Aquiclude

2. Which of the following formations neither contains water nor transmits it?
   (a) Aquiclude (b) Aquifer
   (c) Aquifuse (d) Aquitard

3. Which of the following formations may contain water but cannot transmit it?
   (a) Aquifuse (b) Aquifer
   (c) Aquiclude (d) Aquitard

4. In case of a gravity well, the piezometric surface is
   (a) Above the groundwater table
   (b) Below the groundwater table

(c) Coincides with the groundwater table
(d) Between the groundwater table and ground level

5. An unconfined aquifer is also known as
   (a) A leaky aquifer (b) A water table aquifer
   (c) An artesian aquifer (d) A perched aquifer

6. An aquifer that is underlain by an impermeable layer and not confined at the top is called as
   (a) A confined aquifer (b) An unconfined aquifer
   (c) A perched aquifer (d) A leaky aquifer

7. Porosity is considered large if it is more than
   (a) 50% (b) 40%
   (c) 20% (d) 15%

8. Porosity has the dimensions as
   (a) m/s (b) m/s^2
   (c) m^2/s (d) Dimensionless

9. Transmissibility has the dimensions as
   (a) m/s (b) m^2/s
   (c) m/s^2 (d) Dimensionless

10. A flowing well has the piezometric surface
    (a) Above the ground level
    (b) Below the ground level
    (c) Below the water surface level in the well
    (d) Between the water surface in the well and the ground level

11. Water available in the capillary fringe is a part of
    (a) Groundwater zone (b) Zone of aeration
    (c) Soil–water zone (d) Intermediate zone

12. The surface joining the water levels in different wells in a confined aquifer represents
    (a) Groundwater table surface (b) Piezometric surface of the aquifer
    (c) Capillary fringe (d) None of the above

13. The ratio of volume of water drained divided by volume of soil sample is
    (a) Porosity (b) Specific yield
    (c) Storage coefficient (d) Specific retention

14. The dimensions of specific yield are
    (a) m^3/s (b) m^2/s
    (c) m/s (d) Dimensionless

15. The ratio of volume of water retained divided by the volume of soil sample is
    (a) Porosity (b) Specific retention
    (c) Storage coefficient (d) Specific yield

16. The dimensions of specific retention are

(a) m^3/s (b) m^2/s
(c) Dimensionless (d) m/s

17. The drawdown at which stage dislodging of soil particles starts is known as

(a) Critical depression head (b) Safe depression head
(c) Working head (d) All the above

18. Normally, the ratio of safe depression head and critical depression head is taken to be

(a) $\frac{2}{3}$ (b) $\frac{1}{3}$
(c) $\frac{1}{2}$ (d) $\frac{1}{4}$

19. The dimensions of specific capacity are

(a) m^3/s (b) m/s
(c) T^{-1} (d) m

20. Which of the following formations has poor permeability, but seepage is possible?

(a) Aquiclude (b) Aquifer
(c) Aquifuse (d) Aquitard

ANSWERS TO MULTIPLE CHOICE QUESTIONS

1. (b)	2. (c)	3. (c)	4. (c)	5. (b)	6. (b)	7. (c)	8. (d)
9. (b)	10. (a)	11. (a)	12. (b)	13. (b)	14. (d)	15. (b)	16. (c)
17. (a)	18. (b)	19. (c)	20. (d)				

*** ——— ***

Appendix I

INTERNATIONAL SYSTEM OF UNITS [SI]

The 11th General Conference on Weights and Measures (1960) adopted the name *Système International d'Unités* (International System of Units, international abbreviation SI), for the recommended practical system of units of measurement. The SI is not static but evolves to match the world's increasingly demanding requirements for measurement.

There are seven well-defined units called the *base units*, which by convention are regarded as dimensionally independent. These are:

'The metre, The kilogram, The second, The ampere, The kelvin, The mole, and the Candela.'

The SI list presented here contains only those units that are used in engineering hydrology.

As per the 11th Conference Générale des Poids et Mesures (CGPM), the dimensions of the base units related to engineering hydrology are as follows:

Metre

The metre is the length of the path travelled by light in vacuum during a time interval of 1/299 792 458 of a second.

Kilogram

The kilogram is the unit of mass; it is equal to the mass of the international prototype of the kilogram.

Second

The second is the duration of 9 192 631 770 periods of the radiation corresponding to the transition between the two hyperfine levels of the ground state of the caesium 133 atom.

The **units** formed by combining base units according to the algebraic relations linking the corresponding quantities are called the *derived units*.

The details of the basic and the derived units related to engineering hydrology are as follows:

BASE UNITS AND DERIVED UNITS

Sr. No.	Item	Symbol	Definition	Dimension	Units	Important Relations	Value for Water
FUNDAMENTAL UNITS							
1	Mass	M	Matter contained in a body	$M^1 L^o T^o$	ton; kg; gram	1 ton = 10^3 kg = 10^6 gram	
2	Length	L	Separation between two defined points	$M^o L^1 T^o$	km; m; cm	1 km = 10^3 m = 10^5 cm	
3	Time	T	An expression of separation among events occurring in the same physical location	$M^o L^o T^1$	hour; minute; second	1 hour = 60 minutes = 3600 seconds	

Sr. No.	Item	Symbol	Definition	Dimension	Units	Important Relations	Value for Water
DERIVED UNITS							
4	Area	A	Extent of a closed perimeter figure on a plane	$M^\circ L^2 T^\circ$	km^2; hect; are; m^2; cm^2	1 km^2 = 10^6 m^2, 1 hect = 10^4 m^2, 1 are = 10^2 m^2	
5	Volume	∀	Space occupied by a body	$M^\circ L^3 T^\circ$	km^3; m^3; litre; c.c.	1 km^3=10^9 m^3, 1 m^3 = 10^6 c.c 1 litre = 10^3 c.c.	
6	Discharge	Q	Volume flowing per unit time	$M^\circ L^3 T^{-1}$	m^3/s; litre/s; c.c./s.	1 m^3/s = 10^6 c.c./s 1 litre/s = 10^3 c.c./s	
7	Discharge intensity	q	Discharge per unit width	$M^\circ L^2 T^{-1}$	m^3/s/m		
8	Velocity	V	Distance travelled per unit time	$M^\circ L^1 T^{-1}$	km/hour; m/s; cm/s		
9	Acceleration	a	Change in velocity per unit time	$M^\circ L^1 T^{-2}$	m/s^2; cm/s^2	Acceleration due to gravity = 'g' = 9.81 m/s^2	
10	Angle	°	Arc divided by the radius of the circle	$M^\circ L^\circ T^\circ$	Radian; degree	π radian = 180°	
11	Mass density [Specific mass]	ρ	Mass per unit volume	$M^1 L^{-3} T^\circ$	kg/m^3; gram/c.c.		1000 kg/m^3, 1 gram/c.c.
12	Force	F	Produces or tends to produce or destroys or tends to destroy motion	$M^1 L^1 T^{-2}$	Newton; dyne	1 N = 1 kg × 1 m/s^2 1 dyne = 1 gram × 1 cm/s^2 1 N = 10^5 dyne	
13	Weight	W	Force due to gravity	$M^1 L^1 T^{-2}$	Newton; dyne		
14	Weight density [Specific weight]	γ	Weight per unit volume	$M^1 L^{-2} T^{-2}$	Newton/m^3; dyne/c.c.	$\gamma = \rho \times g$	9.81×10^3 N/m^3
15	Pressure	P	Normal compressive force per unit area	$M^1 L^1 T^{-2}$	Pascal: bar, N/m^2, m of fluid	1 Pa = 1 N/m^2, 1 bar = 10^5 N/m^2, 'm' of fluid = P/γ	P atm = 1.02×10^5 N/m^2 = 10.33 m of water
16	Momentum	M	Mass × velocity	$M^1 L^1 T^{-1}$	kg m/s		
17	Specific gravity	S	Ratio of specific weight of fluid and water	$M^\circ L^\circ T^\circ$		s = ρf/ρw = γf/γw	1.00
18	Specific volume	∀s	Volume per unit weight	$M^{-1} L^2 T^2$	m^3/N		1.02×10^{-4} m^3/N

*** ——— ***

Appendix II

CONVERSION TABLE FOR SI UNITS

Earlier, the British or the Imperial System of units was in use in India. Now, however, the SI units are used.
A conversion between these two systems is given below

Length

1) 1 in	=	2.54 cm	i) 1 cm	=	0.394 in		
2) 1 ft	=	0.305 m	ii) 1 m	=	3.281 ft	=	1.093 yd
3) 1 yd	=	0.914 m	iii) 1 km	=	0.621 mile		
4) 1 mile	=	1.609 km					

Area

1) 1 in^2	=	6.452 cm^2	i) 1 cm^2	=	0.155 in^2		
2) 1 ft^2	=	0.093 m^2	ii) 1 m^2	=	10.764 ft^2	=	1.195 yd^2
3) 1 $mile^2$	=	2.588 km^2	iii) 1 km^2	=	0.386 $mile^2$		
4) 1 acre	=	0.405 ha	iv) 1 ha	=	2.469 acre		

Volume

1) 1 in^3	=	16.393 cc	i) 1 cc	=	0.061 in^3		
2) 1 c ft	=	0.0283 m^3	ii) 1 m^3	=	35.330 ft^3	=	1.306 yd^3
3) 1 $mile^3$	=	4.165 km^3	iii) 1 km^3	=	0.240 $mile^3$		
4) 1 acre-ft	=	0.123 ha-m	iv) 1 ha-m	=	8.130 acre-ft		

Temperature

1) $\frac{(°F - 32) \times 5}{9} = °C$ i) $\frac{(°C \times 9)}{5} + 32 = °F$

Logarithms

1) $\log_{10} e$	=	0.4343	i) $\ln_e 10$	=	2.303
2) $\log_{10} x$	=	0.4342 $\ln_e x$	ii) $\ln_e x$	=	2.303 $\log_{10} x$

Acceleration due to gravity

g = 32.2 ft/s^2 = 9.81 m/s^2

*** ——— ***

Appendix III

ROMAN NUMERALS

The Romans who were active in trade and commerce right from the time they learnt to write devised a system of symbols to indicate numbers. The difference between the Roman and the Arabic numericals [The one which is used today] is that the Romans did not have a symbol of zero. The following chart gives the several basic Roman numerals and explains how to determine the value of other numbers.

I	The easiest way to note down a number is to make marks. Thus, I means 1, II means 2, III means 3.
V	However, since four strokes seemed too many, the Romans moved on to the symbol V for 5. To obtain 4, they decided to place I before V—IV. Thus, placing any small number before a basic Roman number indicates subtraction and placing a number after the basic Roman number indicates addition. Following this pattern, VI, VII and VIII indicate 6, 7 and 8 respectively.
X	The symbol X indicated the number 10. Placing I before X—IX—gives 9 and placing I after X—XI— gives 11. Numbers in the teens, twenties and thirties follow the same pattern. In these sets, the number of X indicates the number of tens. So, XXVI is 26 and XXXIV is 34.
L	L stands for 50. To obtain 40, which is 10 subtracted from 50, it will be written as XL. And thus 60, 70, and 80 are LX, LXX and LXXX.
C	C stands for *centum*, the Latin word for 100. This is used in words like "century" and "cent". As per the rule, 90 will be written as XC. Like the X's and L's, the C's are tacked on to the beginning of numbers to indicate how many hundreds are there. CCCLXIX is 369.
D	D stands for 500. As can be guessed, CD means 400. So CDXLVIII is 448.
M	M is 1,000. This symbol finds a lot of use when Roman numerals are used to indicate dates. For instance, 2007 will be written as MMVII.
—	Large numbers were indicated by putting a horizontal line over them. It means to multiply the number by 1000. Thus $\overline{V}$ means 5000.

*** ——— ***

Appendix IV

GREEK ALPHABET

The Greek alphabet is used as mathematical symbol in most of the scientific studies. The following table gives the upper- as well as the lower-case letters.

Upper-case	Lower-case	Greek Name	English
A	α	Alpha	a
B	β	Beta	b
Γ	γ	Gamma	g
Δ	δ	Delta	d
E	ϵ	Epsilon	e
Z	ζ	Zeta	z
H	η	Eta	h
Θ	θ	Theta	th
I	ι	Iota	i
K	κ	Kappa	k
Λ	λ	Lambda	l
M	μ	Mu	m
N	ν	Nu	n
Ξ	ξ	Xi	X
O	o	Omicron	o
Π	π	Pi	p
P	ρ	Rho	r
Σ	σ	Sigma	s
T	τ	Tau	t
Υ	ϑ	Upsilon	u
Φ	ϕ	Phi	ph
X	χ	Chi	ch
Ψ	ψ	Psi	ps
Ω	ω	Omega	o

*** ——— ***

Appendix V

METRIC PREFIXES

In practice, several parameters which are very large or very small are required to be referred to. These are mentioned as multiples of 10. In SI, a number of **metric prefixes** are used to mention their largeness or smallness. These are as follows.

		Prefix	***Symbol***
million million [Trillion]	10^{12}	tera-	T
thousand million [Billion]	10^{9}	giga-	G
million	10^{6}	mega-	M
thousand	10^{3}	kilo-	k
hundred	10^{2}	hecto-	h
ten	10	deca-	da
tenth	10^{-1}	deci-	d
hundredth	10^{-2}	centi-	c
thousandth	10^{-3}	milli-	m
millionth	10^{-6}	micro-	μ
thousand millionth	10^{-9}	nano-	n
million millionth	10^{-12}	pico-	p
thousand million millionth	10^{-15}	femto-	f
million million millionth	10^{-18}	atto-	a

Since hecto-, deca-, deci-, and centi- are not multiples of powers of three (10^3), these prefixes are not approved and should be avoided in formal texts.

*** ——— ***

Appendix VI

KEY TO SURVEY MAPS

The Survey of India was set up as the National Survey and Mapping Organization in 1767 under the Ministry of Science and Technology, Government of India, with headquarters at Dehradun. This organization publishes various types of maps.

The maps published by the Survey of India are of two series.

A) India and Adjacent Countries (I A C) series
B) International Map of the World (I M W) series

A) The I. A. C. series

In this series, the maps are numbered. Each numbered sheet covers 4° in latitude and 4° in longitude. The numbering starts from NW corner 40° Latitude North and 44° Longitude and increases from north to south. Sheets covering sea are not numbered.

The Indian Union is located between 8°-4' to 37°-6' north latitude and 68°-7' to 97°-25' east longitude. Sheets covering India are numbered from 39 to 88.

These sheets are known as **'I/M or million sheets'**. The representative scale of these sheets is 1:1,000,000. The map of India showing this division is as shown in Fig. AVI.1.

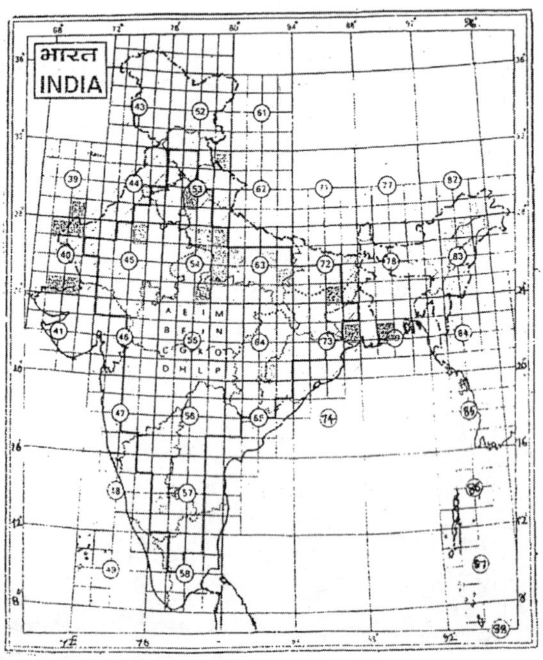

Fig. AVI.1 The map of India showing the I/M sheets
Courtesy: Survey of India

Each I/M sheet is divided into 16 equal parts and are designated from A to P. Each part therefore covers 1° in latitude and 1° in longitude and hence is called **'Degree Sheet'**. For example, the I/M sheet 47 is divided into 16 degree sheets as shown in Fig. AVI.2.

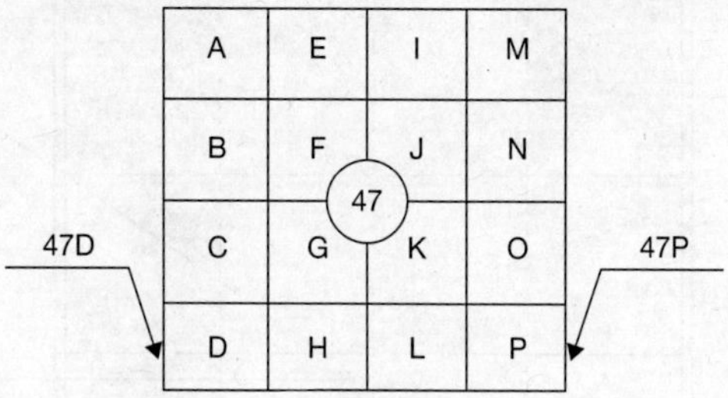

Fig. AVI.2

An accurate survey is done and each degree sheet is plotted to a scale of 1 : 250 000. All the details such as rivers, roads, railways, buildings; and administrative boundaries such as taluka, district, state; and also forest area, tanks, reservoirs and other important landmarks located in that area are shown in different colours. Contours at an interval of 100 m are normally shown.

Each degree sheet is further divided into 16 equal parts numbered from 1 to 16. For instance, the degree sheet 47 I is divided as shown in Fig. AVI.3.

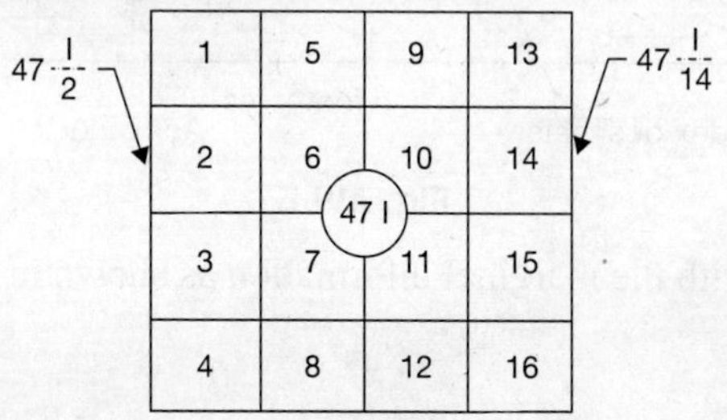

Fig. AVI.3

Here each part covers 0°-15' in latitude and 0°-15' in longitude and is known as **Topo-Sheet** and is plotted to a scale of 1:50,000. All the details shown on a degree sheet are also shown on a topo-sheet. The contours are at an interval of 20 m.

A topo-sheet to a scale of 1:50000 is further divided into 4 equal parts each 0°-7.5' in latitude and 0°-7.5' in longitude and are plotted to a scale of 1:25000. All possible details as shown on the degree sheet as well as on the topo-sheet are also shown in this map. Contours at an interval of 10 m are shown. If the ground is relatively plain, then the contour interval is maintained as 5 m.

The topo-sheet '47A/4', for instance, is divided as shown in Fig. AVI.4.

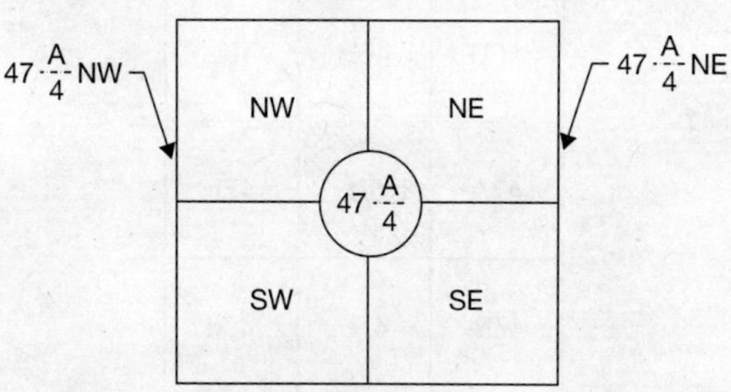

Fig. AVI.4

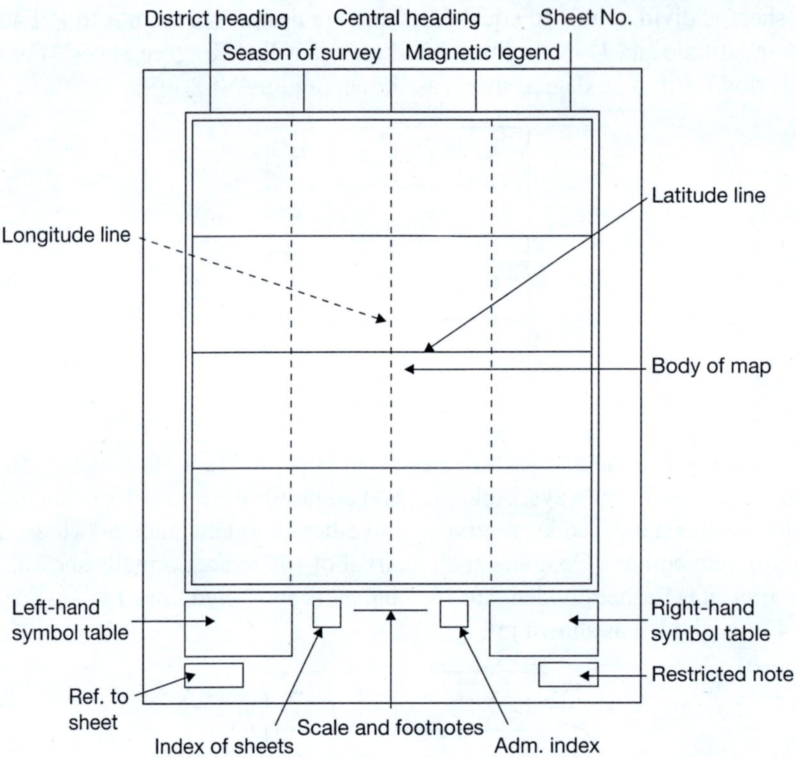

Fig. AVI.5

All maps are provided with the marginal information as shown in Fig. AVI.5 as follows:

1) Title/central heading.
2) District heading.
3) Edition: Year of publication.
4) Sheet number.
5) Legend. [This table explains the various symbols used.]
6) Footnotes: Any information which cannot be shown on the map is given here.
7) Index to sheets: Sometimes, the adjoining sheet has to be referred to. In such case, the adjoining sheet can be easily located from the 'Index to sheets' shown on each of the sheet as shown in Fig. AVI.6. The central shaded sheet is the one under reference and the adjoining ones are indicated in the index.
8) Administrative index.

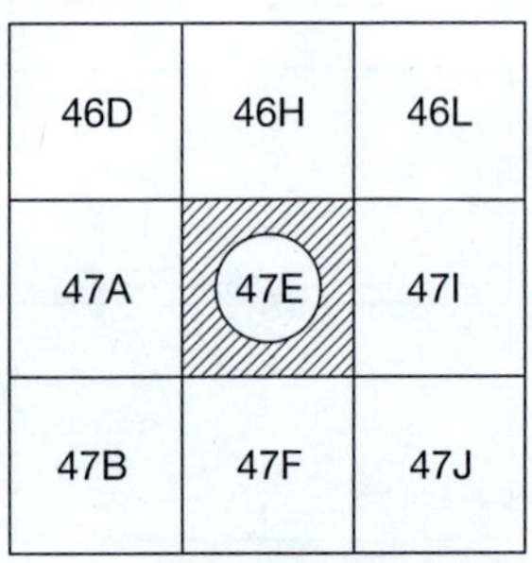

Fig. AVI.6

Some of the I/M maps, degree sheets and toposheets are classed as **RESTRICTED** for security purposes, and cannot be obtained directly. These can be obtained only after permission from the 'Ministry of Defense, Government of India'.

This I A C series is normally used in India.

B) The I. M. W series

The system is applicable globally (to whole of the world).

The entire area of the world is divided into sheets each covering an area of 4° latitude and 6° longitude.

The geographical positions are defined by two letters 'N' for the northen hemisphere and 'S' for the south hemisphere.

The 90° latitudes in each of the hemisphere are divided into horizontal bands, each of 4° Latitudes and are identified by alphabetical letters from 'A' onwards. Thus the first band from 0° to 4° as 'A'; 4° to 8° as 'B'; 8° to 12° as 'C' and so on.

There are two sets of brands 'A. B. C. -----', on either side of 0° Latitude i.e. the equator.

Similarly, the 360° Longitudes are divided into 60 vertical bands each of 6° Longitudes and are numbered from 1 to 60. However the numbering starts from 180° Longitude and goes from west to east.

Thus there will be two vertical bands '30 and 31' on either side of 0° Longitude i.e. Greenwich.

The *Mercator projection* of the Earth is shown in Fig. AVI.7.

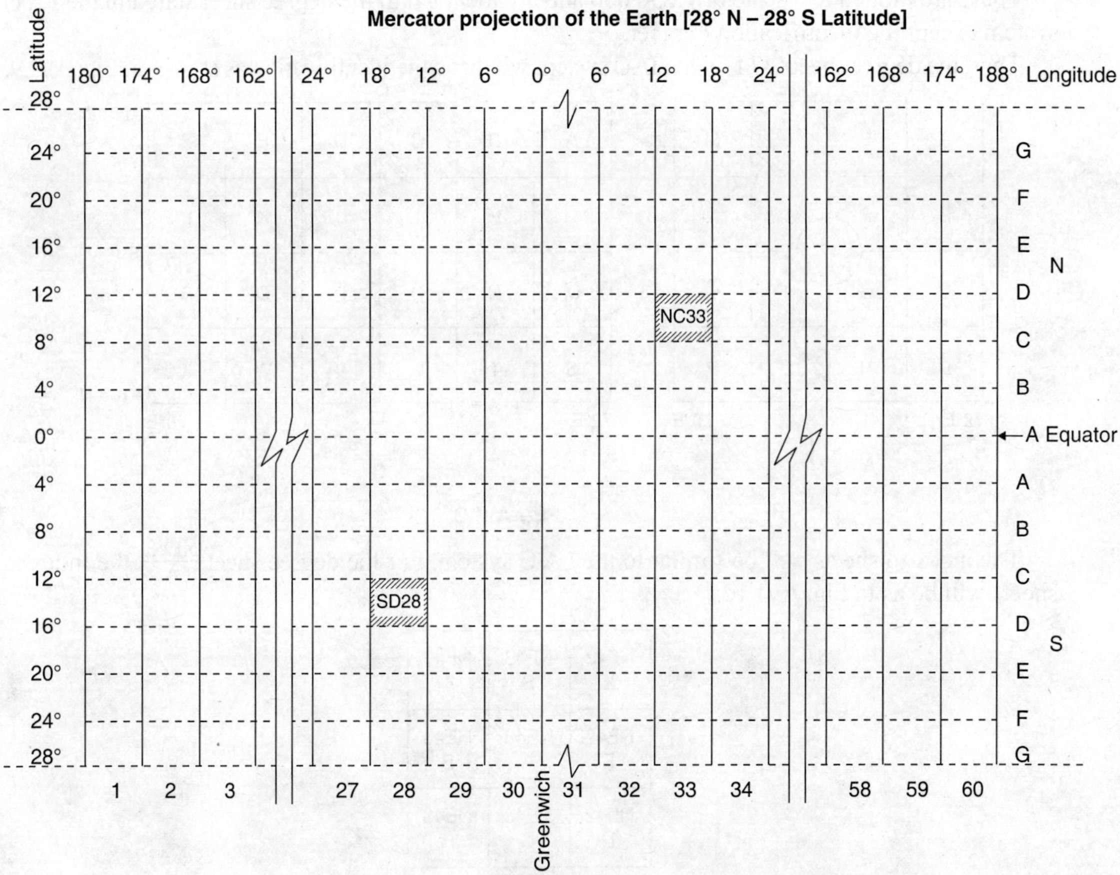

Fig. AVI.7

Thus, each sheet will cover an area of 4° Latitude and 6° Longtitude.
The Indian Union is located between 8°-4' to 37°-6' N Latitude and 68°-7' to 97°-25' Longitude.
Further, each sheet (4° Lat. and 6° Long.) is divided into 24 parts as shown in Fig. AVI.8.

A	B	C	D	E	F
G	H	I	J	K	L
M	N	O	P	Q	R
S	T	U	V	W	X

Fig. AVI.8

Each part will cover an area of 1° Latitude and 1° Longitude and hence it is also termed as **'Degree Sheet'**. The degree sheet in this International System will be identified as $\frac{\text{ND43}}{\text{H}}$. This means that the degree sheet is in the north hemisphere, in the 'D' band of Latitudes i.e. in the vertical band '43' i.e. between 12° to 16° of Longitudes, and 'H' sub-part.

These are plotted to a scale of 1:250 000 and are identical to the degree sheet stated in the I A C system except the identification number.

Thus, the degree sheet '47 I' in the IAC system will have the identification as shown in Fig. AVI.9.

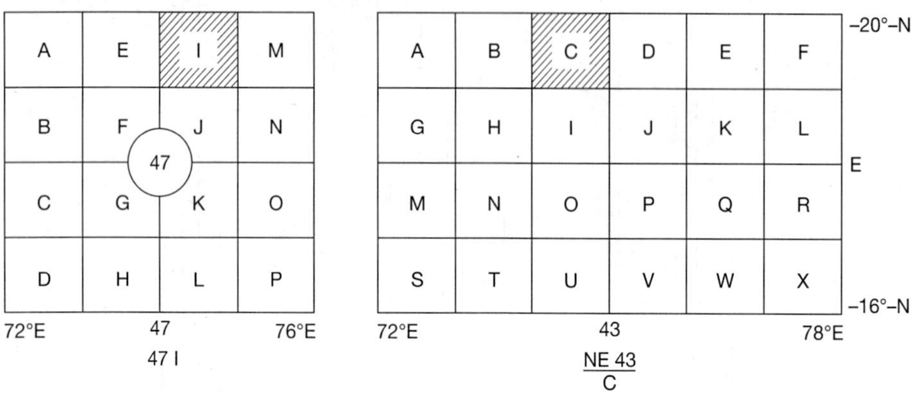

Fig. AVI.9

The index to sheets will be similar to the I A C system. For the degree sheet $\frac{\text{NE43}}{\text{A}}$, the Index to sheets will be as in Fig. AVI.10.

NF42 X	NF43 S	NF43 T
NE42 F	NE43 A	NE43 B
NE42 L	NE43 G	NE43 H

Fig. AVI.10

In all the maps, the north direction will be upwards.

All the maps are available on payment with the
Survey of India
P. O. box No:–18
Hathibarkala estate
Dehradun [Uttaranchal]
PIN 248001, INDIA

These are also available with the regional offices of the Survey of India or with authorized maps sales agents.

*** ——— ***

Appendix VII

MULTIFACETED WATER

Ballast water: Huge tanks from oil-rich countries are used to transport oil. These tanks are emptied in different countries. After emptying the oil, during their return journey, the tanks are filled with the available seawater to maintain the balance of the tanks. These tanks are emptied in the sea at the loading oil-rich country to refill them again with oil. This sea water which is used for balancing purpose and emptied in the sea at the oil-rich country is called *ballast water*. It is noticed that this ballast water causes some environmental problems.

Black water: Waste water from toilets is called *black water*.

Blue water: Water in rivers, flowing ground water, etc. is called *blue water*.

Gray water: Waste water from bathrooms, bathtubs, showers, wash basins, laundry, kitchen sinks, etc. is called *gray water*.

Green water: Water from soil/land utilized for bio mass generation by way of evapo-transpiration, etc. is called *green water*.

Hard water: Water from rivers, springs, wells, etc. contain a certain amount of dissolved salts such as calcium, magnesium salts. Such water that will not produce good amount of lather with soap is called *hard water*.

Heavy water: The oxide of heavy hydrogen (Deuterium) 'D_2O' is called *Heavy water*. It is like normal water in appearance. This was discovered by 'Uray' in 1932. It is used in Atomic energy studies.

Magnetized water: It is noticed that after magnetization, some of the physical and chemical properties of water change. The extent of magnetization depends on the magnetic poles used, intensity of magnetic field, duration, amount of water used, etc.

Some believe that this *magnetic water* is useful for healing some diseases.

Soft water: Water with a very low content of dissolved salts is called *soft water*. Some people claim that their thirst is not satisfied by drinking this type of water.

Virtual water: When any item/commodity is imported by a nation, the water required for growing or producing that item is not accounted for in water balance e.g., when one kg of wheat is imported by a nation, then 1.2 m^3 of water required for growing that one kg of wheat is indirectly imported. This 1.2 m^3 of water is called *virtual water*.

*** ——— ***

Appendix VIII

JOURNALS AND PERIODICALS ON HYDROLOGY AND WATER RESOURCES

A number of journals/periodicals on hydrology and water resources are published. A list of some Indian and foreign periodicals is given below.

A – INDIAN PERIODICALS

Sr. No	Title	Publisher
1.	ISH Journal of Hydraulic Engineering	The Indian Society for Hydraulics, Central Water Power Research Station, Khadakwasla, Pune – 411024
2.	Journal of Applied Hydrology	Association of Hydrologists of India & Andhra Pradesh, Visakhapatnam – 530003
3.	Journal of Indian Water Resources Society	Indian Water Resources Society; Indian Institute of Technology, Roorkee – 247667
4.	Indian Journal of Power and River Valley Development	Books & Journals Pvt. Ltd., 6/2 Madan Street, Kolkatta – 700072
5.	Journal of Indian Water Works Association	Indian Water Works Association, Pipeline Road, Vakola, Santakruz (East), Mumbai – 400055
6.	Hydrology Journal	Indian Water Resources Society, Indian Institute of Technology, Roorkee – 247667
7.	Water and Energy **Research Digest**	CBIP, Malcha Marg, Chanakyapuri, New Delhi – 110021
8.	Journal of Institution of Engineers – Civil Engineering	Institution of Engineers, 8, Gokhale Road, Kolkatta – 700020
9.	Dams, Rivers and People	Editor, 86 D, AD Block, Shalimar Bagh, Delhi – 110088
10.	Jal Vikas	National Water Development Agency, 18-20 Saket Community Centre, New Delhi

B – FOREIGN PERIODICALS

Sr. No.	Title	Publisher
1.	Journal of the European Water Association (EWA)	European Water Association, Theodor-Heuss-Allee, 17, D-53773 Hennef
2.	Journal of Hydrologic Engineering	American Society of Civil Engineers (A.S.C.E.), Subcription Orders, P.O.Box 79342, Baltimore, MD, 21279-0342, USA
3.	Journal of Water Resources Planning & Management	American Society of Civil Engineers (A.S.C.E.), Subcription Orders, P.O.Box 79342, Baltimore, MD, 21279-0342, USA
4.	Journal of Spatial Hydrology	Department of Environmental Studies, Florida, International University, 11200 SW, 8th Street, Miami FL 33199
5.	Ground Water	Ground Water, 601 Dempsey Road, Westerville, OH 43081
6.	Hydrology & Earth System Sciences	European Geophysical Society, Max/Planck- Str., 13, 371, Katlenburg, Lindau, Germany
7.	Journal of Hydrology	R. Krzysztafoweiz, University of Verginia, Thorntok Hall, SE, Charlottsville, VA 22903, USA (Elsevier, 3 Killiney Road, # 08-01 Winsland House I, Singapore 239519)
8.	Nordic Hydrology	ISVA Technical University of Denmark, Building 115 Dk 2800, Lyngby, Denmark
9.	Journal of Contaminant Hydrology	Elsevier, 3 Killiney Road, # 08-01 Winsland House I, Singapore 239519
10.	Journal of Environmental Hydrology	International Association of Environmental Hydrology, 2607 Hopeton, Dr. San
11.	Hydrological Processes	Journals Administration Dept., John Wiley & Sons Ltd., 1 Oldlands Way, Bognar Regis, West Sussex, Po. 229SA, England
12.	Hydrological Sciences Journal	Centre for Ecology & Hydrology, Welingford, Oxfordshire, OX 10 8BB, U.K.
13.	Advances in Water Resources	Elsevier, 3 Killiney Road, # 08-01 Winsland House I, Singapore 239519
14.	International Journal of Water Resources Development	Taylor & Francis, Customer Services T&F Informa UK. Ltd. Sheepen Place, Colchester, CO3 3LP, UK
15.	Water Resources Research	American Geophysical Union (AGU), 2000 Florida Avenue N. W. Washington DC, 20009-12-77, USA
16.	Journal of American Water Resources Association	American Water Resources Association, 4 West Federal Street, P.O. Box. 1626, Middleburg, VA 200118-16126
17.	Journal of Water Resources Planning & Management	American Society of Civil Engineers (A.S.C.E.), Subcription Orders, P.O.Box 79342, Baltimore, MD, 21279-0342, USA
18.	Florida Water Resources Journal	Buena Vista Publishing, 1402 Emerald Lakes Dr. Clermont, FL 34711, U.S.A.
19.	Stochastic Hydrology & Hydraulics	Springer-Verlag Wien, Journals Department, Sachsenplatz, 4-6, 1200 Vienna, Austria
20.	Water Resources	Springer-Verlag Wien, Journals Department, Sachsenplatz, 4-6, 1200 Vienna, Austria

Sr. No.	Title	Publisher
21.	Canadian Water Resources Journal	Canadian Water Resource Association, P.O. Box. 1329, 400 Clyde Road, Cambridge, Ontario N1 R7 GB, Canada
22.	Journal of American Water Resources Association	American Water Resources Association, 4 West Federal Street, P.O. Box. 1626, Middleburg, VA 20158-16126, USA
23.	Journal of Hydrologic Processes	Wiley Publications, Corporate Quarters, 111, River Street, Hobaken, NJ 07030-5774 USA
24.	Water Resources Journal	The Economic & Social Commision for Asia & the Pacific, United Nations Building, Rajadamnern Avenue, Bankok 10200, Thailand
25.	Water Resources Management	European Water Resources Association (EWRA), Springer, The Netherlands

*** ———— ***

Appendix IX

WATER RESOURCES OF INDIA

The Indian Union lies between 8°-4' to 37°-6' North latitude and 68°-7' to 97°-25' East longitude.

India is also known as 'Bharat, Hindustan, Bharat-varsh, Arya-varta'.

The Indian Union consists of 28 states, six centrally governed union territories and one national capital territory.

States: Andhra Pradesh, Arunachal Pradesh, Assam, Bihar, Chhatisgarh, Goa, Gujrat, Haryana, Himachal Pradesh, Jammu and Kashmir, Jharkhand, Karnataka, Kerala, Madhya Pradesh, Maharashtra, Manipur, Meghalaya, Mizoram, Nagaland, Orissa, Punjab, Rajasthan, Sikkim, Tamil Nadu, Tripura, Uttar Pradesh, Uttarakhand, West Bengal.

Centrally governed union territories: Andaman and Nicobar, Chandigarh, Dadra and Nagar Haveli, Daman and Diu, Lakshadweep, Pondicherry.

National capital territory: Delhi.

Physiographically India is divided into seven divisions 1) The Northern mountains, 2) The Great plains, 3) The Central highlands, 4) The Peninsular plateau, 5) The East coast belt, 6) The West coast belt, 7) The islands.

Total land area	:	328 million hectares 3.28 million km^2 [2.11% of total world area]
Land frontier	:	15,200 km
Coastline	:	6,100 km
Average annual precipitation	:	1140 mm
Maximum average annual precipitation	:	10860 mm [At Mawsyram in Assam near Cherrapunji]
Minimum average annual precipitation	:	30 mm [At Churu dist. Bikaner Rajasthan]
Annual volume of water received over the country	:	3700 km^3 370×10^6 hectare m
Evaporation losses [33%]	:	1230 km^3 123×10^6 hectare m
Seepage in subsoil [22%]	:	800 km^3 80×10^6 hectare m
Run-off in the rivers [45%]	:	1670 km^3 167×10^6 hectare m

Surface water resources basin wise in km^3:

	Available	Utilizable
1) Indus basin	72.24	46
2) Ganga basin	478.48	250
3) Brahamaputra basin	506.63	24
4) East flowing rivers	326.49	151
5) West flowing rivers	286.16	219
Total:	1670.00	690

Per capita availability of water:

Year	:	1955	1990	2025
Amount in m^3	:	5277	2464	1496

Ultimate irrigation potential	:	140×10^4 km^2 140×10^6 hectare
Cumulative irrigation potential achieved	:	99×10^4 km^2 99×10^6 hectare
Annual utilizable ground water resource	:	422.86 km^3 42.28×10^6 hectare m
Annual utilized ground water resource	:	100 km^3 10×10^6 hectare m
Economical hydroelectric potential of the country [From 845 schemes]	:	84044 MW at 60% load factor 150000 MW at Average load factor
Share of run-off of river plants [256 schemes]	:	40%
Reservoir type projects	:	60%
Hydroelectric power developed so far	:	16032 MW at 60% load factor
Flood prone area in the country	:	40×10^4 km^2 40×10^6 hectare
Flood affected area on an average every year	:	7.5×10^4 km^2 7.5×10^6 hectare

*** ——— ***

Appendix X

INDIA'S NATIONAL WATER POLICY*

India adopted a national water policy (NWP) in September 1987 and revised it in 2002, under the aegis of National Water Resources Council, established in 2000. It recognized water as one of the crucial elements in the developmental planning. The policy document, *inter alia*, lays down that planning and development of this natural resource, i.e., water, needs to be governed by a national perspective. Resource planning, in case of water, has to be done for hydrological unit such as drainage basin as a whole or for a sub-basin. All individual developmental projects and proposals should be formulated by states and considered within the framework of such an overall plan for a basin or a sub-basin so that the best possible combinations of options can be made.

The policy document lays down that water should be made available to water-short areas by transfer from other areas including from one river basin to another, based on a national perspective, after taking into account the requirements of areas/basins. Transfer of water from one river to another, especially if it involved inter-state transfer, has always been a sensitive issue amongst people and the states.

On the issue of equity and social justice, the policy lays down that water allocation in an irrigation system should be done so that disparities in the availability between the head-reach and tail-end farms and between large and small farms should be obviated by adoption of a rotational water distributional system and supply of water on a volumetric basis, subject to certain ceilings.

For implementing the above programme, the policy document lays down that appropriate organizations should be established for planned development and management of a river basin as a whole.

The policy recognizes water as a prime natural resource, a basic human need and a precious national asset. The policy has laid down that, in planning and operation of systems, water allocation priorities should be as follows.

Drinking water

Irrigation

Hydro-power

Navigation

Industrial and other uses

However, these priorities might be modified, if necessary, in particular regions with reference to area-specific considerations.

The policy stipulates that adequate drinking water facilities should be provided to the entire population, both in urban and rural areas. Irrigation and multi-purpose projects should invariably include a drinking water component, wherever there is no alternative source of drinking water. Drinking water needs of human beings and animals should be provided for all irrigation projects.

---

*Reproduced with permission from Government of India.

The policy further provides that economic development and activities including agricultural, industrial and urban development should be planned with due regard to the constraints imposed by the configuration of water availability. There should be a water zoning of the country and the economic activities should be guided and regulated in accordance with such zoning.

The implementation of the policy guidelines is overseen by the National Water Resources Council in which all chief ministers are members and the prime minister is the chairman. The National Water Board headed by the Secretary, Union Ministry of Water Resources, and the representatives from all the states assist the Council. The National Water Board has so far finalized a number of policy papers on water-related issues for the consideration of the National Water Resources Council.

*** ——— ***

Appendix XI

THE BUREAU OF INDIAN STANDARDS

The 'Indian Standards Institution' was established in the year 1947 for (i) formulation of standards and (ii) promotion and implementation of standards.

It was reconstituted as 'Bureau of Indian Standards' in 1986 by an Act of the Parliament. The standards are formulated on various topics by a committee taking into consideration the consciousness of all interested through a wide circulation of the draft. About 18000 standards on various groups have been finalized.

The B.I.S. has its headquarters in Delhi (Manak Bhawan 9, Bahadur Shah Jafar Marg, New Delhi, 110002, India) and regional and branch offices all over the country.

The B.I.S. publications are priced publications and are available at all these offices.

The B.I.S. has recommended criteria for various parameters related to hydrology. Some of them are listed below:

IS Number/DOC Number	Title
IS 1191:2003 | ISO 772:1988	Hydrometric determination—Vocubulary
IS 1192:1981 | ISO 748:1979	Velocity area Methods for measurement of flow of water in open channels (first revision)
IS 1194:1960	Forms for recording measurement of flow of water in open channels
IS 2912:1998 / ISO 1070:1992	Liquid flow measurement in open channels—Slope-area method
IS 3910:1992 | ISO 2537:1988	Requirements for rotating element current meters (cup type) for water flow measurement (first revision)
IS 3911:1994	Surface floats—Functional requirements (first revision)
IS 3912:1993 | ISO 3454:1983	Sounding rods—Functional requirements (first revision)
IS 3918:1966	Code of practice for use of current meter (cup type) for water flow measurement
IS 4073:1967	Specification for fish weights
IS 4080:1994	Vertical staff gauges—Functional requirements (first revision)
IS 4858:1968	Specification for velocity rods
IS 4986:2002	Code of practice for installation of rain-gauge (non-recording type) and measurement of rain (second revision)
IS 4987:1994	Recommendations for establishing network of raingauge stations
IS 5542:2003	Guide for storm analysis (first revision)

(*continued*)

IS Number/DOC Number	Title
IS 6062:1971	Method of measurement of flow of water in open channels using standing wave flume-fall
IS 6063:1971	Method of measurement of flow of water in open channels using standing wave flume
IS 6064:1971	Specification for sounding and suspension
IS 6330:1971 | ISO 3847	Recommendation for liquid flow measurement in open channels by weirs and flumes-end depth method for estimation of flow in rectangular channels with a free overfall (approximate method)
IS 8389:2003	Code of practice for installation and use of raingauges, recording (second revision)
IS 9108:1979 | ISO 1438-1	Liquid flow measurement in open channels using thin plate weirs
IS 9116:2002	Specification for water stage recorder (float type) (first revision)
IS 9119:1979	Method for flow estimation by jet characteristics (approximate method)
IS 9163(Part 1):1979 | ISO 9555-1:1973	Dilution methods of measurement of steady flow Part 1 constant rate injection method
IS 9922:1981 | ISO 8363:1980	Guide for selection of method for measuring flow in open channels
IS 12752:1989 | ISO 8368:1980	Guidelines for the selection of flow gauging
IS 13083:1991 / ISO 4377:1990	Liquid flow measurement in open channels—Flat-V
IS 13084:1991 / ISO 4374:1990	Liquid flow measurement in open channels—Round nose horizontal crest weirs
IS 13371:1992 | ISO 3455:1976	Code of Practice for calibration (rating) of rotating element current meters instraight open tank
IS 14359:1996 | ISO 4366:1979	Echo sounders for water depth measurements
IS 14371:1996 / ISO 9826:1982	Measurement of liquid flow in open channels—Parshall and SANIIRI flumes
IS 14573:1998 / ISO 1088:1985	Liquid flow measurement in open channels—Velocity area methods—Collection and processing of data for determination of errors in measurement
IS 14574:1998 / ISO 4371:1984	Measurement of liquid flow in open channels by weirs and flumes- end depth method for estimation of flow in non rectangular channels with a free overfall (approximate method)
IS 14615 (Part 1):1999 / ISO 5167-1:1991	Measurement of fluid flow by means of pressure differential devices Part 1: Orifice plates, nozzles and venturi tubes inserted in circular cross-section conduits running full
IS 14673:1999 | ISO 4360:1984	Liquid flow measurement in open channels by weirs and flumes—Triangular profile weirs
IS 14869:2000 / ISO 4359:1983	Liquid flow measurement in open channels—Rectangular, trapezoidal and U-shape flumes
IS 14973:2001 / ISO 3966:1997	Measurement of fluid flow in closed conduits—Velocity area method using Pilot Static Tubes
IS 14974:2001 / ISO 3846:1989	Liquid flow measurement in open channels by weirs and flumes—Rectangular broad-crested weirs
IS 14975:2001 / ISO 9827:1994	Measurement of fluid flow in open channels—Stream lind triangular profile weirs
IS 15117:2002 / ISO 4375:2000	Hydrometric determination—Cable way system for stream gauging

IS Number/DOC Number	Title
IS 15118:2002 / ISO 4373:1995	Measurement of liquid flow in open channel—Water level measuring devices
IS 15119 (Part 1):2002 / ISO 1100-1:1996	Measurement of liquid flow in open channels—Part 1 Establishment and operation of a gauging station (superseding IS 2914:1964)
IS 15119 (Part 2):2002 / ISO 1100-2:1998	Measurement of liquid flow in open channels—Part 2 Determination of the stage-discharge relation (superseding IS 2914:1964)
IS 15123:2002 / ISO 4362:1999	Hydrometric determination—Flow measurement in open channels using structures-Trapezoidal broad-crested weirs (superseding IS/ISO 4362:1992)
IS 15352:2003 / ISO 6420:1984	Liquid flow measurement in open channels—Position fixing equipment for hydrometric boats
IS 15353:2003 / ISO 8333:1985	Liquid flow measurement in open channels by weirs and flumes—V-shaped broad-crested weirs
IS 15362:2002 | 114139:2000	Liquid flow measurement in open channels—Flow measurements in open channels using structures-compound gauging structure (Adoption of ISO 14139:2000)
IS 15454:2004	Liquid flow measurement in open channels—Velocity-area method using a restricted number of verticals (based on ISO/TR 9823:1990)
IS 15527:2004	Measurement of liquid flow in open channels—Measurement in meandering rivers and in streams with unstable boundaries (based on ISO/TR 9210:1992)
DOC.WRD 1(338)	Measurement of liquid flow in open channels—Field measurement of discharge in large rivers and floods (based on ISO 9825:1994)
DOC.WRD 1(358)	Measurement of total discharge in open channels—Stage-fall-discharge relationship (adoption of ISO 9123:2001)
DOC.WRD 1(389)	Measurement of liquid flow in open channels—Method of specifying performance of hydrometric equipment' (adoption of ISO
DOC.WRD 1(333)	Liquid flow measurement in open channels velocity-area methods—investigation of total error (based on ISO/TR 7178:1983)
DOC.WRD 1(335)	Measurement of liquid flow in open channel—Determination of the wet-line correction (based on ISO/TR 9209:1989)
DOC.WRD 1(390)	Measurement of liquid flow in open channels—Electromagnetic current meters (adoption of ISO/TR 11974:1987)
DOC.WRD 1(337)	Draft Indian Standard Measurement of fluid flow—Evaluation of uncertainties (based on ISO/TR 5168:1998)
DOC.WRD 1(361)	Hydrometric determinations—Unstable channels and ephemeral streams (based on ISO/TR
IS 4410 (Part 1):1991	Glossary of terms relating to river valley projects: Part 1 Irrigation practice (first revision)
IS 4410 (Part 11/ Sec 1):1972	Glossary of terms relating to river valley projects Part 11 Hydrology Section 1 General terms
IS 4410 (Part 11/ Sec 2):1972	Glossary of terms relating to river valley projects: Part 11 Hydrology Section 2 Precipitation and run-
IS 4410 (Part 11/ Sec 3):1973	Glossary of terms relating to river valley projects: Part 11 Hydrology: Section 3: Infiltration and water
IS 4410 (Part 11/ Sec 4):1973	Glossary of terms relating to river valley projects: Part 11 Hydrology Section 4 Hydrographs
IS 4410 (Part 11/ Sec 5): 1977	Glossary of terms relating to river valley projects: Part 11 Hydrology Section 5 Floods

(continued)

IS Number/DOC Number	Title
IS 4410 (Part 11/ Sec 6):1994	Glossary of terms relating to river valley projects: Part 11 Hydrology Section 6 Ground water
IS 14476 (Part 1):1998	Test pumping of water wells—Code of practice Part 1 General
IS 14476 (Part 2):1998	Test pumping of water wells—Code of practice Part 2 Hydrogeological considerations
IS 14476 (Part 3/ Sec 1):1998	Test pumping of water wells—Code of practice Part 3 Pre-test planning Sec 1 General aspects
IS 14476 (Part 5):1998	Test pumping of water wells—Code of practice Part 5 Pumping test
IS 14476 (Part 8):1998	Test pumping of water wells—Code of practice Part 8 Water level and discharge measuring devices
DOC.WRD 3(370)	Guidelines for artificial recharge to ground water
DOC.WRD 3(456)	Manual method for measurement of a ground water level in a well
IS 6939:1992	Methods for determination of evaporation from reservoirs (first revision)
DOC.WRD 10(354)	Determination of volume of water & water level in lakes & reservoirs
DOC.WRD 13(378)	Guidelines for adopting coefficient of friction (Rugosity coefficient) for design of canals

*** ——— ***

Appendix XII

WATER IN INDIAN LANGUAGES

India is a multi-linguistic country. Equivalent words for water in these languages are given here.

1. Assamese:	i) পানী	(pani)
	ii) জল	(jal)
	iii) নীৰ	(neer)
2. Bengali:	i) বারি	(bari)
	ii) নীর	(neer)
	iii) জল	(pani)
3. Dogri:	i) पानी	(pani)
	ii) नीर	(neer)
4. Gujrati:	i) પાણી	(pani)
	ii) જળ	(jal)
5. Hindi:	i) पानी	(pani)
	ii) जल	(jal)
6. Kannada:	i) ನೀರು	(niiru)
	ii) ಉದಕ	(udak)
	iii) ತೀರ್ಥ	(teertha)
7. Kashmiri:	i) پۆنۍ	(poni)
	ii) آب	(aab)
8. Konkani:	i) उदक	(udak)
	ii) उदीक	(udeek)
	iii) उदोक	(udok)

9. Maithili: i) পানি (pani)

10. Malayalam: i) വെള്ളം (vallam)
ii) നീര് (niira)
iii) ജലമ (jelam)

11. Manipuri: i) ꯏꯁꯤꯡ (isin)

12. Marathi: i) जल (jal)
ii) पाणी (paanee)
iii) उदक (udak)

13. Nepali: i) पानी (paani)
ii) जल (jal)
iii) नीर (neer)

14. Oriya: i) ପାଣି (paani)
ii) ଜଳ (jal)
iii) ନୀର (neer)
iv) ବାରି (baari)
v) ପୟ (paya)

15. Punjabi: i) ਪਾਣੀ (paani)
ii) ਜਲ (jal)

16. Rajasthani: i) पानी (pani)
ii) जल (jal)

17. Sindhi: i) آبُ (aab)
ii) جَلُ (jalu)
iii) پاڻِي (paani)

18. Tamil: i) நீர் (neer)
ii) ஜலம் (jalam)
iii) தண்ணீர் (thanneer)

19. Telugu: i) నీరు (neeru)
ii) జలమ్ (jalam)
iii) నీళ్లు (neellu)

20. Urdu: i) پانی (pani)

ii) آب (aab)

21. Sanskrit: There are as many as 70 words for water. Each one being originally Sanskrit and not derived from any other language. These are as follows.

वाः	Vãha	सलम्	Salam
वारि	Vãri	ऊर्जम्	Urjam
सलिलम्	Salilam	घृतम्	Ghrutam
कमलम्	Kamalam	वाजम्	Vãjam
पयः	Payaha	आपः	Aapaha
कीलालम्	Keelãlam	सरिलम्	Sarilam
अमृतम्	Amrutam	साम्बः	Sambaha
भुवनम्	Bhuvanam	अन्धम्	Andham
वनम्	Vanam	कुशम्	Kusham
कबन्धम्	Kabandham	कर्बुरम्	Karburam
उदकम्	Udakam	क्षीरम्	Ksheeram
पाथः	Pãthaha	तामरम्	Tãmaram
पुष्करम्	Pushkaram	ह्रीवेरम्	Hriveram
सवरम्	Savaram	सञ्चलम्	Sanchalam
अम्भः	Ambhaha	सम्वरम्	Sambaram
अर्णः	Arnaha	जडम्	Jadam
तोयम्	Toyam	जीवनम्	Jeevanam
दकम्	Dakam	कोमलम्	Komalam
कम्बलम्	Kambalam	तामरम्	Tãmaram
स्यन्दनम्	Syandanam	घनीसासम्	Ghanisãsam
आपः	Aapaha	कृप्पेरम्	Krupperarn
सदनम्	Sadanam	रेपालम्	Repãlam
इरा	Irã	शम्बरम्	Shambaram
अम्बु	Ambu	घनरसः	Ghanarasaha
कम्	Kam	जलपीथम्	Jalapeetham
जडः	Jadaha	कपन्धम्	Kapandham
सोमम्	Somam	चन्द्रोरसम्	Chandrorasam

व्योम	Vyoma	गोकलनम्	Gokolanam
नारम्	Nãram	सर्वतोमुखम्	Sarvatomukham
क्षरम्	Ksharam	मेघपुष्पम्	Meghapushpam
सरः	Saraha	अभ्रपुष्मप्	Abbhrapushpam
नीरम्	Neeram	पीप्पलम्	Peeppalam
उदम्	Udam	काण्डम्	Kãndam
सरम्	Saram	कृपीटम्	Krupeetam
ऋतम्	Rutam	सम्बलम्	Sambalam

*** ——— ***

Appendix XIII

JAL SUBHASHITANI

In Sanskrit, there is an unlimited treasure of short elegant compositions [Subhashitani सुभाषितानी], wherein qualities of good or bad are quoted, discussed and compared. In some cases the qualities of water are referred to.

A collection of some of such subhashitani is given below

१) 'निम्नमनुसरति जलं'

Water flows towards the lowland.

२) आपूर्यमाणमचलप्रतिष्ठं समुद्रमापः प्रविशन्ति यद्वत्।

(भगवद्गीता २.७०)

As the waters (of different rivers) enter the ocean, which though full on all sides, remains undisturbed.

३) ॐ आपो हिष्ठा मयोभुवः

(अथर्ववेद काण्ड - १ सूक्त - ५)

Oh! water deity you give pleasure to us.

४) तडागोदरसंस्थानां परीवाह इवाम्भसाम्।

(कालिदास द्वात्रिंष्पुत्तलिका)

Spill way is the guaranty of safety for a reservoir of waters which is a pool having plenty of water.

५) आस्वाद्यतोयाः प्रभवन्ति नद्यः समुद्रमासाद्य भवन्त्यपेयाः

The rivers having delicious water at the source became unpotable when they merge into the ocean.

६) महामेघः क्षारं पिबति कुरूते वारि मधुरम्

A huge cloud consumes salty water and turns the same into sweet water.

७) गतोदके सेतुबन्धो

Building a bridge after the bed becomes dry (is of no use).

८) जलसेकेन वर्धन्ते तरवः

The trees grow by watering.

९) क्षारभावमपनीय गृहणते वारिधेः सलिलमेव वारिदाः

Clouds collect the water of the ocean keeping away its salinity.

१०) तरूमूलादिषु निहितं जलमाविर्भवति पल्लवाग्रेषु

Water given at the roots of trees manifests itself high up at the tips of the leaves.

११) नदी वेगेन शुध्यति

(वृद्धचाणक्य ६ - ३)

A river purifies due to the speed of its flow.

१२) नलिनीदलगतजलमतितरलं तद्वत् जीवितमतिशयचपलम्

Drops of water on the leaves of lotus are very transient. Same is the case of human life.

१३) परोपकाराय वहन्ति नद्यः

Rivers flow for benevolence.

१४) कुण्डे कुण्डे नवं पयः।

Novel is the water in every reservoir.

१५) पृथिव्यां त्रीणि रत्नानि जलमन्नं सुभाषितम्

On this earth there are three precious things; Water, food and a witty word.

१६) संभूयाम्भोधिमभ्येति महानद्यः नगापगाः।।

Great rivers go away from the mountains and collectively flow into the sea.

१७) मूलसिक्तस्य वृक्षस्य फलं शाखासु दृश्यते।

A tree which is watered at the roots shows the fruits up in the branches.

१८) मृदुना सलिलेन हन्यमानानि अवधृष्यन्ति गिरेः अपि स्थलानि

Continuously striken, even by soft water, regions of rugged mountains give away, hard hills are eroded.

१६) यथा भूमिः तथा तोयं

Water has the quality of soil from which it flows.

२०) तस्य विस्तारिता बुद्धिः तैलबिन्दुः यथा जले॥

His mental orbit grows like a drop of oil in the water.

२१) व्रजन्ति न निवर्तन्ते स्त्रोतांसि सरितां यथा।

Currents of rivers flow away. They do not return.

२२) अजीर्णे भेषजं वारि जीर्णे वारि बलप्रदम्

Water works like a medicine, when the food is not digested, but it is invigorating after the food is digested.

२३) तस्मान्नरो वह्निविवर्धनार्थं मुहुर्मुहुर्वारि पिबेद् अभूरि

Hence, for the strengthening of appetite, man should drink a small quantity of water, a number of times.

२४) आयुष्यं जलबिन्दुलोलचपलं फेनोपमं जीवितम्।

Life is fickle like a drop of water which is ever unsteady.

२५) पर्वतानां जलं जरा

For the mountains water works like an old age. (As it decreases human energy.)

२६) कृत्वा ज्ञानम् स्वयं नस्येज्जलं केतकरेणुवत्

(चि. वि. १५६ शंकराचार्य आत्मबोध)

The clearing nut, when dropped into turbid water, removes the turbidity of water and also vanishes itself.

२७) अल्पतोयश्चलत्कुम्भः

A pitcher with scanty water is ever unsteady.

२८) आयुरूदञ्चनयन्त्रोपममिति

Life is like a revolving water wheel.

२६) एकदेशेन चावर्तः संवर्तः सर्वतो जलम्।
पुष्करे दुष्करं वारि द्रोणे बहुजला मही॥

(कृषिपाराशर)

Avarta (cloud) is confined to a particular locality. Under Samvarta cloud water is everywhere, under Puskara cloud water is scarce and under Drona cloud the water is abundant on earth.

३०) गुरूत्वं जलभूम्योः पतनकर्मकारणम्
अप्रत्यक्षं पतनकर्मानुमेयम्

(न्याय कंदलि Sci. S Tech. in Ancient India. - Page no. 98)

Gravity is the cause for falling liquids and solids. It is invisible and is inferred by the falling motion.

३१) जलबिन्दुनिपातेन क्रमशः पूर्यते घटः

A pitcher gets gradually filled due to the falling water drops.

३२) किमावरीवः कुह कस्य शर्मन् अंभः किमासीद्‌गहनं गभीरम्॥

(नासदीयसूक्त चरम खंड)

Was there plenty of water so deep?

३३) नास्ति मेघसमं तोयं

(वृद्धचाणक्य - संत्रिका)

No water is comparable to the water from the cloud.

३४) यंत्रेणावनतादिनाच निपुणं यद्वांबुसंपूरणे
नोर्वी चारूसमीकरोत्यथ दृढंस्यात्पादसंपूरणं।।

Foundation should either be levelled by the instrument, or the foundation should be filled with water (and the level should be checked according to water level).

३५) संस्थाप्य मृण्मये पात्रे ताम्रपत्रं सुसंस्कृतम्
छादयेच्छिखिग्रीवेन चार्द्राभिः काष्ठपांसुभिः
दस्तालोष्टोनिधातव्यः पारदाच्छादिदस्ततः
संयोगाज्जायते तेजो मैत्रावरूणसंज्ञितम्
अनेन जलभंगोऽस्ति प्राणोदानेषु वायुषु।

Place copper plates in an earthen pot; cover it with powdered coal and moistened saw dust spread zinc powder and cover it with mercury. Due to chemical interaction positive and negative electricity is produced. Due to this, water is decomposed into oxygen and hydrogen.

३६) अंधं जलं बलं प्राहुः।

Stored water is potential energy.

३७) शं नो वरुणः

Let God Varun protect us.

३८) 'अग्रेव धूम्रो जायते धूम्राद भ्रम भ्राद् वृष्टिः'

Fire produces smoke, smoke clouds and clouds rain.

*** ——— ***

Appendix XIV

WISDOM OF WATER

Scientists, philosophers and statesmen, have realized the importance of 'water' and have quoted it extensively in their speeches and writings.

A few of them have been quoted here.

If we could ever competitively, at a cheap rate, get fresh water from saltwater, ... (this) would be in the long-range interests of humanity which could really dwarf any other scientific accomplishments.

—John F. Kennedy

Water and air, the two essential fluids on which all life depends, have become global garbage cans.

—Jacques Cousteau (1910–1997)

There is nothing softer and weaker than water, and yet there is nothing better for attacking hard and strong things.

For this reason there is no substitute for it.

—Lao-Tzu (C.B.C. 550)

Water that has been begged for does not quench the thirst.

—Geographical Origin Uganda

All earth's full rivers cannot fill
The sea that drinking thirsteth still

—Christina Georgina Rossetti (1830–1894), By the Sea

Water is H_2O, hydrogen two parts, oxygen one, but there is also a third thing that makes water and nobody knows what that is.

—D. H. Lawrence (1885–1930), Pansies, 1929

Water is life's matter and matrix, mother and medium. There is no life without water.

—Albert Szent-Gyorgyi quotes (Hungarian Biologist, 1937 Nobel Prize for Medicine 1893–1986)

Filthy water cannot be washed.

—West African proverb

All streams flow into the sea, yet the sea is never full. To the place the streams come from, there they return again.

—Ecclesiastes 1:7 from New International Version of *The Bible*

Muddy water, let stand–becomes clear

—Lao Tzu quotes [Chinese Taoist philosopher, Founder of Taoism, wrote "*Tao Te Ching*" (also "*The Book of the Way*") 600 BC–531 BC]

Nothing is softer or more flexible than water, yet nothing can resist it

—Lao Tzu quotes [Chinese Taoist philosopher. Founder of Taoism, wrote "*Tao Te Ching*" (also "*The Book of the Way*") 600 BC–531 BC]

Water, like religion and ideology, has the power to move millions of people. Since the very birth of human civilization, people have moved to settle close to it. People move when there is too little of it. People move when there is too much of it. People journey down it. People write, sing and dance about it. People fight over it. And all people, everywhere and every day, need it.

—Mikhail Gorbachev, President of Green Cross International quoted in Peter Swanson's *Water; The Drop of Life*, 2001.

Water has no taste, no color, no odor; it cannot be defined, art relished while ever mysterious. Not necessary to life, but rather life itself. It fills us with a gratification that exceeds the delight of the senses.

—Antoine De Saint-Exupery (1900–1944), *Wind, Sand, and Stars*, 1939

Water is the one substance from which the earth can conceal nothing; it sucks out its innermost secrets and brings them to our very lips.

—Jean Giraudoux (1882–1944), *The Madwomen of Chaillot*, 1946

If there is magic on this planet, it is contained in water.

—Loran Eisley (Anthropologist), *The Immense Journey*, 1957

Throughout the history of literature, the guy who poisons the well has been the worst of all villains.

—Author unknown

Thousands have lived without love, not one without water.

—W.H. Auden

Fire, water and government know nothing of mercy.

Wetlands have a poor public image.... Yet they are among the earth's greatest natural assets... mankind's waterlogged wealth.

—Edward Maltby, *Waterlogged Wealth*, 1986

Life originated in the sea, and about eighty percent of it is still there.

—Isaac Asimov, Isaac Asimov's *Book of Science and Nature Quotations*, 1988

The estuary is the point where man, the sea—his immemorial ally and adversary—and the land meet and challenge each other.

—U.S. Department of the Interior, National Estuarine Pollution Study, November 1969

In every glass of water we drink, some of the water has already passed through fishes, trees, bacteria, worms in the soil, and many other organisms, including people.... Living systems cleanse water and make it fit, among other things, for human consumption.

—Elliot A. Norse, in R.J. Hoage, ed., *Animal Extinctions*, 1985

Between earth and earth's atmosphere, the amount of water remains constant; there is never a drop more, never a drop less. This is a story of circular infinity, of a planet birthing itself.

—Linda Hogan, *Northern Lights*, Autumn 1990

The stone in the water knows nothing of the hill which lies parched in the sun.

—African Proverb

The highest good is like water. Water gives life to the ten thousand things and does not strive. It flows in places men reject and so is like the Tao.

—Excerpt from the *Tao Te Ching*, chapter 8

By means of water, we give life to everything.

—Koran, 21:30

It is a fascinating and provocative thought that a body of water deserves to be considered as an organism in its own right.

—Lyall Watson, Supernature

Water links us to our neighbor in a way more profound and complex than any other.

—John Thorson

Water flows uphill towards money.

—Famous saying by Anonymous, saying in the American West, quoted by Ivan Doig in Marc Reisner, Cadillac Desert, 1986

Still waters run no mills.

—quoted by Aglionby, *Life of Bickerstaff* (p. 5)

A river is the report card for its watershed.

—Alan Levere

Smooth runs the water where the brook is deep.

—William Shakespeare (1564–1616), *Henry VI*, Part II

We must begin thinking like a river if we are to leave a legacy of beauty and life for future generations.

—David Brower quoted by *E-Wire*, 7 Apr 2000

Rain is grace; rain is the sky condescending to the earth; without rain, there would be no life.

—John Updike, *Self-Consciousness: Memoirs*, 1989

Water helped ancient man learn those first lessons about the rights of others and responsibility to a larger society… . It became part of the moral and mental legacy parents passed on to their children.

—M. Meyer, *Water in the Hispanic Southwest*

A river is more than an amenity, it is a treasure.

—Justice Oliver Wendell Holmes (quoted by the Supreme Court in its decision in U.S. v. Republic Steel, 1960)

Truths are first clouds; then rain, then harvest and food.

—Henry Ward Beecher

Any river is really the summation of the whole valley. To think of it as nothing but water is to ignore the greater part.

—Hal Borland, *This Hill, This Valley*

Water is the basis of life and the blue arteries of the earth! Everything in the non-marine environment depends on freshwater to survive.

—Sandra Postel, *Sandra Postel, Global Water Policy Project,* Grist Magazine 26 Apr 04

Clean water is not an expenditure of Federal funds; clean water is an investment in the future of our country.

—Bud Shuster, U.S. Representative,
quoted in *The Washington, Post*, 1 Sep 1987

No one can see their reflection in running water. It is only in still water that we can see.

—Taoist proverb

As water runs towards the sword, money towards the rich man's hand.

—Proverb

We used to think that energy and water would be the critical issues for the next century. Now we think water will be the critical issue.

—Mostafa Tolba of Egypt,
former heard of the United Nations Environment Program

Like swift water, an active mind never stagnates.

Water flows humbly to the lowest level.
Nothing is weaker than water,
Yet for overcoming what is hard and strong,
Nothing surpasses it.

—Lao Tzu, Tao Te Ching

To trace the history of a river... is to trace the history of the soul, the history of the mind descending and arising in the body.

—Gretel Ehrlich

A good word extinguishes more than a pailful of water.

—Author: Proverb

Water is the only drink for a wise man.

—Henry David Thoreau

The noblest of the elements is water

—Pindar, 476 B.C.

Next to blood relationships come water relationships.

—Stanley Crawford, *Mayordomo*

Water, fire, and soldiers, quickly make room.

—Author: George Herbert

Water is a very good servant, but it is a cruel master.

—C.G.D. Roberts, *Adrift in America*, 1891

The wars of the twenty-first century will be fought over water.

—Ismail Serageldin, World Bank Vice President for Environmental Affairs, quoted in Marq de Villiers' *Water*, 2000

Water is fundamental for life and health. The human right to water is indispensable for leading a healthy life in human dignity. It is a pre-requisite to the realization of all other human rights.

—The United Nations Committee on Economic, Cultural and Social Rights, *Environment News Service*, 27 Nov 02

Rivers are roads which move, and which carry us whither we desire to go

—By Blaise Pascal–famous water sayings

Water is sometimes sharp and sometimes strong, sometimes acid and sometimes bitter, sometimes sweet and sometimes thick or thin, sometimes it is seen bringing hurt or pestilence, sometime health-giving, sometimes poisonous. It suffers change into as many natures as are the different places through which it passes. And as the mirror changes with the colour of its subject, so it alters with the nature of the place, becoming noisome, laxative, astringent, sulfurous, salty, incarnadined, mournful, raging, angry, red, yellow, green, black, blue, greasy, fat or slim. Sometimes it starts a conflagration, sometimes it extinguishes one; is warm and is cold, carries away or sets down, hollows out or builds up, tears or establishes, fills or empties, raises itself or burrows down, speeds or is still; is the cause at times of life or death, or increase or privation, nourishes at times and at others does the contrary; at times has a tang, at times is without savor, sometimes submerging the valleys with great floods. In time and with water, everything changes

—Leonardo da Vinci

When you drink the water, remember of spring

—Chinese Proverb

When the well is dry, we learn the worth of water

—Benjamin Franklin

It is the surplus water that causes danger than the stored.

Too much or too little of water and its quality has always been responsible for the largest number of diseases and deaths.

Civilization is the dialogue between man and nature.

The movement of heavenly bodies despite their distances from earth have presented fewer difficulties to me than the movement of water which is within my reach

—Galileo

Rain is a blessing when it falls gently on parched fields, turning the earth green, causing the birds to sing.

—Donald Worster, *Meeting the Expectation of Land*, 1984

The cure for anything is salt water—sweat, tears, or the sea.

—Tagore–a Bengali poet and novelist

Water should not be judged by its history, but by its quality

—Dr Lucas Van Vuuren, National Institute of Water Research, South Africa

A waster of water is a waster of better.

—Old Irish Adage

The heart of the wise man lies quiet like limpid water.

—Geographical Origin Cameroon

It is really important to solve the problem of rational utilization and distribution of water supplies. I dare say, the shortage of fresh water is the major ecological problem of this moment.

—Mikhail Gorbachev, President of Green Cross International
quoted in Peter Swanson's *Water: The Drop of Life*, 2001

Water is the driver of Nature.

—Leonardo da Vinci

The good rain, like the bad preacher, does not know when to leave off.

—Ralph Waldo Emerson

Water, taken in moderation, cannot hurt anybody.

—Mark Twain

Any water in the desert will do.

—Geographical Origin Saudi Arabia

For the benefit of the flowers, we water the thorns, too.

—Geographical Origin Egypt

Water links us to our neighbor in a way more profound and complex than any other.

—John Thorson

Aquifer: a mysterious, magical and poorly defined area beneath the surface of the earth that either yields or withhold vast or lesser quantities of standing/flowing water, the quantity and/or quality of which is dependent on who is describing it or how much money may be at stake.

—R. Radden, "Watershed Resources", Jan. 2002

Irrigation of the land with seawater desalinated by fusion power is ancient. It's called rain

—Michael McClary

Don't empty the water jar until the falls.

—Philippine proverb

We forget that the water cycle and the life cycle are one.

—Jacques Cousteau

A used plough shines, standing water stinks.

—Author: Proverb

You should respect each other and refrain from disputes; you should not, like Water and oil, repel each other, but should, like milk and water, mingle together.

—Buddha quotes, (Hindu Prince Gautams ***
the founder of Buddhism, 563–483 B.C.)

Muddy water, let stand - becomes clear

—Lao Tzu quotes (Chinese Taoist Philosopher).

The wise adapt themselves to circumstances, as water moulds itself to the pitcher

—Anonymous (Chinese Proverb) quotes

Like swift water, an active mind never stagnates.

By means of water, we give life to everything.

—*Koran*, 21:30

Beware of still water, a still dog, and a still enemy.

—Proverb

The wise man of Miletus thus declared the first of things is water

—J.S. Blackie, 1877

Nothing on earth is so weak and yielding as water, but for breaking down the firm and strong it has no equal.

—Lao-Tsze

In sweet water there is a pleasure ungrudged by anyone.

—Ovid, 13 AD

*** ——— ***

Bibliography

Chitale, M. A. 2000. *A Blue Revolution*, Mumbai: Bhavans Book University.

Chow, V. T., D. R. Maidment, and L. W. Mays, 1973. *Applied Hydrology*, New York: McGraw-Hill.

Chow, V. T. 1964. *Handbook of Applied Hydrology*, New York: McGraw-Hill.

Kulandaiswamy, V. C. 1971. *Lecture Notes on Hydrology*, Guindy: ISTE Summer School.

Linsley, R. K., M. A. Kohler, and J. L. H. Paulhus, 1988. *Hydrology for Engineers*, New York: McGraw-Hill.

Manning, J. C. 1987. *Applied Principles of Hydrology*, Princeton: Merrill Publishing Company.

Mutreja, K. N. 1982. *Applied Hydrology*, New Delhi: Tata McGraw-Hill.

National Water Policy. 2002. Ministry of Water Resources, Government of India.

Nemec, J. 1973. *Engineering Hydrology*, New Delhi: Tata McGraw-Hill.

Patra, K. C. 2002. *Hydrology and Water Resources Engineering*, Delhi: Narosa Publishing House.

Ponce, V. M. 1989. *Engineering Hydrology*, Englewood Cliffs: Prentice Hall.

Raghunath, H. M. 1983. *Groundwater*, New Delhi: Wiley Eastern Limited.

Raghunath, H. M. 1985. *Hydrology*, New Delhi: Wiley Eastern Limited.

Reddy, P. J. 1992. *A Textbook of Hydrology*, New Delhi: Laxmi Publications.

Sharma, R. K. 1983. *A Textbook of Hydrology and Water Resources*, New Delhi: Dhanpat Rai & Sons.

Subramanya, K. 1994. *Engineering Hydrology*, 2nd ed., New Delhi: Tata McGraw-Hill.

Todd, D. K. 1959. *Ground Water Hydrology*, Tokyo: Toppan Company Ltd.

Varshney, R. S. 1986. *Engineering Hydrology*, Roorkee: Nem Chand & Bros.

Viessman, W. Jr., G. L. Lewis, and J. W. Knapp, 1989. *Introduction to Hydrology*, New York: Harper & Row, Publishers Inc.

Wanielista, M., R. Kersten, and R. Eaglin, 1997. *Hydrology*, Hoboken: John Wiley & Sons Inc.

Ward, R. C. and M. Robinson, 1989. *Principles of Hydrology*, New York: McGraw-Hill.

Water Resources of India. 1987. Central Water Commission, Government of India.

Wisler, C. O. and E. F. Brater, 1959. *Hydrology*, New York: John Wiley & Sons.

Index

Y